CAMBRIDGE MONOGRAPHS ON
MECHANICS AND APPLIED MATHEMATICS

General Editors
G. K. BATCHELOR, F.R.S.
University of Cambridge
CARL WUNSCH
Massachusetts Institute of Technology
J. RICE
Harvard University

WAVE INTERACTIONS AND FLUID FLOWS

Wave interactions and fluid flows

ALEX D. D. CRAIK

Reader in Applied Mathematics, St Andrews University,
St Andrews, Fife, Scotland

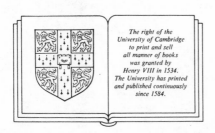

The right of the
University of Cambridge
to print and sell
all manner of books
was granted by
Henry VIII in 1534.
The University has printed
and published continuously
since 1584.

CAMBRIDGE UNIVERSITY PRESS

Cambridge

London   New York   New Rochelle

Melbourne   Sydney

Published by the Press Syndicate of the University of Cambridge
The Pitt Building, Trumpington Street, Cambridge CB2 1RP
32 East 57th Street, New York, NY 10022, USA
10 Stamford Road, Oakleigh, Melbourne 3166, Australia

First published 1985

Printed in Great Britain by the University Press, Cambridge

British Library cataloguing in publication data
Craik, Alex D. D.
Wave interactions and fluid flows.—(Cambridge
monographs on mechanics and applied mathematics)
1. Fluid mechanics
I. Title
532   QC145.2

Library of Congress cataloguing in publication data
Craik, Alex D. D.
Wave interactions and fluid flows.
(Cambridge monographs on mechanics and applied
mathematics)
Bibliography: p.
Includes index.
1. Fluid dynamics. 2. Wave-motion, Theory of.
I. Title. II. Series.
TA357.C73   1985   532′.593   85-7803
ISBN 0 521 26740 4

'A wave is never found alone, but is mingled with as many other waves as there are uneven places in the object where the said wave is produced. At one and the same time there will be moving over the greatest wave of a sea innumerable other waves proceeding in different directions.'

Leonardo da Vinci, *Codice Atlantico*, c. 1500. (Translation by E. MacCurdy, *The Notebooks of Leonardo da Vinci*.)

'Since a general solution must be judged impossible from want of analysis, we must be content with the knowledge of some special cases, and that all the more, since the development of various cases seems to be the only way of bringing us at last to a more perfect knowledge.'

Leonhard Euler, *Principes généraux du mouvement des fluides*, 1755.

'Notwithstanding that...the theory is often not a little suspect among practical men, since nevertheless it rests upon the most certain principles of mechanics, its truth is in no way weakened by this disagreement, but rather one must seek the cause of the difference in the circumstances which are not properly considered in the theory.'

Leonhard Euler, *Tentamen theoriae de frictione fluidorum*, 1756/7.
(Translations by C. A. Truesdell, *Leonhardi Euleri Opera Omnia*, Ser. 2, vol. 12.)

CONTENTS

PREFACE

When, over four years ago, I began writing on nonlinear wave interactions and stability, I envisaged a work encompassing a wider variety of physical systems than those treated here. Many ideas and phenomena recur in such apparently diverse fields as rigid-body and fluid mechanics, plasma physics, optics and population dynamics. But it soon became plain that full justice could not be done to all these areas – certainly by me and perhaps by anyone.

Accordingly, I chose to restrict attention to incompressible fluid mechanics, the field that I know best; but I hope that this work will be of interest to those in other disciplines, where similar mathematical problems and analogous physical processes arise.

I owe thanks to many. Philip Drazin and Michael McIntyre showed me partial drafts of their own monographs prior to publication, so enabling me to avoid undue overlap with their work. My colleague Alan Cairns has instructed me in related matters in plasma physics, which have influenced my views. General advice and encouragement were gratefully received from Brooke Benjamin and the series Editor, George Batchelor.

Various people kindly supplied photographs and drawings and freely gave permission to use their work: all are acknowledged in the text. Other illustrations were prepared by Mr Peter Adamson and colleagues of St Andrews University Photographic Unit and by Mr Robin Gibb, University Cartographer. The bulk of the typing, from pencil manuscript of dubious legibility, was impeccably carried out by Miss Sheila Wilson, with assistance from Miss Pat Dunne.

My wife Liz, who well knows the traumas of authorship, deserves special thanks for all her understanding and tolerance; as do our children Peter and Katie, for their welcome distractions.

xi

Many have instructed and stimulated me by their writing, lecturing and conversation: I hope that this book may do the same for others. I hope, too, that errors and serious omissions are few. But selection of material is a subjective process, and I do not expect to please everyone!

Such writing as this must often be set aside because of other commitments. But for two terms of study leave, granted me by the University of St Andrews, this book would have taken longer to complete. Things were ever so: in 1738, Colin Maclaurin wrote to James Stirling as follows –

> '...it is my misfortune to get only starts for minding those things and to be often interrupted in the midst of a pursuit. The enquiry, as you say, is rugged and laborious.'

St Andrews, September 1984

Chapter one

INTRODUCTION

1 Introduction

Waves occur throughout Nature in an astonishing diversity of physical, chemical and biological systems. During the late nineteenth and the early twentieth century, the linear theory of wave motion was developed to a high degree of sophistication, particularly in acoustics, elasticity and hydrodynamics. Much of this 'classical' theory is expounded in the famous treatises of Rayleigh (1896), Love (1927) and Lamb (1932).

The classical theory concerns situations which, under suitable simplifying assumptions, reduce to linear partial differential equations, usually the wave equation or Laplace's equation, together with linear boundary conditions. Then, the principle of superposition of solutions permits fruitful employment of Fourier-series and integral-transform techniques; also, for Laplace's equation, the added power of complex-variable methods is available.

Since the governing equations and boundary conditions of mechanical systems are rarely strictly linear and those of fluid mechanics and elasticity almost never so, the linearized approximation restricts attention to sufficiently small displacements from some known state of equilibrium or steady motion. Precisely how small these displacements must be depends on circumstances. Gravity waves in deep water need only have wave-slopes small compared with unity; but shallow-water waves and waves in shear flows must meet other, more stringent, requirements. Violation of these requirements forces abandonment of the powerful and attractive mathematical machinery of linear analysis, which has reaped such rich harvests. Yet, even during the nineteenth century, considerable progress was made in understanding aspects of weakly-nonlinear wave propagation, the most notable theoretical accomplishments being those of Rayleigh in acoustics and Stokes for water waves.

1

Throughout the present century, development of the linear theory of wave motion in fluids and of hydrodynamic stability has been steady and substantial: much of this is described in the books of Lin (1955), Stoker (1957), Chandrasekhar (1961), Lighthill (1978) and Drazin & Reid (1981). In contrast, the present vigorous interest in nonlinear waves and stability in fluids dates mainly from the 1960s. Particularly deserving of mention are the monographs of Eckhaus (1965), Whitham (1974), Phillips (1977) and Joseph (1976) and the collections edited by Leibovich & Seebass (1974) and Swinney & Gollub (1981). Related works by Weiland & Wilhelmsson (1977) on waves in plasmas and Nayfeh (1973) on perturbation methods are also of interest to fluid dynamicists.

The great scope, and even greater volume, of recent work on nonlinear waves and stability pose a daunting task for any student entering the field and a continuing, time-consuming challenge to all who try to keep abreast of recent developments. Comprehensive, yet broad, surveys of research in this area become increasingly difficult to write as the subject expands. But collections of more narrowly-focused reviews by groups of specialists often fail to emphasize the many similarities which exist between related areas; similarities which can reveal fresh insights and generate new ideas.

The underlying theme of the present work is that of wave interactions, primarily in incompressible fluid dynamics. But similar mathematical problems arise in a variety of other disciplines, especially plasma physics, optics, electronics and population dynamics: accordingly, some of the work cited derives from the latter fields of study.

Many fascinating and unexpected wave-related phenomena occur in fluids. For instance, water-wave theory has experienced a revolution in the last two decades: solutions are now available, for waves modulated in space as well as time, which exhibit properties as diverse as solitons, side-band modulations, resonant excitation, higher-order instabilities and wave-breaking. Recent progress has been no less dramatic in nonlinear hydro-dynamic stability: the role of mode interactions in the processes leading towards fully-developed turbulence in shear flows is now fairly well understood, and the discovery of low-dimensional 'chaos' in certain fluid flows and in corresponding differential equations is of great current interest. Throughout the history of mathematical analysis, fluid mechanics has provided a challenge and source of inspiration for new theoretical developments: there is every indication that this situation will persist for generations to come.

Chapter 2 is devoted to linear wave interactions, but the remainder of this work concerns aspects of nonlinearity. The underlying assumptions are usually those necessary for development of a *weakly* nonlinear theory: that is to say, linear theory is considered to provide a good starting point in the search for better, higher-order, approximations. However, the nonlinear evolution equations which result from such approximations are sometimes amenable to exact solution: when this is so, an account of their properties is given.

Nonlinear problems are treated in broad categories, on the basis of mathematical rather than physical similarity. Chapter 3 provides a general theoretical introduction; then Chapter 4 treats wave-driven mean flows and waves modified by weak mean flows. Chapter 5 deals with cases of three-wave resonance driven by nonlinearities which are quadratic in wave amplitudes; Chapter 6 concerns nonlinear evolution of a single dominant wave-mode which experiences cubic nonlinearities and Chapter 7 mainly considers interaction of several (typically three or four) wave-modes coupled by cubic nonlinearities. Chapter 8 briefly considers local secondary instabilities and aspects of turbulence. Included in most categories are problems concerning surface waves, internal waves in stratified or rotating fluids and wave-modes in thermal convection and shear flows. Inviscid, and so in some sense conservative, systems are treated side by side with dissipative ones, in order to demonstrate similarities and differences. Typically, the resulting nonlinear evolution equations are soluble analytically in conservative cases, but have rarely been solved other than numerically in dissipative ones. Numerical work which attempts to encompass high-order nonlinearities beyond the range of present analytical techniques is discussed where appropriate.

The use of non-rigorous, sometimes non-rational, procedures – most notably series truncation – is a feature of much work of undoubted interest and value. Unlike Joseph (1976), I have not scrupled to give a full account of the 'state of the art': but it must firmly be borne in mind that the connection between a theoretical model so derived and physical reality is often unclear and perhaps less close than the original author's enthusiasm led him to believe. It is also true that many of the physical configurations so readily envisaged by theoreticians can be rather intractable for experimentalists: even the most obvious restriction to channels of finite length, width and depth immediately causes difficulties! The tendency to make comparisons between theories and experiments which are not strictly comparable is natural and widespread. Theories which are rationally deduced, for some limiting case, have restricted domains of validity which may not overlap with available experimental evidence: comparisons made outwith this range of validity are no more rational – indeed may be less so – than those based on less rigorous theories. Throughout this work, the existing experimental evidence is discussed.

Mechanical systems normally vibrate when displacements from equilibrium are resisted by restoring forces. Examples in fluid mechanics are sound waves, surface gravity and capillary waves, and internal waves sustained by density-stratification, uniform rotation or electromagnetic fields. Such waves may exist in fluid otherwise at rest and they are usually damped by diffusive processes associated with viscosity, thermal or electromagnetic conductivity. But doubly or triply diffusive systems are known to support other instabilities, such as 'salt fingering'.

Relative motion of parts of the fluid, maintained by moving boundaries or applied stresses, modifies wave properties and admits new, possibly unstable, modes. The (Kelvin–Helmholtz) instability of waves at a velocity discontinuity and the centrifugal (Rayleigh–Taylor) instability of differentially rotating flows were among the first to be successfully analysed by linear theory. In unstable rotating flows, the centrifugal force is analogous to the destabilizing body force due to buoyancy in fluid layers heated from below: the latter causes convective (Bénard) instability.

Surface tension provides a restoring force on plane surface waves; but it causes instability of cylindrical columns or jets of liquid. This occurs for geometrical reasons related to the total curvature of the deformed surface, and is analogous to certain instabilities of magnetic flux tubes. Variations in surface tension, due to gradients of temperature or concentration of adsorbed contaminants, may also enhance or inhibit instabilities.

The linear instability of parallel and nearly-parallel flows in channels, boundary layers, unbounded jets and wakes is profoundly influenced by the presence of one or more 'critical layers' where the local flow velocity is close to the phase velocity of a wavelike perturbation. When the primary velocity profile has no inflection point, there are no unstable inviscid modes. But viscosity plays a dual role: as well as providing dissipation, it can also admit new unstable modes which continually absorb energy from the primary flow at the critical layer. Such viscous instability has similarities with Landau damping of plasmas.

Density stratification and the presence of boundaries also play dual roles. A gravitationally-stable density distribution may suppress shear-flow instability; but it can also admit new modes which may interact linearly or nonlinearly to give instability. Likewise, a boundary may enhance viscous dissipation, largely due to the intense oscillatory boundary layer in its vicinity; but it can also reflect wave energy generated elsewhere within the flow and so encourage wave growth.

These few examples serve to illustrate the variety and subtlety of instability mechanisms in fluids. Excellent detailed accounts of linear stability theory are presently available, which it is pointless to duplicate here. The existence of linear instability of a particular flow indicates that this flow cannot normally persist, but will evolve into another type of motion if given an arbitrary small disturbance. However, it is sometimes possible to stabilize a flow by eliminating potentially unstable modes: the party trick of inverting a gauze-covered glass of water is an example, for the gauze prevents growth of the longer wavelength gravitationally-unstable modes not already stabilized by surface tension. Of more practical interest are recent attempts to suppress boundary-layer instability by artificially creating a wave with phase such as to 'cancel' the spontaneously-growing mode. Such stabilization by controlled vibration is effective in dynamical systems with just a few degrees of freedom – for instance the inverted pendulum – but may also induce new parametric instabilities.

If interest is restricted to a finite region of space, say the surface of an aeroplane wing or turbine blade, the mere existence of instability is not the only important aspect. One needs to know whether a disturbance of certain size initiated at some location, say part of the leading edge, will attain significant amplitudes within the region of interest; and, if so, where the greatest amplitudes will occur. Hence, consideration of spatial, as well as temporal, growth is important.

Though linear theory may successfully yield criteria for onset of instability to small disturbances (and sometimes may not!) a finite disturbance can assume a form remote from that of the most unstable linear mode. It *may* happen that nonlinear effects stabilize the disturbance at some small fixed amplitude and that its form broadly resembles the single linear mode from which it evolved.

An instance of this is the toroidal-vortex motion in Taylor–Couette flow between concentric rotating cylinders, at Taylor numbers marginally above the critical one for onset of linear instability. Other examples are near-critical Bénard convection and wind-generated ripples in rather shallow water at just above the critical wind speed. In all such cases, there is a stable solution of the nonlinear equations in the immediate vicinity of the critical conditions for onset of linear instability: this solution bifurcates at the critical point from the trivial zero-amplitude solution.

But, when nonlinear terms have a destabilizing influence, there is no *stable* small-amplitude solution near the critical point and large enough disturbances typically evolve to more complex states. As one moves further from the linear critical conditions, even those constant-amplitude solutions which were stable may lose their stability and support spontaneous growth of other modes. In a similar way, water waves, which are neutrally-stable according to linear theory, exhibit nonlinear instability and modulation.

When a flow becomes very irregular, it is normally described as being turbulent. In fully-developed turbulence, there is no discernible regularity of spatial or temporal structure: Fourier spectra in both space and time are then continuous and broadband, without distinct peaks. When not fully developed, turbulence may be intermittent, confined to localized regions which propagate within an otherwise laminar (though disturbed) flow. A weaker sort of turbulence is found in certain flows which retain a dominant periodic structure amid the broadband 'noise': an example is Taylor–Couette flow at very large Taylor numbers, where spatially-periodic toroidal vortices persist.

Still weaker apparently chaotic motions may occur due to the mutual interaction of a small number of modes: though the temporal structure may be broadband, usually with a few dominant peaks, the spatial structure remains highly organized. Behaviour of this kind, indicative of a 'strange attractor' in the solution space of the governing equations, has deservedly received much recent attention. Both Bénard convection and Taylor–Couette flow can exhibit such behaviour. However, frequent use of the word 'turbulence' in this connection seems misplaced: although the motion is certainly 'chaotic' in time, it remains highly organized in space.

Sometimes, instability and subsequent nonlinear growth have no connection whatever with turbulence. The capillary instability of liquid jets leads to breaking into discrete droplets, usually of regular size; other interfacial instabilities also lead to droplet formation and entrainment. Low Reynolds-number flow of thin liquid films, down an incline under gravity or horizontally under an airflow, may support large-amplitude but still periodic waves or may break up to form dry patches.

Throughout most of this work, the governing equations are the incompressible Navier–Stokes equations,

$$\left.\begin{array}{l} (\partial/\partial t + \mathbf{u}\cdot\nabla)\,\mathbf{u} = -\rho_0^{-1}\,\nabla p + \mathbf{f} + \nu\,\nabla^2\mathbf{u}, \\ \nabla\cdot\mathbf{u} = 0. \end{array}\right\} \tag{1.1a, b}$$

Here, $\mathbf{u}(\mathbf{x}, t)$ and $p(\mathbf{x}, t)$ respectively denote the velocity vector and pressure at each point $\mathbf{x}$ and instant t and $\mathbf{f}$ is a body force per unit mass. The fluid density ρ_0 is taken to be constant, though this constant may differ in different fluid layers; also, continuous changes in density, assumed small compared with ρ_0, may be incorporated into the gravitational body force (the so-called Boussinesq approximation). The kinematic viscosity ν is also assumed constant and is related to the dynamic viscosity coefficient μ by $\nu = \mu/\rho_0$. Equation (1.1a) yields three scalar momentum equations, one for each co-ordinate direction, and (1.1b) is the continuity equation.

Equations (1.1) are frequently expressed in dimensionless form, relative to characteristic scales of mass, length and time. If the latter are defined by a length L, velocity V and the density ρ_0, dimensionless counterparts of (1.1) are

$$\left.\begin{array}{l} (\partial/\partial T + \mathbf{U}\cdot\mathbf{\nabla}_1)\mathbf{U} = -\mathbf{\nabla}_1 P + \mathbf{F} + R^{-1}\mathbf{\nabla}_1^2\mathbf{U}, \\ \mathbf{\nabla}_1\cdot\mathbf{U} = 0, \end{array}\right\} \qquad (1.1\mathrm{a,b})'$$

with the new variables related to the old by $\mathbf{U} = \mathbf{u}/V$, $P = p/\rho_0 V^2$, $\mathbf{F} = \mathbf{f}L/V^2$. The new space co-ordinates, if Cartesian, and dimensionless time T are respectively

$$(X, Y, Z) = (x/L, y/L, z/L), \quad \mathbf{\nabla}_1 \equiv (\partial/\partial X, \partial/\partial Y, \partial/\partial Z), \quad T = tV/L.$$

Viscosity is now represented by the *Reynolds number* $R \equiv VL/\nu$. In the following chapters, lower-case symbols are sometimes used to denote these dimensionless variables: there should be no risk of confusion.

The choice of scales for non-dimensionalization is to some extent arbitrary, but strong conventions exist. For example, plane Poiseuille flow through a plane channel is usually characterized by the half-width of the channel and the maximum flow velocity at mid-channel, yielding the dimensionless velocity profile

$$U(Z) = 1 - Z^2 \quad (-1 \leqslant Z \leqslant 1). \qquad (1.2)$$

Similarly, boundary-layer flows may be non-dimensionalized relative to the (local) free-stream velocity and displacement thickness.

When there occur variations of temperature θ, and so of density, (1.1) must be supplemented by the thermal equation and by an equation of state expressing variation of density with θ. In the Boussinesq approximation, the former becomes

$$(\partial/\partial t + \mathbf{u}\cdot\mathbf{\nabla})\theta = \kappa\,\mathbf{\nabla}^2\theta \qquad (1.3)$$

where κ is thermal diffusivity, and consequent density variations from ρ_0 are considered sufficiently small to be retained only in the gravitational body force $\rho\mathbf{g}$ per unit volume. The dimensionless counterpart of (1.3) has κ replaced by $Pr^{-1}R^{-1}$ where $Pr \equiv \nu/\kappa$ is the *Prandtl number*.

A steady state $\mathbf{u} = \mathbf{u}_0(\mathbf{x})$, $p = p_0(\mathbf{x})$ which satisfies (1.1) may experience a perturbation to

$$\mathbf{u} = \mathbf{u}_0 + \epsilon\mathbf{u}'(\mathbf{x}, t), \quad p = p_0 + \epsilon p'(\mathbf{x}, t),$$

where ϵ is a small parameter characteristic of the initial magnitude of the perturbation. From (1.1),

$$(\partial/\partial t + \mathbf{u}_0 \cdot \nabla)\,\mathbf{u}' + (\mathbf{u}' \cdot \nabla)\,\mathbf{u}_0 = -\rho_0^{-1}\,\nabla p' + \mathbf{f}' + \nu\,\nabla^2\mathbf{u}' - \epsilon(\mathbf{u}' \cdot \nabla)\,\mathbf{u}', \\ \nabla \cdot \mathbf{u}' = 0, \qquad\qquad\qquad\qquad\qquad\qquad\qquad\qquad \text{(1.4a, b)}$$

where $\epsilon\mathbf{f}'$ denotes any perturbation of the body force from its steady-state value. When the disturbance is sufficiently small, it may be justifiable to neglect the term $\epsilon(\mathbf{u}' \cdot \nabla)\,\mathbf{u}'$ in (1.4a): if so, the resultant set of equations for the disturbance is linear and may be solved to find a first approximation to the true perturbed solution. Weakly-nonlinear theory then builds on this by constructing the solution as a series in ascending powers of ϵ.

When viscosity is negligible, equations (1.1) reduce to Euler's equations. If the body force $\mathbf{f}$ is conservative (say $\mathbf{f} = -\nabla\Omega$), these greatly simplify for irrotational flows: for then the vorticity $\nabla \times \mathbf{u}$ remains zero at all times if zero initially. Accordingly, the velocity is expressible as $\mathbf{u} = \nabla\phi$ in terms of a scalar velocity potential $\phi(\mathbf{x}, t)$ and (1.1b) immediately yields Laplace's equation. Integration of (1.1a) along any line element within the fluid gives

$$\partial\phi/\partial t + p/\rho_0 + \Omega + \tfrac{1}{2}(\nabla\phi)^2 = f(t), \\ \nabla^2\phi = 0 \qquad\qquad\qquad\qquad\qquad\qquad \text{(1.5a, b)}$$

and the arbitrary function $f(t)$ may be absorbed into ϕ without loss. Here, the nonlinear Euler's equations have reduced exactly to the linear Laplace's equation, without restriction on any disturbance amplitude, and p is given directly by (1.5a) once ϕ is known. However, in many cases to be discussed, the boundary conditions remain nonlinear and so solution is not straightforward.

The physical condition at solid boundaries is that the velocity of the fluid immediately adjacent to the boundary equals that of the boundary: i.e. $\mathbf{u}(\mathbf{x}, t) = \mathbf{u}_b(\mathbf{x}, t)$ on the boundary surface $B(\mathbf{x}, t) = 0$. Here, $\mathbf{u}_b$ denotes the velocity of material particles of the boundary. The boundary itself must satisfy a kinematic condition connecting $\mathbf{u}_b$ with the boundary position $B = 0$. However, for inviscid flows, the 'no-slip' boundary condition must be discarded and only the velocity component normal to the boundary is prescribed: i.e. $(\mathbf{u} - \mathbf{u}_b) \cdot \hat{\mathbf{n}} = 0$ where $\hat{\mathbf{n}}$ is the unit normal to the boundary.

At free surfaces and fluid interfaces, there are both kinematic and dynamical boundary conditions. Continuity of velocity (or, for inviscid flows, the normal component of velocity) across interfaces is required; also the location of the interface is related to the velocity of particles comprising it by a kinematic condition. In addition, dynamical boundary conditions express the force balance at the interface. In Cartesian form, the stress tensor σ_{ij} $(i, j = 1, 2, 3)$ within either fluid (designated by superscripts 1, 2) and the unit normal $\hat{\mathbf{n}} = \hat{n}_j$ at the interface satisfy

$$(\sigma_{ij}^{(1)} - \sigma_{ij}^{(2)})\hat{n}_j = T_i \quad (i = 1, 2, 3) \tag{1.6}$$

with summation over j. Here, $T_i = \mathbf{T}$ represents interfacial forces per unit area; when these derive solely from surface tension, $\mathbf{T}$ equals $\gamma(\mathbf{\nabla} \cdot \hat{\mathbf{n}}) \hat{\mathbf{n}}$ where γ is the coefficient of interfacial surface tension. The stress tensor σ_{ij} is related to $\mathbf{u} = u_i$ and p by

$$\sigma_{ij} = -p\delta_{ij} + \mu(\partial u_i/\partial x_j + \partial u_j/\partial x_i) \tag{1.7}$$

where $\mathbf{x} = x_i$ denote Cartesian co-ordinates and δ_{ij} is the Krönecker delta. At a free surface, σ_{ij} is zero for the absent fluid.

Since these boundary conditions apply at the moving interface, the position of which may be unknown, approximations valid for small displacements from some known location are usually employed. The boundary conditions applicable to inviscid water-wave theory are set out in §§11 and 14. Both the kinematic equation and the pressure boundary condition are inherently nonlinear; further nonlinearities result from constructing the approximate boundary conditions at the mean level of the water surface.

On nomenclature, note that Figures are numbered by Chapter but equations by section. For instance, Figure 6.1 is in Chapter 6 and equation (6.1) is in §6, Chapter 2.

Chapter two

LINEAR WAVE INTERACTIONS

2 Flows with piecewise-constant density and velocity

2.1 *Stability of an interface*

We begin by considering the flow shown in Figure 2.1. Two inviscid incompressible fluids of effectively unlimited extent have respective constant densities ρ_1, ρ_2 and horizontal velocities U, 0. Their common interface is situated at $z = 0$. Gravitational acceleration g acts downwards, in the $-z$ direction, and there may be an interfacial surface tension γ.

Figure 2.1. Kelvin–Helmholtz flow configuration.

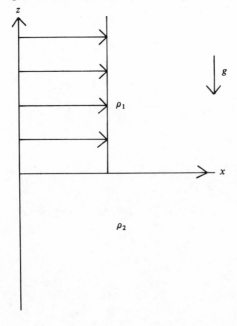

We envisage that, superimposed on this flow, there is an irrotational disturbance such that the interface is displaced to $z = \eta(x, y, t)$ where t is time and y the horizontal co-ordinate perpendicular to the flow U. The associated velocity perturbations have the form $\mathbf{u} = \nabla\phi$ where ϕ is a velocity potential satisfying Laplace's equation in either fluid. The disturbances are assumed to decay to zero as $|z| \to \infty$. The pressure p is known, in terms of the velocity field, from equation (1.5a).

At the interface, kinematic boundary conditions relate the velocity in either fluid to the displacement $\eta(x, y, t)$. There is also a dynamical boundary condition expressing the force balance across the interface, between pressure, gravity and surface tension. These relationships yield three nonlinear boundary conditions, in ϕ and η, to be satisfied at the interface.

For sufficiently small disturbances, these nonlinear boundary conditions may be replaced by linear approximations applied at the undisturbed surface-level, $z = 0$. A typical Fourier component of the displacement $\eta(x, y, t)$ has the form

$$z = \eta(x, y, t) = \epsilon \operatorname{Re}[\exp i(kx + ly - \omega t)]$$

where k, l are horizontal wavenumber components, assumed real, and $\omega(k, l)$ is a possibly complex frequency. The associated velocity potential which decays as $|z| \to \infty$ is

$$\phi = \begin{cases} \epsilon \operatorname{Re}[A_1 e^{-mz} \exp i(kx + ly - \omega t)] & (z > 0) \\ \epsilon \operatorname{Re}[A_2 e^{mz} \exp i(kx + ly - \omega t)] & (z < 0) \end{cases}$$

where $m \equiv (k^2 + l^2)^{\frac{1}{2}}$ denotes the modulus of the wavenumber vector (k, l). The kinematic conditions at $z = 0$ yield

$$A_1 = i m^{-1}(\omega - kU), \quad A_2 = -i m^{-1}\omega$$

and the dynamical boundary condition yields the linearized eigenvalue relationship for $\omega = \omega(k, l)$ as

$$(\rho_2 - \rho_1)gm + \gamma m^3 = \rho_2 \omega^2 + \rho_1(\omega - kU)^2. \tag{2.1}$$

Full details of the derivation of these results are given in Lamb (1932 art. 232) (see also Drazin & Reid 1981 and §11.2 following).

This quadratic equation for ω has either two real roots or a complex-conjugate pair. For real roots, there are two wave-modes which propagate with constant amplitude. For complex-conjugate roots, $\omega = \omega_r \pm i\omega_i$ ($\omega_i > 0$) subscripts 'r' and 'i' indicating real and imaginary, the mode with the positive imaginary part grows exponentially with time t while that with negative imaginary part decays. When there exists such an exponentially-growing mode for some wavenumber pair (k, l), the primary flow is unstable. If there is no such mode for any (k, l), the flow is regarded as stable to linearized disturbances. Note that stability, in this sense, does not imply the decay of all disturbances as $t \to \infty$, but merely the absence of growing modes with the chosen form: modes with real ω are neutrally stable. Formal definitions of stability in the sense of Liapunov are discussed, for example, by Drazin & Reid (1981, p. 9) and Knops & Wilkes (1966) but these need not be considered here.

On defining the horizontal co-ordinate $x' \equiv m^{-1}(kx + ly)$ in the direction of the wavenumber vector (k, l) (i.e perpendicular to wave crests), it is seen that waves propagate in the direction of increasing x' with phase speed ω_r/m. The value of ω/m, as given by (2.1), is dependent on k and l: the waves are therefore dispersive and (2.1) is called the (complex) *dispersion relation* for $\omega(k, l)$. Various special cases deserve attention.

When $U = 0$, the primary state is one of rest and

$$\omega = \pm \left[\left(\frac{\rho_2 - \rho_1}{\rho_1 + \rho_2} \right) gm + \frac{\gamma m^3}{\rho_1 + \rho_2} \right]^{\frac{1}{2}}, \tag{2.2}$$

which are real roots whenever $(\rho_2 - \rho_1)g + \gamma m^2$ is positive. When $\rho_2 > \rho_1$, they represent interfacial capillary–gravity waves. When $\rho_2 < \rho_1$, the heavier fluid is on top and there is gravitational instability of all wavenumbers with $m^2 < (\rho_1 - \rho_2)g/\gamma$: that is, sufficiently long waves are unstable but short waves are stabilized by surface tension if present. An identical instability exists when g is replaced by an acceleration in the direction of increasing density. This is usually known as *Rayleigh–Taylor instability*.

When $\rho_1 = \rho_2$, $\gamma = 0$ and $U \neq 0$,

$$2(\omega/kU)^2 - 2(\omega/kU) + 1 = 0 \tag{2.3}$$

giving roots

$$\omega/kU = \tfrac{1}{2}(1 \pm i).$$

Here, there exists an unstable mode for all non-zero values of k, with exponential growth rate $\omega_i = \tfrac{1}{2}|kU|$. This is the well-known Helmholtz instability of a vortex sheet.

For $\rho_2 > \rho_1$ and non-zero γ and g, the combined restoring forces of gravity and surface tension prevent this instability whenever the discriminant of the quadratic equation (2.1) is positive. The condition for instability, with given k and l, is therefore

$$\frac{\rho_1 \rho_2}{\rho_1 + \rho_2} k^2 U^2 > (\rho_2 - \rho_1)gm + \gamma m^3. \tag{2.4}$$

There exists instability, for *some* k and l, if and only if

$$|U| > \left[\frac{2(\rho_1 + \rho_2)}{\rho_1 \rho_2}\right]^{\frac{1}{2}} [(\rho_2 - \rho_1) g\gamma]^{\frac{1}{4}} \equiv U_c.$$

As $|U|$ is progressively increased from zero, the first unstable mode appears on exceeding the critical value U_c, this mode having wavenumber components $(k_c, 0)$ with $k_c = [g(\rho_2 - \rho_1)/\gamma]^{\frac{1}{2}}$. The instability criterion for oblique wave-modes with wavenumber vector (k, l) is the same as that for two-dimensional modes $(m, 0)$ but with the reduced velocity kU/m replacing U. This is just the component $U \cos\theta$ of the primary flow in the direction of the wavenumber vector (k, l), θ being the angle between (k, l) and the flow direction $(1, 0)$. This instability is known as *Kelvin–Helmholtz instability*.

The energy associated with a mode with real frequency ω is transmitted with the horizontal group velocity

$$\mathbf{c}_g = (\partial\omega/\partial k, \partial\omega/\partial l).$$

Figure 2.2. Typical dispersion curves ω *vs.* k of Kelvin–Helmholtz flow. (*a*) $U = 0$, (*b*) $U < U_c$, (*c*) $U > U_c$. Complex conjugate roots occur along the dashed portion of (*c*).

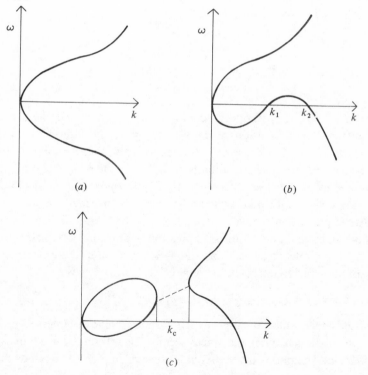

This is also the velocity of propagation of the envelope of a slowly-modulated 'almost periodic' wave-train (see §11). When U is zero, c_g is parallel to the wavenumber vector; but this is not so in the presence of a primary flow. When $U = 0$, the dispersion relation describes capillary–gravity waves. For sufficiently short waves, $|c_g|$ then exceeds the phase speed $|\omega/m|$; but long waves dominated by gravity have $|c_g| < |\omega/m|$. Equality of $|c_g|$ and ω/m occurs at the wavenumber $m = k_c$ defined above, where the phase speed is a minimum. When ω is complex, so also is the group velocity and its close connection with energy propagation is lost.

The various possible forms of the dispersion curves $\omega(k, 0)$ for two-dimensional wave-modes, as U varies, are shown schematically in Figure 2.2. The upper and lower branches intersect when $U = U_c$ and $k = k_c$. For $U > U_c$, there are complex conjugate roots ω with real parts lying along the dashed line in Figure 2.2(c). Instability associated with the appearance of complex-conjugate roots is a common occurrence when different branches of the dispersion curve approach one another. But this condition is neither necessary nor sufficient for instability, as is shown in the next section.

2.2   A three-layer model

We now consider a fluid with

$$\rho = \begin{cases} \rho_1 \\ \rho_2, \\ \rho_3 \end{cases} \quad U = \begin{cases} U_1 & (z > h) \\ U_2 & (|z| < h) \\ U_3 & (z < -h) \end{cases} \tag{2.5}$$

where $\rho_1 < \rho_2 < \rho_3$. Gravity acts along the $-z$-axis and there may be interfacial surface tensions γ and γ' at the respective interfaces $z = h$ and $z = -h$. When $U_1 = U_2 = U_3 = 0$, the fluid can support capillary–gravity waves on either interface, suitably modified by the presence of the other. A wave-mode with periodicity $\exp i(kx + ly - \omega t)$ centred on one interface is only weakly influenced by the other interface provided $mh \gg 1$. In such cases, equation (2.2) yields a good first approximation for waves on the upper interface; and a similar expression with ρ_2, ρ_3, γ' replacing ρ_1, ρ_2, γ is applicable to modes on the lower one. But these approximate dispersion curves intersect, at some value of m, whenever

$$\gamma'/\gamma < (\rho_2 + \rho_3)/(\rho_1 + \rho_2), \quad \rho_1 \rho_3/\rho_2^2 > 1:$$

this is shown schematically in Figure 2.3(a). Near the intersections, the linear coupling between the modes becomes significant. Unlike the example of Kelvin–Helmholtz instability, it is obvious on physical grounds that no instability can occur here; but the structure of the exact dispersion curves is altered near the intersections.

A complete linear analysis of this problem (Cairns 1979; Craik & Adam 1979) shows that the roots do indeed remain real near the 'intersections': the curves do not actually intersect but exchange their identities as shown in Figure 2.3(b). The alternative type of behaviour, exemplified by Kelvin–Helmholtz instability, is shown schematically in Figure 2.3(c), the range of unstable wavenumbers being indicated.

Exact dispersion curves for three-layer flows of type (2.5) are given by Craik & Adam (1979) for cases with $U_2 = U_3 = \gamma' = 0$ and various values of U_1. Two of these are reproduced in Figure 2.4(a), (b). Modes labelled 1 and 2 are gravity waves on the lower interface; those labelled 3 and 4 are capillary–gravity waves on the upper interface, which are modified by the vortex sheet. In Figure 2.4(a), there are four real roots for all values of k and it is seen that modes 2 and 3 exchange their identities near $k = 1.2$. This is close to the intersection point of the dispersion curves for waves on either interface, treated independently. In Figure 2.4(b) modes 1 and

Figure 2.3. (a) Typical approximate dispersion curves for the three-layer model (2.5) with $U_1 = U_2 = U_3 = 0$, treating the interfaces as independent of one another. (b) Nature of the exact solution near the mode 'intersections', showing interchange of identities. (c) Alternative behaviour near 'intersections', characteristic of Kelvin–Helmholtz instability.

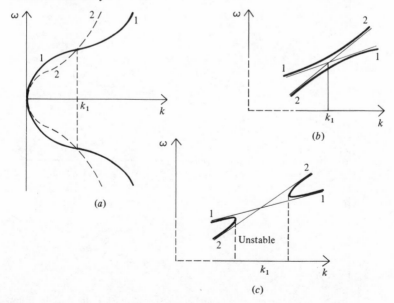

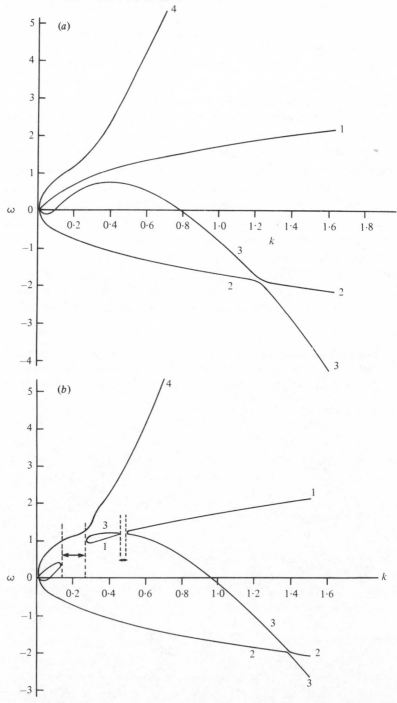

Figure 2.4. Frequency *vs.* wavenumber dispersion curves for the three-layer flow (2.5), with $U_2 = U_3 = \gamma' = 0$, $\gamma = 74 \text{ g s}^{-2}$, $h = 8 \text{ cm}$, $g = 981 \text{ cm s}^{-2}$, $\rho_1 = 1.015$, $\rho_2 = 1.020$, $\rho_3 = 1.026 \text{ g cm}^{-3}$. Case (*a*) shows $U_1 = 8.0 \text{ cm s}^{-1}$ and case (*b*) $U_1 = 8.7 \text{ cm s}^{-1}$ (from Craik & Adam 1979).

3 have 'crossed' giving rise to two bands of instability associated with complex-conjugate roots. The interaction of modes 2 and 3 is therefore of the type shown in Figure 2.3(b) while that of modes 1 and 3 is of the type shown in Figure 2.3(c). Craik & Adam further show that the crossing of modes 3 and 4 at a still greater value of U_1 yields another band of instability, this simply being the Kelvin–Helmholtz instability of the upper interface.

Note that when $\rho_2 = \rho_3$, the flow remains stable at larger values of U than when $\rho_3 > \rho_2$. The introduction of the apparently 'more stable' density distribution with heavier fluid below promotes instability by supporting additional wave-modes, one of which interacts with a wave on the upper interface to give earlier instability: this phenomenon was first remarked on by Taylor (1931).

2.3   *An energy criterion*

A useful criterion for predicting the nature of mode interactions, without calculating the exact dispersion curves, is given by Cairns (1979). His account is based upon considerations of wave energy analogous to those in plasma physics. He envisages the generation of a particular wave-mode, centred on one interface, by suitable (notional) periodic forces applied at some location $z = z_0 + \eta_1(x, t)$.

The amount of work (per unit horizontal area) expended in driving up the wave from zero amplitude is regarded as defining the wave energy $\mathscr{E}$ per unit horizontal area.

Cairns shows that this wave energy $\mathscr{E}$ is related to a suitably-defined dispersion function $D(\omega, k)$ by

$$\mathscr{E} = \tfrac{1}{4}\omega(\partial D/\partial \omega)\,|A|^2.$$

Here, A is the amplitude of the periodic vertical displacement of fluid particles centred at $z = z_0$, the dispersion relation is $D(\omega, k) = 0$ and attention is restricted to two-dimensional waves with $l = 0$. In fact, the function $D(\omega, k)$ is equal to $P(z_0+) - P(z_0-)$, where $AP(z_0+)$ and $AP(z_0-)$ are the amplitudes of the pressure fluctuations just above and below $z = z_0$, evaluated by considering the periodic flows in $z > z_0$ and $z < z_0$ respectively. For natural modes of the system, these must be equal. In plasma physics, the function $D(\omega, k)$ is identified with the plasma dielectric 'constant'. There is a close connection between this treatment and the averaged Lagrangian formulation of Whitham (1965): see §11.4.

For the Kelvin–Helmholtz flow of Figure 2.1, $D(\omega, k)$ is found to be (cf. (2.1) and Cairns' equation 5)

$$D(\omega, k) = (\rho_1 - \rho_2)g - k^2\gamma + k^{-1}\rho_1(\omega - kU)^2 + k^{-1}\rho_2\omega^2$$

when z_0 is chosen to be just below the interface $z = 0$. It follows that

$$\mathscr{E} = \tfrac{1}{2}|A|^2 k^{-1}\omega[(\rho_1+\rho_2)\omega-\rho_1 kU] \qquad (2.6)$$

where A is the interfacial wave amplitude. Substitution from the dispersion relation $D(\omega,k) = 0$ yields

$$\mathscr{E} = \pm\omega(\rho_1+\rho_2)\,U\left[\frac{c_0^2}{U^2}-\frac{\rho_1\rho_2}{(\rho_1+\rho_2)^2}\right]^{\tfrac{1}{2}},$$

$$c_0^2 \equiv \left(\frac{\rho_2-\rho_1}{\rho_1+\rho_2}\right)\frac{g}{k}+\frac{\gamma k}{\rho_1+\rho_2},$$

the $+$ and $-$ signs corresponding to the greater and lesser roots ω respectively and the positive square root being taken. Kelvin–Helmholtz instability occurs when the expression in square brackets is negative, in agreement with the result found above: here we suppose that this expression is positive.

When $U \geqslant 0$, the greater root of ω is positive for all $k > 0$ and the energy $\mathscr{E}$ of this mode is therefore positive. The lesser root ω is certainly negative for sufficiently small positive values of U and then carries positive energy $\mathscr{E}$. But this root ω may become positive, for some k, at large values of U. An instance of this is shown in Figure 2.2(b) with $k_1 < k < k_2$; another is shown in Figure 2.4(a) for mode 3 with $0.1 < k < 0.8$. In such cases, the wave energy $\mathscr{E}$ is *negative*.

The realization that waves may possess either positive or negative energy $\mathscr{E}$ is the key to understanding the two types of mode interaction near 'intersection points' of the approximate dispersion curves obtained by regarding the interfaces as uncoupled. When two approximate dispersion curves cross, and the respective modes have energies of the *same* sign, then the exact dispersion curves 'exchange identities' when they approach one another and both roots remain real. But when two modes cross and have energies of *opposite* sign, the coupling causes complex conjugate roots to appear when the modes are sufficiently close. The form of the dispersion curves is then as shown in Figure 2.3(c), with instability near the 'intersection' point.

The reason for this behaviour is easily demonstrated. The exact dispersion relation for the three-layer flow (2.5) with $U_2 = U_3 = 0$ has the form (cf. Cairns 1979)

$$D_1(\omega,k)\,D_2(\omega,k) = (\rho_2^2\,\omega^4/k^2)\,\text{cosech}^2\,2kh \qquad (2.7)$$

where $D_1 = 0$ and $D_2 = 0$ are the respective dispersion relations for waves on either interface, with the other replaced by a plane rigid wall. The right-hand side, an exponentially-small quantity when kh is large, represents the coupling between the modes. This has negligible effect except when the roots of $D_1 = 0$ and $D_2 = 0$ are close. In this case, let

$$D_1(\omega_1, k) = D_2(\omega_2, k) = 0$$

for some k where $\omega_2 - \omega_1 = \delta$, a real quantity with $|\delta/\omega_1|$ small. On setting $\omega = \omega_1 + \Delta$ in (2.7) one obtains the leading-order approximation

$$(\Delta\, \partial D_1/\partial\omega)_{\omega=\omega_1} [(\Delta - \delta)\, \partial D_2/\partial\omega]_{\omega=\omega_2} = (\rho_2^2 \omega_1^4/k^2)\, \text{cosech}^2\, 2kh.$$

This has the form

$$\Delta^2 - \Delta\delta - K = 0$$

with K positive if the energies $\mathscr{E}$ of the waves have the same sign and K negative if the signs of $\mathscr{E}$ differ. In the former case, the roots Δ are real and no instability occurs. In the latter, they are complex conjugates whenever $|\delta/K| < 2$: instability then occurs whenever the roots ω_1 and ω_2 are sufficiently close.

Application of the above energy criterion remains valid for all multi-layered flows with piecewise-constant density and velocity. For continuous velocity and density profiles, though, the situation is complicated by the occurrence of critical layers within the flow where energy may be exchanged between waves and mean flow. The situation is then somewhat analogous to Landau damping in plasmas (Briggs, Daugherty & Levy 1970).

2.4 *Viscous dissipation*

The above consideration of wave energy also provides some insight into the dissipative rôle of viscosity. We illustrate this for the vortex-sheet configuration (Figure 2.1), following Cairns (1979): (see also Landahl 1962; Benjamin 1963; Ostrovsky & Stepanyants 1983). Let the lower fluid have kinematic viscosity ν, while the upper fluid remains inviscid. Since the lower fluid is at rest apart from the periodic disturbance, viscosity must continuously extract energy from the system. In this case, a positive-energy wave must gradually decay as its energy diminishes. In contrast, a negative-energy wave must *grow* to accommodate the gradual decrease in energy!

For example, in Figure 2.2(b) which has a flow velocity below that for Kelvin–Helmholtz instability, mode 2 has negative energy for $k_1 < k < k_2$. Such wavenumbers must therefore be destabilized by the viscosity of the lower fluid, while all other waves are damped. The dispersion relation then has the form

$$D(\omega, k) = -4\mathrm{i}\rho_2 \nu\omega k \tag{2.8}$$

where $D(\omega, k) = 0$ for inviscid flow. When $\omega_0(k)$ is a root of the latter equation, a good approximation to the solution of (2.8), when ν is sufficiently small, is $\omega = \omega_0 + \delta\omega$ where

$$\delta\omega = \frac{-4i\rho_2 \nu\omega k}{(\partial D/\partial\omega)_{\omega=\omega_0}}. \tag{2.9}$$

Since $\delta\omega$ is imaginary, this approximation denotes growth or decay according as the sign of the wave energy, given by (2.6), is negative or positive.

This instability is of a different sort from the Kelvin–Helmholtz type. In the language of plasma physics (e.g. Bekefi 1966) the latter is a 'reactive' instability where one mode reacts on another to produce complex conjugate roots, while the present type is an example of 'resistive' instability. Growth rates of the resistive instability are typically much smaller than for the reactive.

The effect of dissipation by viscosity within a particular layer of fluid possessing uniform velocity U may be inferred by determining the sign of the wave energy in the reference frame moving with this velocity U. If the wave appears in this frame to have positive energy, then the local viscous dissipation tends to diminish the wave amplitude; if negative, the amplitude will tend to increase. Of course, the envisaged flow is a rather artificial one; and one cannot normally regard viscosity as zero in one part of the flow and non-zero in another. For instance, in the rest frame of the *upper* fluid of the Kelvin–Helmholtz flow of Figure 2.1, the wave which the lower fluid perceived as having negative energy appears to have positive energy: viscosity in the upper fluid therefore tends to cause this mode to decay and the influences of the upper and lower fluid viscosities are in conflict.

Furthermore, the above treatment assumes that the major part of the viscous dissipation is accomplished by the straining of the irrotational flow field. While this is so for wave motion with interfaces which are virtually free from tangential stress and in the absence of (or with sufficiently distant) rigid horizontal boundaries, this is not always the case. A nearby upper or lower rigid boundary, or an interface between two viscous fluids, usually has a rather strong periodic viscous boundary layer in its vicinity and this may account for the bulk of the dissipation (see §3.2 below), thereby contradicting result (2.9). Nevertheless, the present account serves to demonstrate the rather unexpected, but frequent, destabilizing rôle of viscosity. A further example of resistive instability is that of a uniform shear flow with free surface above and rigid boundary below (Miles 1960; Smith & Davis 1982).

Benjamin (1963) proposed a three-fold classification of unstable disturbances. His class C instability is of the Kelvin–Helmholtz, or reactive, type. His class A instability is the resistive type in which negative-energy waves grow by a net extraction of energy from the system by dissipation. His class B instability corresponds to the instability of a wave of positive energy by the net addition of energy from an external source. However, categories A and B are not entirely distinct: the choice of a different reference frame may change the energy sign of a disturbance and so the category to which it belongs. Unfortunately, this classification is not particularly helpful for shear flows with critical layers, where energy may be transferred between mean flow and wave.

Finally, we observe that bounded flows of homogeneous viscous fluid, with prescribed velocities at the boundaries, are globally stable at sufficiently small Reynolds numbers (Synge 1938; Joseph 1976). But this is not necessarily so for flows with free surfaces or internal interfaces. Examples of small-R instability are those of Benjamin (1957), Craik (1966, 1969), Yih (1967) and Hooper & Boyd (1983). Variable tangential stresses at the mean interface level typically play an important rôle in causing such instability at small dimensionless wavenumbers.

3   Flows with constant density and continuous velocity profile
3.1   *Stability of constant-density flows*

Vortex-sheet profiles cannot persist, being eroded by viscosity into continuously-varying shear layers. The study of such discontinuous profiles is therefore based on the expectation that they retain characteristic features of continuous shear-layer profiles, while allowing simpler mathematical treatment. This is indeed so, but care is required in interpreting the results.

A fundamental difference between vortex-sheet profiles and continuous ones is the appearance, in the latter, of one or more critical layers whenever the phase velocity of the wave lies between the maximum and minimum flow velocities. In the linear inviscid approximation, the governing equation is singular at such locations. For a primary parallel shear flow $U(z)$ of constant-density fluid with kinematic viscosity ν, a small two-dimensional wavelike disturbance has velocity perturbations

$$u' = \partial\psi/\partial z, \quad w' = -\partial\psi/\partial x,$$

$$\psi = \epsilon \operatorname{Re}\{\phi(z)\exp[ik(x-ct)]\}, \quad c = c_{\mathrm{r}}+ic_{\mathrm{i}}.$$

This dependence of u' and w' on the perturbation stream function ψ ensures that the continuity equation (1.1b) is identically satisfied. The eigenfunction $\phi(z)$ may be shown to satisfy the *Orr–Sommerfeld equation*

$$(U-c)(\phi''-k^2\phi)-U''\phi = (\nu/ik)(\phi^{\mathrm{iv}}-2k^2\phi''+k^4\phi) \qquad (3.1a)$$

where the prime denotes d/dz. This is just the vorticity equation, obtained by eliminating the pressure p from the momentum equations (1.1a).

The complex frequency $\omega(k)$ equals kc and the wavenumber k is real for purely temporal growth or decay. At rigid plane boundaries, say at $z = z_1$ and z_2, the appropriate boundary conditions are

$$\phi = 0, \quad \phi' = 0 \quad (z = z_1, z_2). \tag{3.2a, b}$$

The latter of these must be ignored for inviscid flows, $\nu = 0$, when the order of the equation (3.1) is reduced from four to two. Equation (3.1a) with $\nu = 0$,

$$(U - c)(\phi'' - k^2\phi) - U''\phi = 0, \tag{3.3a}$$

is known as *Rayleigh's equation*. When $U'' = 0$, Rayleigh's equation in turn reduces to $\phi'' = k^2\phi$, which is equivalent to Laplace's equation for ψ.

The customary introduction of dimensionless variables, relative to chosen velocity and length scales V and h, leads to a similar equation with U, c, k, z replaced by their dimensionless counterparts and ν^{-1} replaced by the Reynolds number $R \equiv Vh/\nu$. The dimensionless wavenumber kh is customarily denoted by α and the dimensionless form of (3.1a) is

$$(U - c)(\phi'' - \alpha^2\phi) - U''\phi = (i\alpha R)^{-1}(\phi^{iv} - 2\alpha^2\phi'' + \alpha^4\phi) \tag{3.1b}$$

where U, c and z now represent dimensionless quantities. Similarly, the dimensionless form of Rayleigh's equation for inviscid flow is

$$(U - c)(\phi'' - \alpha^2\phi) - U''\phi = 0. \tag{3.3b}$$

The eigenvalue problem for $c = c(\alpha, R)$, posed by (3.1b) and (3.2) or alternative boundary conditions, with $U(z)$ given, has received much attention, the most up-to-date account being that of Drazin & Reid (1981). When αR is sufficiently large, the general solution of Rayleigh's equation (3.3) for inviscid flow gives good approximations, over most but not all of the flow domain, to *two* of the four independent solutions of (3.1). The remaining two solutions depend explicitly on the Reynolds number R. The inviscid solutions normally have singular derivatives at locations $z = z_c$ where $U(z_c) = c$, and they do not satisfy the no-slip condition (3.2b) at rigid boundaries. Accordingly, modification of the inviscid solutions takes place near critical layers and walls.

Much effort has been devoted to the development of asymptotic techniques, incorporating the viscous terms, to obtain acceptable approximations for $\phi(z)$ and c. Indeed, this was for long the only means of progress; but the development of high-speed computers and improvements in numerical techniques now provide a ready means of solving (3.1) directly. A comprehensive account of the asymptotic theory, including recent successes in achieving uniformly-valid representations, is given by

Drazin & Reid (1981), who also review the better-known computational methods.

For the present, we confine attention to results for the particular dimensionless shear-layer profile

$$U = \tanh z,$$

which illustrate the connection with the vortex-sheet model. The inviscid problem, with unbounded flow, was studied by Gotoh (1965). He found two modes with imaginary parts c_i of c which approach ± 1 as the dimensionless wavenumber α approaches zero. The real parts c_r of both modes approach zero as $\alpha \to 0$. This long-wave limit is in complete accord with result (2.3) for the vortex sheet, on taking account of the change of reference frame. However, as α increases, the imaginary parts c_i of both roots decrease, approaching asymptotes $c_i/\alpha = -0.376$ and -2.22 respectively as $\alpha \to \infty$ (Drazin & Reid p. 238). The mode which is unstable at small α has $c_i = 0$ at $\alpha = 1$ and is damped for all $\alpha > 1$. The shear profile therefore exerts a stabilizing influence on the flow in the absence of viscosity.

The corresponding viscous problem was treated by Betchov & Szewczyk (1963) who found the curve of neutral stability in the α–R plane. At every Reynolds number, no matter how small, there exist unstable modes with sufficiently small wavenumbers α, the band of unstable wavenumbers becoming ever smaller as R decreases (Drazin & Reid p. 239). In dimensional terms, the Helmholtz instability persists for waves long compared with both the shear-layer thickness h and the viscous length-scale $(\nu/Vk)^{\frac{1}{2}}$.

3.2   *Critical layers and wall layers*

The critical layer plays an important rôle in the instability of parallel flows of homogeneous fluid: see §9 and, for a fuller account, Drazin & Reid (1981). Inviscid flows satisfy the Rayleigh equation (3.3) and the presence of lateral boundaries at $z = z_1, z_2$ requires $\phi(z_1) = \phi(z_2) = 0$; alternatively, for unbounded and 'semi-bounded' flows, $\phi \to 0$ as $z \to -\infty$ or $+\infty$. For a neutrally-stable wave ($c_i = 0$), the mean dimensionless Reynolds stress $\tau = -(\overline{u'w'}) = -\frac{1}{2}\alpha \operatorname{Im}\{\phi'\phi^*\}$ is constant except at any critical layer $z = z_c$, where it has a discontinuity

$$\tau(z_c+) - \tau(z_c-) = \tfrac{1}{2}\alpha\pi(U_c''/U_c')|\phi_c|^2,$$

the subscript c denoting evaluation at z_c. Here, * denotes complex conjugate and the overbar an x-average. But the above boundary conditions require that τ vanishes at z_1 and z_2. If there is just one critical layer, neutral and amplified disturbances may exist if and only if U'' vanishes somewhere in the flow domain. This and other general theorems are reviewed by Drazin & Reid. If there is more than one critical layer, the sum of all the discontinuities in τ must vanish for neutral disturbances.

For continuous, monotonically increasing profiles of boundary-layer form, there is one critical layer when $U_{min} < c < U_{max}$. For plane Couette–Poiseuille channel flows, there may be one or two. If U'' does not change sign, there can be no unstable or neutrally-stable inviscid modes. However, viscosity admits additional modes, the *Tollmien–Schlichting waves*, some of which may be unstable. The mathematical reason for the appearance of these additional modes is the increased order of the governing equation, from the two of Rayleigh's equation to the four of the Orr–Sommerfeld equation. In physical terms, when $\alpha R \gg 1$ and c is $O(1)$, there are thin oscillatory viscous boundary layers with thickness $O[(\alpha R)^{-\frac{1}{2}}]$ adjacent to the wall; also viscous modifications in a layer of thickness $O[(\alpha R)^{-\frac{1}{3}}]$ about $z = z_c$. These may permit neutral or growing waves with finite values of τ in the inviscid region between critical layer and wall, since τ is able to decrease rapidly to zero within the viscous wall layer.

Near the walls, the viscous modes resemble the flow induced by oscillating a rigid boundary in its own plane, the wavelength $2\pi/\alpha$ typically being large compared with the $O[(\alpha R)^{-\frac{1}{2}}]$ wall-layer thickness within which the viscous modes rapidly decay. In place of an oscillating boundary, it is the inviscid flow just beyond the wall layer which oscillates: the inviscid and viscous modes must combine to satisfy the no-slip condition at the stationary wall. Similar viscous modes also occur at deformable boundaries where different boundary conditions must be met (see §§9–10).

For the Blasius boundary layer on a flat plate, unstable modes exist for all $R = U_0 \delta(x)/\nu$ greater than the critical value $R_c = 520$ (Jordinson 1970). Here U_0 is the free-stream velocity and $\delta(x)$ the displacement thickness. This result was obtained by solving the Orr–Sommerfeld equation, neglecting the weak non-parallelism of the flow. The pioneering experiments of Schubauer & Skramstad (1947, but first issued in 1943) revealed at least approximate agreement with quasi-parallel linear theory and more complete data were obtained by Ross, Barnes, Burns & Ross (1970). Despite some as yet unresolved points of detail concerning non-parallelism, the agreement between linear theory and experiment is gratifyingly good.

The influence of flow divergence has been variously treated. Bouthier (1973), Gaster (1974) and Saric & Nayfeh (1975) developed approximate procedures for flow at near-critical Reynolds numbers. Their results show some reduction of R_c due to non-parallelism and agree even better with experiment than do those of quasi-parallel theory. Rational asymptotic analyses of Smith (1979a) and Bodonyi & Smith (1981) determine the asymptotes to the lower and upper branches of the neutral stability curve as $R \to \infty$. The influence of non-parallelism then leads to rather complicated flow structures. Extrapolation of their results to lower $O(10^2)$ Reynolds numbers at which experimental data are available again yields quite satisfactory agreement. However, such extrapolation is 'non-rational', for the respective analyses employ $R^{-\frac{1}{4}}$ and $R^{-\frac{1}{20}}$ as small parameters and the latter equals 0.7 when $R = 10^3$, for instance.

For plane Poiseuille flow, satisfactory agreement between linear theory and experiment was only recently achieved, by the experimental work of Nishioka, Iida & Ichikawa (1975). The critical Reynolds number for this flow is $R_c = 5772$ and, at such high Reynolds numbers, great care is necessary to achieve conditions free from nonlinearity and the influence of side walls and entry region. Without such care, subcritical instability and transition to turbulence occurs for all $R > 10^3$ or so. In fact, as R increases, the range of validity of linear theory is confined to ever-smaller wave amplitudes. The neutral curve in the α–R plane, which separates regions of stability and instability, is shown in Figure 6.3 of §20.

For plane Couette flow (Romanov 1973) and for Poiseuille flow in a pipe of circular cross-section, linear theory indicates no normal mode instability; but a rigorous proof for the latter flow is still lacking. In practical terms, nonlinear mechanisms will govern the stability of these flows.

The rôle of the critical layer is also crucial in the generation of water waves by wind. This was first elucidated by Miles (1957a). There is a critical layer in the airflow above downwind-propagating water waves provided their phase velocity does not exceed the maximum wind speed. On assuming the airflow to be 'quasi-laminar' (though, in fact, turbulent fluctuations ought not to be ignored), Miles showed that this induces a mean Reynolds stress τ much as described above. In the absence of such Reynolds stress, the pressure fluctuation p' at the air–water interface is in exact antiphase with the upwards surface displacement η and the only possible instability mechanism is the Kelvin–Helmholtz one discussed in §2.1. A non-zero Reynolds stress is associated with a phase-shift of p' relative to η, and this phase-shift is responsible for instability at wind speeds far smaller than those required for Kelvin–Helmholtz instability. The rate of working of p' on a neutrally-stable wave is

$$\overline{W} = -\overline{p' \, \partial \eta / \partial t} = \tfrac{1}{2} c k \, \mathrm{Im} \, \overline{p' \eta^*}$$

per unit area, which is positive when p' has a component in phase with the wave-slope $\partial \eta / \partial x$. Instability, and so wave generation, occurs when $\overline{W}$ exceeds the rate of energy dissipation by viscosity within the water.

Miles' original inviscid theory was extended by Miles (1959), Benjamin

(1959) and others to viscous airflows. Agreement is good with subsequent direct computations (Caponi, Fornberg *et al.* 1982) of the airflow over small-amplitude waves. However, as expected, the calculated airflow over larger waves, and the corresponding pressure distribution, are rather different from linear estimates: then, separation eddies form in the troughs of the waves.

The linear quasi-laminar theory has also been integrated with another, complementary, theory of Phillips (1957) which dealt with waves generated in response to imposed (turbulent) pressure variations convected by the airflow (Phillips 1977; Barnett & Kenyon 1975). The influence of small-scale turbulence in the vicinity of the critical layer was discussed heuristically by Lighthill (1962) and a more comprehensive theory of turbulent airflow was developed by Davis (1972, 1974). However, such theories remain far from complete.

Experiments in laboratory channels by Cohen & Hanratty (1965) and Plant & Wright (1977) – see also the reviews by Barnett & Kenyon (1975) and Hanratty (1983) – show satisfactory agreement with the quasi-laminar theory only for rather short waves. In thin layers of water, the energy transfer is insufficient to overcome increased viscous dissipation: in this

Figure 2.5. Experimental results of Craik (1966) on onset of wind-generated waves in thin horizontal films of water.

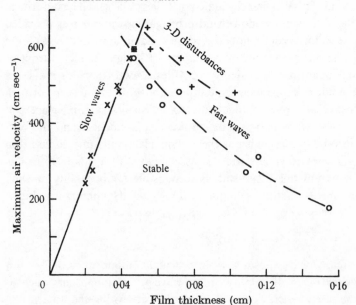

case, there is a range of depths for which the interface is stable. But a new type of instability appears in sufficiently thin films (typically, with depths d of order 0.3 mm or less) with small Reynolds numbers. In this, waves grow because the restoring forces of gravity and surface tension are overcome by a combination of normal and *tangential* stresses at the interface, the latter becoming increasingly important as $kd \rightarrow 0$ where k is the dimensional wavenumber (Craik 1966): see Figure 2.5.

Of less practical concern, but of interest in the context of mode interactions, is a possible coupling between Tollmien–Schlichting waves of a laminar airflow and the water-wave mode. Under appropriate circumstances, the real phase velocities c_r of the two modes may nearly coincide for a range of wavenumbers k. This leads to a rather abrupt increase in the growth rates of the surface-wave (Miles 1962; Blennerhassett 1980; Akylas 1982).

Many surface additives, particularly oils and detergents, have a remarkable capacity to inhibit wave generation by wind and to damp out waves already present. Varying surface concentrations of such additives produce changes in surface tension and the interface acquires elastic or elastico-viscous properties whereby it resists local extension or contraction (see, for example, Miles 1967; Smith & Craik 1971). In effect, the interface now possesses additional natural modes of oscillation, comprising periodic extensions and compressions of the free surface. These are strongly coupled to the underlying water motion, through modified surface boundary conditions. The strength of the viscous boundary layer, and so the rate of dissipation, just beneath the surface is greatly enhanced (see § 10.1 below). The limiting case of an inextensible surface is particularly easy to treat mathematically. In this case, the damping rate of deep-water gravity waves is just half the maximum possible damping rate when the extensional mode and water wave are most strongly coupled, and much greater than for a clean surface. Scott (1979) gives a comprehensive bibliography of such work.

4   Flows with density stratification and piecewise-constant velocity

4.1   *Continuously-stratified flows*

Flows in which the density varies continuously with height are of particular interest in meteorology and oceanography. For these, it is customary to simplify matters by making the so-called Boussinesq approximation. In this, the influence of density variations upon the gravitational force is retained but such variations are ignored in the fluid inertia. This is an acceptable procedure whenever the maximum density variation is small compared with the mean density: this is certainly so in the ocean and is often, but not always, a satisfactory approximation in the atmosphere.

In these circumstances, a small two-dimensional wavelike disturbance has the form

$$u' = \partial \psi / \partial z, \quad w' = -\partial \psi / \partial x,$$

$$\psi = \epsilon \, \mathrm{Re} \{\phi(z) \exp [ik(x-ct)]\} \tag{4.1}$$

where ϵ is small and the eigenfunction $\phi(z)$ satisfies the equation

$$(U-c)(\phi'' - k^2\phi) - U''\phi + N^2\phi/(U-c) = 0 \tag{4.2}$$

for inviscid flow. This was first derived by Taylor (1931), Goldstein (1931) and Haurwitz (1931) and is normally named after the first two authors. Here $N(z) \equiv (-g\rho_0^{-1} \, d\rho_0/dz)^{\frac{1}{2}}$ is the buoyancy (Brunt–Väisälä) frequency; $\rho_0(z)$ is the density distribution in the undisturbed state, stably-stratified so that $d\rho_0/dz \leqslant 0$ everywhere, with z measured vertically upwards; g is gravitational acceleration and $U(z)$ is the primary velocity-profile in the horizontal x-direction.

On setting $U(z) = 0$ and $\phi(z) =$ constant, it is seen that there exist solutions corresponding to vertical oscillations with frequencies $\pm N$. More generally, in regions where $U(z)$ and $N(z)$ are constant, there exist propagating internal-gravity waves with $\phi \propto \exp(\pm im z)$ where m is real and non-negative. The possible phase speeds are

$$c = U \pm N(k^2 + m^2)^{-\frac{1}{2}}$$

(where signs need not correspond to those in the exponent of ϕ), which are respectively faster and slower than the free stream U. The corresponding frequency is $\omega = kc$ and the group velocity of the modes $\phi \propto \exp(+im z)$ has horizontal (in the x-direction) and vertical components

$$(u_{\mathrm{g}}, w_{\mathrm{g}}) = (\partial \omega / \partial k, \partial \omega / \partial m).$$

This gives

$$(u_{\mathrm{g}} - U, w_{\mathrm{g}}) = \pm Nm(k^2 + m^2)^{-\frac{3}{2}} (m, -k),$$

showing that, relative to the free stream U, the group velocity is perpendicular to the wavenumber vector (k, m) and so parallel to the wave crests. Since energy is transmitted with the group velocity, it is clear that wave energy is transported upwards by the mode with $c < U$ and downwards by the mode with $c > U$. Similarly, the modes $\phi \propto \exp(-im z)$ have $w_{\mathrm{g}} \gtrless 0$ according as $c \lessgtr U$.

4.2   *Vortex sheet with stratification*

As an illustrative example, we consider the Helmholtz profile

$$U(z) = \begin{cases} U & (z > 0) \\ 0 & (z < 0) \end{cases} \tag{4.3}$$

with constant positive N and U. The corresponding density distribution $\rho_0(z)$ decreases exponentially with height z. As boundary conditions, we relax the requirement that $\psi \to 0$ as $|z| \to \infty$ in order to admit the propagating modes, and require instead that ψ remains finite as $|z| \to \infty$. We let $\phi(z)$ equal $\phi_1(z)$ and $\phi_2(z)$ in the respective regions $z > 0$ and $z < 0$ and denote the disturbed interface by $z = \eta(x, t)$ where

$$\eta(x, t) = \epsilon \, \mathrm{Re}\{\exp \mathrm{i}k(x - ct)\}$$

and k is known.

Appropriate solutions are

$$\left.\begin{aligned} \phi_1(z) &= A \, \mathrm{e}^{\mathrm{i}m_1 z} + B \, \mathrm{e}^{-\mathrm{i}m_1 z} \quad (z > 0) \\ \phi_2(z) &= C \, \mathrm{e}^{\mathrm{i}m_2 z} + D \, \mathrm{e}^{-\mathrm{i}m_2 z} \quad (z < 0) \end{aligned}\right\} \tag{4.4}$$

$$m_1 \equiv \left(\frac{N^2}{(U - c)^2} - k^2\right)^{\frac{1}{2}}, \quad m_2 \equiv \left(\frac{N^2}{c^2} - k^2\right)^{\frac{1}{2}},$$

the positive roots for m_1 and m_2 being taken. Also A, B, C and D are related by the kinematic conditions

$$\frac{A + B}{c - U} = \frac{C + D}{c} = 1$$

while a further boundary condition is furnished by the pressure-balance across the interface. When m_1 and m_2 are real, ϕ remains bounded at infinity and these three boundary conditions are clearly insufficient to determine the four constants A, B, C, D and the phase-speed $c(k)$; in fact, two of these constants may be assigned arbitrary values.

In the region $z < 0$, the wave modes have frequencies

$$\omega = \pm Nk(k^2 + m_2^2)^{-\frac{1}{2}}.$$

Suppose that ω, as well as k, is a given positive quantity: then m_2 is uniquely determined. When m_2 is real, the terms of ϕ_2 in C and D then represent waves with downwards and upwards-propagating energies respectively. Then C may be regarded as a reflected wave. Likewise, in $z > 0$, A and B represent waves whose energy propagates downwards and upwards respectively whenever $c > U$ and m_1 is real; but these directions are reversed when $c < U$.

If the frequency ω is such that $\omega^2 > N^2$, m_2 is imaginary. Since ϕ_2 remains bounded at infinity, it is then necessary to set $C = 0$: the resultant wave amplitude ϕ_2 decays exponentially as $\exp{(|m_2|z)}$ away from the interface and the wave is said to be 'evanescent' in $z < 0$. Likewise, in $z > 0$, m_1 is imaginary whenever $(\omega - Uk)^2 > N^2$; in this case one must set $B = 0$ and ϕ_1 decays as $\exp{(-|m_1|z)}$ as z increases. Both ϕ_1 and ϕ_2 are evanescent when $\omega < -N$, $N < \omega < Uk - N$ or $\omega > N + Uk$. The three interfacial conditions determine whether there are permissible eigenvalues $\omega(k)$ in these ranges.

When $\omega(k)$ is such that both m_1 and m_2 are real and $c = \omega/k > U > 0$, we may consider a given wave D, incident on the vortex sheet from $z = -\infty$, as giving rise to reflected and transmitted waves C and B. The wave A incident from $z = +\infty$ is meantime chosen to be zero. The reflection and transmission coefficents $|C/D|$ and $|B/D|$ are determined by the interfacial conditions to be (cf. Lindzen 1973; Acheson 1976)

$$\frac{C}{D} = \frac{1-Q}{1+Q}, \quad \frac{B}{D} = \frac{2(c-U)}{c(1+Q)} \tag{4.5a, b}$$

where

$$Q = \frac{m_1(c-U)^2}{m_2 c^2}.$$

Since $|C/D| < 1$ for all $Q > 0$, partial reflection occurs. When $Q = 1$, there is no reflection and the transmission coefficient attains its greatest value $(m_2/m_1)^{\frac{1}{2}}$. Qualitatively similar results are obtained for propagating modes with $c < 0$.

A rather different picture emerges for propagating modes with $0 < c < U$. Then, the transmitted wave in $z > 0$ is A, while B may be chosen as zero. The reflection and transmission coefficients (with A replacing B) have the same form as (4.5), but now Q is *minus* the above expression. This means that the reflection coefficient always *exceeds* unity in such cases. Moreover, a singularity occurs at $Q = -1$, when both the reflection and transmission coefficients become infinite. This phenomenon has become known as *over-reflection*, the singular case being called *resonant over-reflection*. The latter case arises when the system can support outgoing 'reflected' and 'transmitted' waves in the absence of an incident wave. This intriguing phenomenon was first noted by Miles (1957b) and Ribner (1957) in the context of acoustics, but was not recognized for internal gravity waves until the work of Jones (1968).

In the absence of incoming wave energy, the only possible neutral modes are those which correspond to resonant over-reflection. For these,

$$m_1(U-c)^2 = m_2 c^2$$

where $0 < c < U$ and m_1, m_2 are as defined above. Substitution for m_1 and m_2 yields a cubic equation for c,

$$(c - \tfrac{1}{2}U)[c^2 - Uc + \tfrac{1}{2}(U^2 - N^2/k^2)] = 0 \qquad (4.6)$$

with the three roots

$$c = \begin{cases} \tfrac{1}{2}U, \quad m_1 = m_2 = \left(\dfrac{4N^2}{U^2} - k^2\right)^{\frac{1}{2}} > 0, \\[2ex] \tfrac{1}{2}U \pm \left(\dfrac{N^2}{2k^2} - \dfrac{U^2}{4}\right)^{\frac{1}{2}}, \quad m_1 = \dfrac{ck}{U-c}, \quad m_2 = \dfrac{(U-c)k}{c}. \end{cases}$$

The first of these roots exists only for $k^2 < 4N^2/U^2$, there being no solution with pure imaginary m_1 and m_2 which remains bounded at infinity. The latter pair of roots exists provided $k^2 > N^2/U^2$, since c must lie in the range $0 < c < U$ for over-reflection. They are real for $N^2/U^2 < k^2 < 2N^2/U^2$ and become complex conjugates when $k^2 > 2N^2/U^2$. The appearance of conjugate roots signals the onset of a Kelvin–Helmholtz instability: sufficiently short waves are always unstable, however strong the stratification.

4.3   Over-reflection and energy flux

An understanding of over-reflection is acquired by examining the energy flux. Unlike the evanescent wave-modes in homogeneous fluid layers, internal waves which propagate vertically to infinity have infinite wave energy per unit horizontal area. One must calculate the *rate* of work done in maintaining such a wave, as it continuously transmits energy to infinity. To do this, consider just half the flow, either above or below the interface $z = \eta(x, t)$, and suppose that an applied periodic pressure distribution $p(x, t)$ at the interface maintains an outgoing wave of constant amplitude. The rate of working $\dot{W}$ by p is just

$$\dot{W} = \int p\mathbf{u} \cdot \hat{\mathbf{n}} \, ds$$

where s denotes arc-length along the interface, $\hat{\mathbf{n}}$ is the unit normal directed into the region considered and $\mathbf{u}$ is the fluid velocity vector at the interface. For the upper region, we have the linear approximations

$$\eta = \epsilon \operatorname{Re}\{\exp[ik(x-ct)]\}, \quad ds = dx, \quad \hat{\mathbf{n}} = (-\partial\eta/\partial x, 1),$$

$$\mathbf{u} = (U, 0) + \epsilon \operatorname{Re}\{iK(m, -k)\exp i(kx + mz - kct)\} \quad (z > 0)$$

where $K = c - U$ by virtue of the kinematic boundary condition.

For an outgoing wave, $m = -m_1$ for $c > U$ and $+m_1$ for $c < U$, where

$$c = U \pm \frac{Nk}{(k^2 + m^2)^{\frac{1}{2}}}.$$

It follows that

$$\dot{W} \approx \int p(w - U \partial \eta / \partial x)\, dx = c(c - U)^{-1} \int pw\, dx$$

where w is the $O(\epsilon)$ z-velocity component evaluated at $z = 0$. The average of $\dot{W}$, per unit length of the interface, is therefore

$$\bar{W} = \left(\frac{c}{c - U}\right) \overline{pw} = \rho_0 \overline{cu'w} = -\tfrac{1}{2}\epsilon^2 \rho_0 (c - U)^2\, c\, \mathrm{Re}\,\{m^*k\} \qquad (4.7)$$

where * denotes complex conjugate and the overbar signifies an x-average. Here, we have used the result relating p to the $O(\epsilon)$ x-velocity u' (cf. Acheson 1976), that

$$p = \rho_0(c - U)\,u'$$

which follows directly from the x-momentum equation.

When $c < 0$ or $c > U$ and m is real, $\bar{W}$ is positive for an outgoing wave. The positive upwards component of group velocity w_g then transmits positive wave energy to $z = +\infty$. But, when $0 < c < U$, $\bar{W}$ is negative; in which case there is a flux of *negative* wave energy towards $z = +\infty$.

For the lower fluid $z < 0$, where there is no primary flow,

$$\bar{W} = \tfrac{1}{2}\epsilon^2 \rho_0\, c^3\, \mathrm{Re}\,\{m^*k\}$$

where $m = m_2$ (> 0) for $c > 0$ and $-m_2$ for $c < 0$ for outwards wave propagation towards $z = -\infty$. Since $\bar{W}$ is positive in both cases, these waves transmit positive energy towards $z = -\infty$.

During over-reflection, a positive-energy wave incident on the vortex sheet from $z = -\infty$ with wave speed c either less than zero or greater than U can only generate transmitted and reflected waves with positive energy. Partial reflection therefore takes place. But an incident wave with $0 < c < U$ necessarily generates a transmitted wave with *negative* energy, and the negative energy carried off to $+\infty$ can only be offset by a reflected positive-energy wave with amplitude greater than that of the incident wave.

Resonant over-reflection occurs when the interfacial conditions admit outgoing-wave solutions with equal and opposite energy flux in the upwards and downwards directions. That is, the upward-propagating negative-energy wave and the downward-propagating positive-energy wave together yield zero total energy flux away from the interface. Waves radiate away from the vortex sheet and there is a constant-amplitude interfacial disturbance.

The addition of a jump in density at the interface admits interfacial gravity waves. Then, it is possible for waves to grow by *radiative instability*; for, when an interfacial wave resonates linearly with an outward propagating internal wave-mode, both grow if their respective energies have opposite signs. Examples of such instability are given by Ostrovsky & Tsimring (1981) and Ostrovsky, Stepanyants & Tsimring (1983).

They discuss a Kelvin–Helmholtz configuration with homogeneous upper fluid of density ρ_1 and uniformly (but weakly) stratified lower fluid with constant Brunt–Väisälä frequency N and reference density ρ_2. When the upper fluid has depth h and the lower fluid is unbounded, the linear dispersion relation is then

$$\rho_1(c-U)^2 \coth kh + \rho_2 c[c^2-(N/k)^2]^{\frac{1}{2}} - (\rho_2-\rho_1)g/k = 0$$

if surface tension is ignored. Note that this is consistent with (2.1) as $N \to 0$ and $kh \to \infty$.

At sufficiently large wavenumbers k, Kelvin–Helmholtz instability occurs (though this may be inhibited by surface tension); also, there is a separate unstable range of smaller wavenumbers, owing to downwards radiation of positive-energy internal waves with the same real frequency as negative-energy interfacial waves. This takes place when the *slower* of the two interfacial modes has $0 < c_r < N/k$, the condition $c_r > 0$ ensuring that this mode has negative energy (see §2) while downward-radiating internal waves must have $c_r < N/k$.

It should be noted that not all the $O(\epsilon^2)$ energy resides in the $O(\epsilon)$ periodic fluctuations, but that an $O(\epsilon^2)$ mean-flow modification also contributes. This fact, and also the relationship between energy and 'wave action', are demonstrated in §11.

4.4   *The influence of boundaries*

For the simple Kelvin–Helmholtz flow of Figure 2.1, but with plane rigid boundaries at $z = a$ and $z = -b$ $(a, b > 0)$, the instability criterion differs from (2.4) by the replacement of the factor $\rho_1 \rho_2 (\rho_1 + \rho_2)^{-1}$ of the left-hand side by $\rho_1 \rho_2 \{\rho_1 \tanh(mb) + \rho_2 \tanh(ma)\}^{-1}$ (Rayleigh 1896, vol. 2, p. 378). Since this factor is always greater than that for unbounded flow, the boundaries have a destabilizing effect on all wavenumbers. The phase speed of capillary–gravity waves with $U = 0$ is reduced by the presence of the boundaries and so a lower value of U can cause coalescence of the two modes.

However, in another sense, the presence of boundaries is stabilizing, in that the growth rate of unstable modes is usually decreased. This is exemplified by the Kelvin–Helmholtz profile with $\rho_1 = \rho_2$ and $\gamma = 0$ with boundaries at $z = a$ and $-b$. For this, the eigenvalue relation (2.3) is altered to

$$\left(1 + \frac{\tanh kb}{\tanh ka}\right)\left(\frac{\omega}{kU}\right)^2 - 2\left(\frac{\omega}{kU}\right) + 1 = 0,$$

for which the unstable root has $\mathrm{Im}\{\omega/kU\} < \frac{1}{2}$ for all non-equal value of a and b.

For continuously-stratified flows, the presence of boundaries has a more dramatic effect. This is most simply demonstrated for the Helmholtz profile

$$\left.\begin{array}{l} U(z) = \begin{cases} U & (z > 0), \\ 0 & (-H \leqslant z < 0) \end{cases} \\ N^2 = \text{constant}\,(-H \leqslant z < \infty) \end{array}\right\} \qquad (4.8)$$

with a rigid plane boundary at $z = -H$. Since the vertical velocity at the boundary must vanish, the stream function in $-H \leqslant z < 0$ has the form given in (4.4), but with

$$\frac{C}{D} = -e^{2im_2 H}.$$

The solution in this region therefore comprises both upwards- and downwards-propagating components. It follows directly from (4.5a) that

$$Q \equiv \pm\frac{m_1(c-U)^2}{m_2 c^2} = \mathrm{i}\cot(m_2 H) \qquad (4.9)$$

when no wave is incident from $z = +\infty$. The dispersion relation defined by (4.4) and (4.9) is a special case of that examined by Lalas, Einaudi & Fua (1976).

This relation yields an *infinite* number of modes, for given k, many of which are unstable. A comprehensive discussion and computed dispersion curves for some of these modes are given by Lalas *et al.* These new modes appear because the region between boundary and interface acts as an (imperfect) waveguide. A wave-component incident on the interface from below may be either partially reflected or over-reflected, while a wave incident on the boundary is totally reflected. A 'trapped' mode suffers repeated reflections, its amplitude increasing or decreasing with each reflection at the interface. Over-reflected waves are therefore unstable and partially-reflected waves are damped.

This instability due to repeated reflections is quite distinct from the radiative instability mentioned in §4.3. Flows which can admit both mechanisms together deserve study.

5   Flows with continuous profiles of density and velocity

5.1   *Unbounded shear layers*

For inviscid flows with continuous velocity profiles $U(z)$, the presence of a critical layer significantly modifies the above results. A full discussion of the various classes of disturbance is given by Banks, Drazin & Zaturska (1976) and Drazin, Zaturska & Banks (1979). The former deal with flows between lateral boundaries and unbounded flows such that $N^2(z) \to 0$ as $|z| \to \infty$: for these, no modes propagate at infinity. The latter consider flows with $N^2(z) \to N^2_{\pm}$ (constants) as $z \to \pm\infty$ for which such propagating modes exist.

Any unstable mode of the inviscid velocity profile with $N^2(z) = 0$ is modified by buoyancy and all instabilities disappear if the local *Richardson number* $J(z) \equiv N^2(dU/dz)^{-2}$ is nowhere less than $\frac{1}{4}$. This is the 'Miles–Howard theorem' (Drazin & Howard 1966). But buoyancy sometimes has a destabilizing effect when $J(z) < \frac{1}{4}$ somewhere within the flow (Howard & Maslowe 1973). In addition to any unstable, or marginally-stable, modes there are exponentially-damped modes, all having phase velocities with $U_{min} < c_r < U_{max}$. There also exist classes of constant-amplitude internal gravity waves with $c_r > U_{max}$ and $c_r < U_{min}$ and a continuous spectrum of singular neutral modes associated with algebraic, rather than exponential, decay of disturbances. If $N^2 < 0$ anywhere, gravitationally-unstable modes of course occur.

Many general results, together with those for particular profiles of $U(z)$ and $N^2(z)$, are described by Drazin & Howard (1966), Hazel (1972) and in the above-mentioned papers. Typical of unbounded shear layers is the hyperbolic-tangent velocity profile with constant Brunt–Väisälä frequency, discussed in detail by Drazin, Zaturska & Banks (1979). In dimensionless variables, this is

$$U = \tanh z, \quad N^2 = J \text{ (constant)} \quad (-\infty < z < \infty). \tag{5.1}$$

If $\alpha = kh$ denotes the corresponding dimensionless wavenumber where h is the length-scale which characterizes the shear-layer thickness, the Taylor–Goldstein equation (4.2) yields neutrally stable solutions

$$J = \alpha^2(1 - \alpha^2), \quad c = 0 \tag{5.2}$$

and unstable solutions exist at all points of the α–J plane enclosed by this curve and the axis $J = 0$.

In the unstratified case, $J = 0$, all sufficiently-short waves ($\alpha > 1$) are damped, as described previously. Now buoyancy provides a further restoring force which acts to inhibit Kelvin–Helmholtz instability, especially at longer wavelengths. It is readily seen that instability disappears for all $J > \frac{1}{4}$ in agreement with the general Richardson-number criterion.

In addition to this mode with $c_r = 0$, Drazin, Zaturska & Banks found two other unstable modes with equal and opposite non-zero phase velocities c_r. These exist for $0 < J < 0.13$ approximately and have values of α somewhat less than the smaller root of (5.2). These modes resemble the propagating neutral modes of (4.6) associated with resonant over-reflection at a vortex sheet, but now have $c_i > 0$ for a range of α on account of the rôle played by the critical layer.

Among the various limiting cases discussed by Drazin *et al.* is that with $\alpha^2 \to 0$ and $J/\alpha^2 \to 2$, for which the dispersion relation simplifies to

$$c^3 + (1 - \tfrac{1}{2}J/\alpha^2)c = i\alpha. \tag{5.3}$$

If the right-hand side were zero, this equation would have roots $c = 0$, $c = \pm(\tfrac{1}{2}J/\alpha^2 - 1)^{\frac{1}{2}}$ which precisely correspond to the roots (4.6) for the

Figure 2.6. Dispersion curves of c_r and c_i *vs.* α for the stratified shear layer $U = \tanh z$, $N^2 = J$ (constant): c_i is shown by a continuous line when $c_r \neq 0$ and a dashed line when $c_r = 0$. Case (*a*) corresponds to $J/\alpha^2 = 0.91$, case (*b*) to $J/\alpha^2 = 0.95$ (from Drazin, Zaturska & Banks 1979).

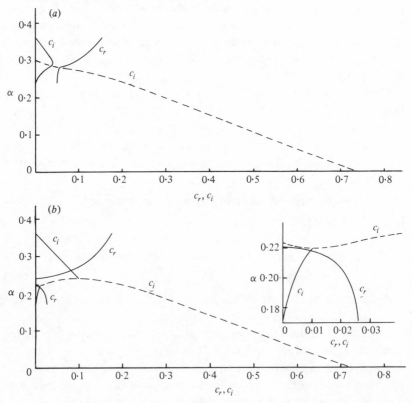

vortex sheet when account is taken of the change of origin. The right-hand side derives from the critical layer and this provides a coupling between the three modes which is absent for the vortex sheet. Unstable roots exist at values of J/α^2 greater than 2, for which the vortex-sheet modes are stable.

Computed dispersion curves of c_r and c_i versus α at various fixed values of J/α^2 are given by Drazin *et al.* Two of these are reproduced in Figure 2.6(a), (b). In Figure 2.6(a) for $J/\alpha^2 = 0.91$, the dashed line denotes the unstable (Kelvin–Helmholtz) mode with $c_r = 0$. Conjugate modes with $c_r \neq 0$ are unstable for a range of α near that of the neutrally-stable mode with $c_r = 0$. In contrast, Figure 2.6(b) for $J/\alpha^2 = 0.95$ shows a double bifurcation of the modes. At $J/\alpha^2 = 1.5$ Drazin *et al.* find a single bifurcation. At $J/\alpha^2 = 3$, there is no bifurcation and no unstable mode with $c_r = 0$. At this last value of J/α^2, the eigenvalues of the corresponding vortex sheet are $c = 0$, $c = \pm 2^{-\frac{1}{2}}$ and there is no Kelvin–Helmholtz instability. Drazin *et al.* find modes with c_r close to the latter values but with small growth rates ($c_i > 0$) for $0 < \alpha < 0.1$. Presumably, a third mode with $c_r \approx 0$ is now damped.

In fact, this rather complicated behaviour may be modelled qualitatively by the root locus of simple cubic equations (see §6 below). Possible forms of the dispersion curves for the three roots are sketched in Figure 2.7(a), (b), (c).

Figure 2.7(a) shows no bifurcation and corresponds to Figure 2.6(a); two conjugate modes with $c_r \neq 0$ pass on either side of the mode with $c_r = 0$. Figure 2.7(b) shows a double bifurcation, as in Figure 2.6(b), the modes with zero, positive and negative c_r now being connected. Figure 2.7(c) shows a single bifurcation, such as occurs for $J/\alpha^2 = 1.5$, the modes with $c_r \neq 0$ having disappeared at small α to be replaced by two more modes with $c_r = 0$. The solid lines denote unstable modes ($c_i > 0$), the dashed curves damped ones.

5.2   *Bounded shear layers*

It was seen above that the presence of boundaries introduces additional unstable modes of the stratified Helmholtz profile (4.3). Not surprisingly, this is also the case for continuous profiles. Einaudi & Lalas (1976) have examined the 'tanh' profile (5.1) with rigid walls situated at $z = a$ and $z = -b$ ($a, b > 0$). Lalas & Einaudi (1976) considered the same profile bounded below at $z = -b$ but unbounded above, thereby allowing upwards wave-propagation towards $z = \infty$. These authors take account of 'non-Boussinesq' inertial terms associated with the variation with height

Figure 2.7. Sketches of possible forms of three-mode dispersion curves in c_i–c_r–α space.

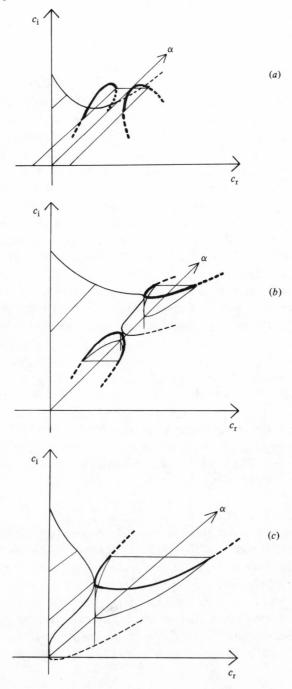

of the mean density profile, such terms being neglected in the Taylor–
Goldstein equation (4.2). The effect of these is represented, in their work,
by a parameter σ which denotes the ratio of shear-layer thickness to a
vertical length-scale of the density profile. The value of σ is typically small
and the inclusion (or neglect) of such terms does not then induce
substantial changes in the results.

Figure 2.8, from Lalas & Einaudi (1976), shows the stability boundaries
of four modes in the J–α plane and Figures 2.9 (a), (b), (c) give corresponding growth rates αc_i and phase velocities c_r versus α for various J. Mode I
is close to result (5.2) except at small α, the phase velocity rather rapidly
decreasing from 0 to -1 as $\alpha \to 0$ and the region of instability being
somewhat greater. This destabilization of long waves accords with the
effect of boundaries, noted earlier, upon Kelvin–Helmholtz instability of
the vortex sheet. The unstable over-reflecting modes II and III are typically
much less strongly amplified than the Kelvin–Helmholtz mode I and are
confined to smaller wavenumbers. The boundary in these cases is situated
at $z = -10$, which is quite distant from the region of strong shear.
Naturally, for smaller values of b, the influence of the boundary on mode
I is more pronounced. For given values of J and α, there is a finite range
of b within which each of the higher modes is unstable. For instance, mode

Figure 2.8. J *vs.* α stability boundaries of the first four modes, for stratified flow
$U = \tanh z$, $N^2 = J$ (constant), $\sigma = 0.1$, with a rigid wall at $z = -10$ but unbounded
as $z \to +\infty$ (from Lalas & Einaudi 1976).

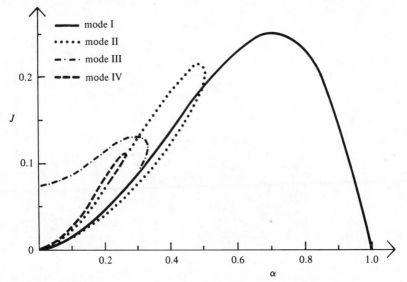

III with $\alpha = 0.01$, $J = 0.1$ and $\sigma = 0.1$ is unstable only for $2.5 < b < 7.5$ approximately. Unstable modes, apart from mode I, are always 'propagating' (i.e. the local vertical wavenumber m is real) for large z. Nearer the boundary, to quote Lalas & Einaudi, 'they are either mostly propagating or mostly evanescent dependent on the actual value of J'.

With both upper and lower boundaries, there can be no energy flux to infinity, wave components being reflected at both walls. Trapped waves impinging on a vortex-sheet profile from *either* side may be partially- or over-reflected and the possibilities for instability are enhanced, but this picture is modified by the shear for continuous profiles. Einaudi & Lalas (1976) have studied the first three modes of the profile (5.1) in such cases.

Figure 2.9. Growth rates αc_i and phase velocities c_r versus α for various J. The flow is as for Figure 2.8: (a), (b) and (c) show modes I, II and III respectively (from Lalas & Einaudi 1976).

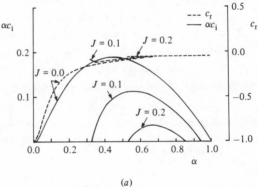

(a)

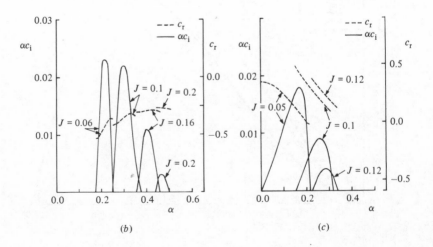

(b)   (c)

For mode I and moderate values of a and b, shorter waves are more stable and longer waves less stable than in unbounded flow. The additional unstable modes appear at smaller wavenumbers than those of mode I, certainly when the boundaries are not too close; but some, and eventually all, of these modes disappear as a and b are reduced. For instance, modes II and III are not unstable for $a = b \leqslant 5.3$ and the instability of mode I is itself suppressed for a and b less than about 1.2. The presence of boundaries therefore plays a dual rôle. Their reflection of waves encourages instability by over-reflection at the shear layer; but they also rule out propagating wave solutions with vertical wavelengths $2\pi/m$, long compared with the channel width $a+b$. As $a+b$ decreases, more such modes are rendered inadmissible until, for sufficiently small values, no instability is possible.

The case of a partially-reflecting, partially-transmitting upper boundary, associated with a sudden increase in N^2, has also been considered by Lalas & Einaudi (1976). For this, they find that the distinct mode structure disappears and there are no longer separate stability boundaries for each mode, as in Figure 2.8. Instead, the transition from one mode to another is continuous, and the neutral curve in the $J–\alpha$ plane is also a continuous one. But the shape of this curve clearly indicates the vestiges of a three-mode structure.

In flows with several layers, separated by partially- or over-reflecting interfaces, a mode may be over-reflected at one interface but only partially reflected at the next. With the added complications of critical layers and reflections from other, more distant, interfaces, some effort would be required to discover whether repeated reflections increase or decrease the wave amplitude.

5.3 *The critical layer in inviscid stratified flow*

The mean Reynolds stress $\tau = -\rho_0 \overline{u'w}$ and rate of work done $\overline{W}$ in maintaining a wave are related, from (4.7), by $\overline{W} = -c\tau$. Since $\tau = \frac{1}{2}\rho_0 \, \mathrm{Im}\{\phi'\phi^*\}$ where $*$ denotes complex conjugate, it follows from (4.2) that $\mathrm{d}\tau/\mathrm{d}z = 0$ when $c_i = 0$, except possibly at any critical layers $z = z_c$ where $U(z_c) = c$. The Reynolds stress, and hence $\overline{W}$, are constant except near such critical layers, where they may suffer discontinuities. In regions where $\overline{W}$ is constant, so also is the vertical energy flux $w_g E$, where E is the net energy density of the disturbance (see §11.5) and w_g the vertical group-velocity component.

The structure of solutions of (4.2) near the critical layer was discussed by Booker & Bretherton (1967) for cases with local Richardson number

$J \equiv N^2(z_c)[U'(z_c)]^{-2}$ greater than $\frac{1}{4}$. Cases with $J < \frac{1}{4}$ were examined by Eltayeb & McKenzie (1975). Near z_c, a Frobenius series solution of (4.2) has leading-order terms

$$\phi \approx \begin{cases} A_1(z_c-z)^{\frac{1}{2}+i\mu} + A_2(z_c-z)^{\frac{1}{2}-i\mu} & (z < z_c) \\ B_1(z-z_c)^{\frac{1}{2}+i\mu} + B_2(z-z_c)^{\frac{1}{2}-i\mu} & (z > z_c) \end{cases}$$

$$A_1 = -iB_1\,e^{\pi\mu}, \quad A_2 = -iB_2\,e^{-\pi\mu},$$

where $\mu \equiv (J-\frac{1}{4})^{\frac{1}{2}}$ is real and positive when $J > \frac{1}{4}$. When $J < \frac{1}{4}$, identical results hold with $\mu = i\nu$ where ν is positive.

For $J > \frac{1}{4}$, let $z = z_1 + \delta$ where z_1 is close enough to z_c for the above approximation to hold and suppose that $|\delta/(z_1-z_c)| \ll 1$. We may then write

$$(z-z_c)^{\pm i\mu} \approx (z_1-z_c)^{\pm i\mu}\exp[im(z_1)\,\delta], \quad m \equiv \pm\mu(z_1-z_c)^{-1}$$

where $m(z_1)$ may be interpreted as a local vertical wavenumber. Wave components with exponent $\frac{1}{2}+i\mu$ have positive vertical group velocity $w_g(z_1)$ and those with exponent $\frac{1}{2}-i\mu$ have negative $w_g(z_1)$. For the former, $m(z_1)$ is negative below and positive above the critical layer; for the latter, vice versa. Since $m(z_1) \to \pm\infty$ as $z_1 \to z_c$, waves are strongly refracted near the critical layer and the group velocity $\mathbf{c}_g(z_1)$ of each component becomes more nearly horizontal. The wave energy density increases without limit as $z_1 \to z_c$. An approximate 'ray-theory' analysis (Bretherton 1966; cf. §11.5) shows that an incident wave-packet takes an infinite time to reach z_c and so becomes 'trapped' in the critical layer. The critical layer may therefore act as an absorber of wave energy which is either dissipated by viscosity or transferred to the mean flow near z_c. The wave component A_1 incident from below yields a 'transmitted' component B_1 with $|B_1/A_1| = \exp(-\pi\mu)$. The transmitted component has much smaller amplitude, since $\exp(-\pi\mu)$ is very small except for J close to $\frac{1}{4}$. Equivalent results hold for downwards-travelling components. Van Duin & Kelder (1982) give some exact solutions for the reflection and transmission coefficients of internal waves incident on the shear layer (5.1). The smallness of waves penetrating a critical level was confirmed experimentally by Thorpe (1981).

The vertical energy flux of a wave incident from below is

$$(\overline{W})_i = -c\tau = \tfrac{1}{2}\rho_0\,c\mu\,|A_1|^2$$

and that of the 'transmitted' wave is

$$(\overline{W})_t = -\tfrac{1}{2}\rho_0\,c\mu\,|B_1|^2 = -(\overline{W})_i\,e^{-2\pi\mu}.$$

Clearly, the energy of the incident wave is positive and that of the transmitted wave is negative. The net rate of accumulation of (positive) wave energy near the critical layer is therefore

$$(\overline{W})_{\mathrm{i}} - (\overline{W})_{\mathrm{t}} = \tfrac{1}{2}\rho_0 c\mu \,|A_1|^2 (1 + \mathrm{e}^{-2\pi\mu}),$$

the outgoing wave being associated with increased, not decreased, absorption of positive energy. The positive-energy wave is totally absorbed and the negative-energy wave is *emitted*, rather than transmitted, at the critical layer.

Eltayeb & McKenzie (1975) considered a wave incident on the shear layer

$$U = \begin{cases} 0 & (z < 0) \\ U_0 z/L & (0 \leqslant z \leqslant L), \quad N^2 \text{ constant.} \\ U_0 & (z > L) \end{cases} \tag{5.4}$$

When $c < 0$ or $c > U_0$ there is no critical layer. The vertical energy flux $\overline{W}$ is therefore constant, the wave components all have positive energy and partial reflections occur at the interfaces $z = 0, L$. When $0 < c < U_0$, $\overline{W}$ is constant on either side of the critical layer $z_{\mathrm{c}} = cL/U_0$. When $J \equiv N^2 L^2/U_0^2 > \tfrac{1}{4}$, the critical layer absorbs energy and a reduced-amplitude negative-energy component is emitted. In addition there are reflections from the interfaces. That from the first interface encountered is of greatest importance, all others being attenuated on each passage through the critical layer. There is no over-reflection for $J > \tfrac{1}{4}$.

For $J < \tfrac{1}{4}$, the above identification of positive and negative-energy waves breaks down. Near z_{c}, the local vertical structure of the wave components is exponential, as $\exp(\pm n\delta)$ where $n = \nu(z_1 - z_{\mathrm{c}})^{-1}$ and $\nu \equiv (\tfrac{1}{4} - J)^{\frac{1}{2}}$. The respective energy fluxes $\overline{W}_-$, $\overline{W}_+$ below and above z_{c} are then

$$\overline{W}_- = \rho_0 \nu \,\mathrm{Im}\{A_1^* A_2\}, \quad \overline{W}_+ = -\rho_0 \nu \,\mathrm{Im}\{B_1^* B_2\} = -\rho_0 \nu \,\mathrm{Im}\{A_1^* A_2 \,\mathrm{e}^{2\mathrm{i}\pi\nu}\}.$$

A wave incident from $z = -\infty$ with $0 < c < U_1$ and $J < \tfrac{1}{4}$ suffers partial reflection at $z = 0$. The transmitted wave is no longer a propagating mode throughout the shear layer, but decomposes into exponentially growing and decaying parts as it approaches z_{c}^- in order to maintain the constant value of $\overline{W}_-$. The values of A_1 and A_2 are known, in terms of the incident wave amplitude, from the matching conditions at $z = 0$. The respective energy fluxes $\overline{W}_{\mathrm{i}}$ and $\overline{W}_{\mathrm{r}}$ of incident and reflected waves in $z < 0$ must satisfy

$$\overline{W}_{\mathrm{i}} - \overline{W}_{\mathrm{r}} = \overline{W}_-;$$

accordingly, over-reflection takes place in $z < 0$ whenever $\overline{W}_-$ is negative. In such cases, the critical layer acts as a *source* of positive energy for the region $z < 0$. For the flow (5.4), Eltayeb & McKenzie show this occurs when $J < 0.1129$. For the region $z > L$, the critical layer acts as a source of negative energy whenever $\overline{W}_+ < 0$, the net rate of accumulation of (positive) energy near z_c being $\overline{W}_- - \overline{W}_+$. In the limit $\nu \to \frac{1}{2} (J \to 0)$, $\overline{W}_+ \to \overline{W}_-$ and over-reflection from the shear layer takes place with no absorption at the critical layer, just as for the vortex sheet (4.8). Instability due to resonant over-reflection may occur with $c_r = \frac{1}{2} U_0$.

Other such flows are discussed by Lindzen & Barker (1985).

5.4   Diffusive effects

The influence of viscosity upon the stability of stratified flows has received comparatively little attention. One might reasonably expect that the eigenfunctions ϕ satisfy an equation with left-hand side as in the Taylor–Goldstein equation (4.2) and right-hand side as in the Orr–Sommerfeld equation (3.1). This is broadly true, *except near the critical layer*, where such an equation remains singular though the Orr–Sommerfeld equation does not. This singularity persists because, in deriving (4.2), use is made of an equation for the density distribution $\rho(x, z, t)$ in the form $D\rho/Dt = 0$. This states that the density of particular fluid particles remains unchanged throughout their motion.

With

$$\rho(x, z, t) = \rho_0(z) + \epsilon \, \mathrm{Re} \{\hat{\rho}(z) \, \mathrm{e}^{\mathrm{i}k(x-ct)}\},$$

this gives

$$(U - c)\hat{\rho} = (\mathrm{d}\rho_0/\mathrm{d}z) \phi \tag{5.5}$$

and it is this approximation for $\hat{\rho}$ which yields the singular term of (4.2) containing N^2. In order to remove this singularity, it is necessary to incorporate diffusive effects. These arise from thermal conductivity, diffusivity of solutes or both. Such terms are significant within the critical-layer region, but are unimportant near isothermal (and iso-solutal) walls where the boundary condition $\hat{\rho} = 0$ is automatically satisfied by (5.5). For other boundary conditions, diffusive effects may be significant in thermal (solutal) boundary layers near the walls. The structure of the solutions in the vicinity of the critical layer was analysed by Baldwin & Roberts (1970) and is essentially the same as for compressible shear flows, studied long before by Lees & Lin (1946).

For the modes presently under discussion, the influence of viscosity and other diffusive agents is broadly similar to the unstratified case. In particular, the rapid change in Reynolds stress across the critical layer persists, still spread over a layer of thickness $O[(\alpha R)^{-\frac{1}{3}}]$ when the ratio of thermal diffusivity to kinematic viscosity is $O(1)$. The jump in τ is the same as that given by inviscid theory on indenting below the singularity z_c in the complex z-plane. However, we note that there are other, essentially diffusive, modes which can lead to instability of a rather different kind (see §6).

Following the discussion of §4.3, it might appear that the negative-energy wave components of over-reflection at a vortex sheet would be susceptible to resistive instability associated with viscous dissipation. However, this does not occur for these propagating modes since, as $z \to \pm \infty$, they always have positive energy in the *rest frame* of the fluid. Their amplitude is therefore diminished by viscous dissipation over most of the flow region. Accordingly, no such resistive instability precedes the Kelvin–Helmholtz instability of an unbounded vortex sheet with constant N^2. But the presence of a density discontinuity at the interface can support interfacial modes which *are* resistively unstable at velocities U below that for Kelvin–Helmholtz instability. This is also likely to be so for stable continuous density distributions such that $N^2(z) \to 0$ as $z \to \pm \infty$. There is then no energy propagation to $z = \pm \infty$ and the energy of bound modes, evanescent as $|z| \to \infty$, may be treated as in §2. Detailed calculations of such cases have yet to be carried out, and it may well turn out that this destabilizing mechanism is normally offset by the stabilizing rôle of the interfacial viscous boundary layers (cf. §2.4).

6   Models of mode coupling

6.1   *Model dispersion relations*

When several modes interact, the dispersion relations exhibit behaviour of seemingly bewildering complexity: the cases discussed above well illustrate this diversity. Nevertheless, the local characteristics of such dispersion relations are frequently well represented by the roots of simple algebraic equations. Examination of simple models provides an aid to better understanding of many exact dispersion relations.

We first consider the quadratic dispersion relationship

$$D(\omega, p) \equiv [\omega - \omega_1(p)][\omega - \omega_2(p)] = \epsilon\, e^{i\phi} \quad (\epsilon \geqslant 0, \ -\pi < \phi \leqslant \pi),$$
(6.1)

where p is a variable real parameter and ϵ, ϕ are real constants. For instance, the parameter p may represent wavenumber α at fixed R or Reynolds number R at fixed α. When ϵ is sufficiently small, good approximations to the two roots are

$$\omega(p) = \omega_j + \frac{\epsilon\, e^{i\phi}}{(\partial D/\partial \omega)_j} + O(\epsilon^2) \quad (j = 1, 2)$$
(6.2)

where $(\ )_j$ denotes evaluation at $\omega = \omega_j$. But this approximation breaks down at any values of p such that $|\omega_1 - \omega_2| \leqslant O(\epsilon^{\frac{1}{2}})$. For instance, the resistive instability (2.8) has ω_1 and ω_2 real, $\phi = -\frac{1}{2}\pi$ and ϵ proportional to viscosity. The $O(\epsilon)$ contribution (2.9) then yields the growth rate $\omega_i(p)$.

More generally, the complex roots $\omega(p) = \omega_r + i\omega_i$ of (6.1) follow continuous curves in three-dimensional (ω_r, ω_i, p) space. If $|\omega_1 - \omega_2| > O(\epsilon^{\frac{1}{2}})$ everywhere, roots 1 and 2 retain their identities as separate curves. Locally-strong interaction takes place where $|\omega_1 - \omega_2|$ is $O(\epsilon^{\frac{1}{2}})$ or less and this may or may not lead to interchange of mode identities. Suppose that $|\omega_1 - \omega_2|$ has a minimum at $p = p_m$ and that $|\omega_1 - \omega_2|^2 \gg \epsilon$ except near p_m. Then the modes interchange identities near p_m if and only if the two curves

$$\Omega(p) = (\omega_1 - \omega_2)^2, \quad \Delta(p) = (\omega_1 - \omega_2)^2 + 4\epsilon\, e^{i\phi}$$

traced out in the complex plane as p varies *enclose a region containing the origin*. Note that $\Delta(p)$ is the discriminant of the quadratic equation (6.1). There is bifurcation only if $\Delta = 0$ for some p: if $|\Delta|$ becomes very small locally, but does not vanish, there will be 'near bifurcations' where the curves almost meet.

Such local behaviour is displayed by the exact (computed) dispersion relations of many flows. Good examples are plane Poiseuille flow (Grosch & Salwen 1968), the Blasius boundary layer (Mack 1976; Antar & Benek 1978), Poiseuille pipe flow (Salwen & Grosch 1972; Salwen, Cotton & Grosch 1980) and stratified plane Couette flows (Davey & Reid 1977a, b; Gustavsson & Hultgren 1980).

Figures 2.10(a), (b) reproduce the results of Salwen, Cotton & Grosch (1980) for the first azimuthal harmonic with periodicity $\exp i[\theta + \alpha(x - ct)]$, when $\alpha = 1.0$ and R varies. At $R = 61$ there is an almost exact bifurcation of modes 2 and 3. At slightly lower values of α, this bifurcation is less perfect. Modes 2 and 4 also experience near-bifurcation around $R = 240$, with identity interchange. In contrast, modes 1 and 2 do not exchange identity: however, they do so at $\alpha = 1.0$ for the third and fourth azimuthal harmonics (Grosch & Salwen 1968).

In Davey & Reid's (1977a, b) results, precise bifurcations occur because of the symmetry of the problem, damped modes having $c_r = 0$ and propagating ones equal positive and negative values c_r. Some of the latter modes have c_r outwith the range of $U(z)$ and so may be identified as internal gravity-wave modes: these are absent in constant-density flow. Others, with $U_{\min} < c_r < U_{\max}$, exist also for constant density and have $c_r \to U_{\min}$ or $U_{\max}$ as $\alpha R \to \infty$. At small R, the eigenvalues αc_i match the decay rates of viscous or thermal (density) diffusion. Along any one continuous curve, the physical character of the mode may change from, say, a gravitational to a diffusive mode.

Figure 2.10. Real and imaginary parts c_r, c_i of the complex phase velocity c *vs.* R at $\alpha = 1.00$ for four modes of the first azimuthal harmonic $\propto \exp i(\theta + x - ct)$ in Poiseuille pipe flow (from Salwen, Cotton & Grosch 1980). Note the 'near-bifurcations' at $R \approx 61$ and 240. These are the four least damped modes as $R \rightarrow 0$; but at $R = 10^6$ they are 1st, 19th, 2nd and 15th respectively.

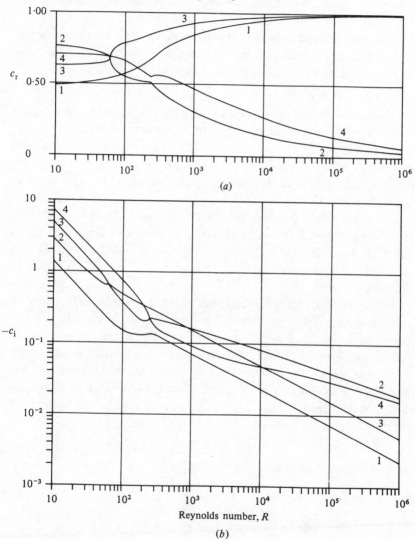

Eigenvalues often group into separate families (see §7.1). Also, the nature of the discrete spectrum may change with R. For boundary-layer flows, there are few discrete modes at low R, but these increase in number as R increases. Discrete modes of the Orr–Sommerfeld equation, as $R \to \infty$, need not correspond to discrete modes of the inviscid Rayleigh equation. Such matters are examined in the following section.

A model of thermal instability in horizontal layers of viscous fluid heated from below (see Drazin & Reid 1981 or Chandrasekhar 1961 for a full account) is given by the algebraic equation

$$(\omega + i\sigma_1)(\omega + i\sigma_2) = \delta$$

for real constants σ_1, σ_2 and δ. In the absence of buoyancy, σ_1 and σ_2 reduce to the decay rates of dynamical and thermal modes, respectively due to viscosity and thermal conductivity. The mode-coupling provided by buoyancy is denoted by δ. Unstable roots exist if and only if $\delta < -\sigma_1 \sigma_2$, a result equivalent to the critical Rayleigh-number criterion for onset of convection.

For 'doubly-diffusive' instability with, say, thermal and solutal diffusivities operating in addition to viscosity, the corresponding model has a cubic left-hand side, the additional factor $(\omega + i\sigma_3)$ being associated with decay of variations in solute concentration (cf. Turner 1973, Chapter 8; Huppert & Turner 1981).

Acheson (1980), among others, has noted that the introduction of a statically-stable density distribution may render a flow unstable by introducing new dynamical modes to the system. Most simply, suppose that a system supports only a decaying mode, $\omega = i\sigma_0$, associated with some dissipative process. The introduction of a statically-stable density distribution admits a new dynamical mode, which in the *absence* of dissipation would be unattenuated; say $\omega = \omega_0$ with ω_0 real. Weak coupling between the modes gives a dispersion relation of the form

$$(\omega - i\sigma_0)(\omega - \omega_0) = \epsilon\, e^{i\phi}$$

and the dynamical mode is approximately

$$\omega = \omega_0 + \epsilon\, e^{i\phi} \left(\frac{\omega_0 + i\sigma_0}{\omega_0^2 + \sigma_0^2} \right) + O(\epsilon^2).$$

If $\phi = \frac{1}{2}\pi$, this mode has a positive growth rate

$$\omega_i = \frac{\epsilon \omega_0}{\omega_0^2 + \sigma_0^2}$$

which is a maximum when $\omega_0 = \sigma_0$. This is a variant of the resistive instability discussed in §2.4. Similar magnetohydrodynamic instabilities occur, with the propagating mode sustained by the magnetic field rather than buoyancy.

Another example of destabilization by mode coupling is given by the cubic equation (5.3), which has the form

$$c(c^2 - c_0^2) = i\alpha$$

when $J/\alpha^2 > 2$, with approximate roots

$$c = -i\alpha c_0^{-2} + O(\alpha^2), \quad c = \pm c_0 + \tfrac{1}{2}i\alpha c_0^{-2} + O(\alpha^2)$$

when α is small. Here the propagating modes are destabilized and the stationary mode stabilized. For $J/\alpha^2 < 2$, the corresponding equation is

$$c(c^2 + \sigma^2) = i\alpha$$

for which the stationary mode $c \approx 0$ is destabilized when α is small.

6.2   *Mode conversion in inhomogeneous media*

Recently, Cairns & Lashmore-Davies (1983a, b) have proposed an interesting model for mode coupling in inhomogeneous media. Let $\omega_1(k, x)$, $\omega_2(k, x)$ be the approximate dispersion relations of two modes, when regarded as uncoupled. These vary slowly with x on account of the inhomogeneity. Suppose that there is a 'mode-crossing point' $x = x_0$ at frequency ω_0 where these modes coalesce. Near x_0, the (coupled) dispersion relation may be approximated by

$$(\omega - \omega_1)(\omega - \omega_2) = \delta. \tag{6.3}$$

Here, δ is a small number, taken as real, and $\omega_1(k_0, x_0) = \omega_2(k_0, x_0) \equiv \omega_0$ say, for some wavenumber k_0. The frequencies ω of the coupled modes are well approximated by ω_1 and ω_2 far from x_0.

A disturbance comprising both modes is regarded as having constant frequency ω_0 and two slowly-varying wavenumbers $k = k_1(x)$, $k_2(x)$ which coalesce into k_0 near $x = x_0$. As the waves pass through the linear resonance region, their interaction causes substantial changes in the amplitudes. Outside the resonance region near x_0, slower amplitude modulations take place owing to the inhomogeneity: the latter, but not the former, are described by Whitham's theory of slowly-varying waves (see §11.3).

At fixed frequency ω_0, let $k = k_0 + \delta$, $x = x_0 + \xi$ where δ, ξ are small in the resonance region. Expanding ω_1, ω_2 about (k_0, x_0) gives the leading-order approximation

$$\omega_1 = \omega_0 + a\delta + b\xi, \quad \omega_2 = \omega_0 + f\delta + g\xi$$

where a, b, f, g denote the appropriate partial derivatives of ω_1, ω_2 at (k_0, x_0).

The relation (6.3) then yields

$$(ak - ak_0 + b\xi)(fk - fk_0 + g\xi) = \delta_0 \tag{6.4}$$

where δ_0 is δ at (k_0, x_0).

The local dispersion relation (6.4) may be associated with a differential equation for the disturbance amplitude, with k identified with the operator $-\mathrm{i}\mathrm{d}/\mathrm{d}\xi$; but straightforward replacement of k in (6.4) to yield a second-order equation is both ambiguous and unjustified as the latter is incompatible with energy conservation. Instead, (6.4) must be converted to a pair of coupled differential equations for two wave amplitudes A_1 and A_2, namely

$$\left.\begin{aligned}
\mathrm{d}A_1/\mathrm{d}\xi - \mathrm{i}(k_0 - ba^{-1}\xi)A_1 &= \mathrm{i}\lambda A_2, \\
\mathrm{d}A_2/\mathrm{d}\xi - \mathrm{i}(k_0 - gf^{-1}\xi)A_2 &= \mathrm{i}\lambda A_1,
\end{aligned}\right\} \tag{6.5}$$

where $\lambda = (\delta_0/af)^{\frac{1}{2}}$. It is readily verified that this satisfies energy conservation,

$$\mathrm{d}(|A_1|^2 + |A_2|^2)/\mathrm{d}\xi = 0$$

when $af > 0$, the amplitudes being normalized so that the $|A_j|^2$ denote energy flux. When $af < 0$, the respective group velocities differ in sign and $\mathrm{i}\lambda$ is real in (6.5): in this case, the energy equation is

$$\mathrm{d}(|A_1|^2 - |A_2|^2)/\mathrm{d}\xi = 0.$$

On eliminating A_2 from (6.5) and making the substitutions

$$A_1(\xi) = \exp[\mathrm{i}k_0\xi - \tfrac{1}{4}\mathrm{i}(ba^{-1} - gf^{-1})\xi^2]\,\psi(\zeta),$$

$$\zeta = (gf^{-1} - ba^{-1})^{\frac{1}{2}}\exp(\tfrac{3}{4}\mathrm{i}\pi\xi),$$

one obtains Weber's equation

$$\mathrm{d}^2\psi/\mathrm{d}\zeta^2 + [\mathrm{i}\delta_0(ag - bf)^{-1} + \tfrac{1}{2} - \tfrac{1}{4}\zeta^2]\psi = 0$$

provided $gf^{-1} > ba^{-1}$. Its solution is a parabolic cylinder function $D_n(\zeta)$ which has known properties. Accordingly, the asymptotic solutions for A_1, and hence A_2, are known as $\xi \to -\infty$ and $\xi \to +\infty$.

These asymptotic solutions may be employed to find the changes in wave amplitude which take place on crossing the resonance region. A wave A_1 incident from $\xi < 0$ is partly transmitted into $\xi > 0$ and partly converted into wave A_2. The energy transmission coefficient is

$$T = \exp[-\pi\delta_0|ag - bf|^{-1}]$$

and that for the converted wave is $1 - T$. Conversion of A_2 to A_1 is of course similar. These remarkably simple results hold for waves with both like and differing signs in the energy equation, and they agree with detailed analyses of particular problems in plasma physics.

This method extends Whitham's theory of slowly-varing waves to encompass direct, linear resonance. Though rigorous justification is still lacking, this may reasonably be expected, perhaps *via* an averaged Lagrangian. Extension to include effects of damping and temporal variation seems feasible, but equations equivalent to (6.5) would not then conserve energy flux.

7 Eigenvalue spectra and localized disturbances
7.1 *The temporal eigenvalue spectrum*

A plane parallel flow $\mathbf{u}^0 = [U(z), 0, 0]$ of homogeneous fluid, supporting small (linearized) wavelike disturbances, has the form $\mathbf{u} = \mathbf{u}^0 + \mathbf{u}'$ where

$$\mathbf{u}' = u_j = \mathrm{Re}\,\{v_j(z)\exp[\mathrm{i}(\alpha x + \beta y) - \mathrm{i}\omega t]\}, \quad (j = 1, 2, 3).$$

The complex frequency, $\omega = \omega_r + \mathrm{i}\omega_i$, depends on the wavenumbers α, β, the Reynolds number R and any other flow parameters. For spatially-periodic wave-modes, α and β must be real.

The velocity components u_j of a given initial disturbance may be represented as a double Fourier integral

$$u_j(\mathbf{x}, 0) = \iint_{-\infty}^{\infty} U_j(\alpha, \beta; z)\exp[\mathrm{i}(\alpha x + \beta y)]\,\mathrm{d}\alpha\,\mathrm{d}\beta, \quad (j = 1, 2, 3).$$

The subsequent evolution of the disturbance with time t is determined by the decomposition of the $U_j(\alpha, \beta; z)$ into a set of eigenfunctions, each of which has its own exponential time-variation, $\exp(-\mathrm{i}\omega t)$. For a given flow and given real wavenumber components (α, β), there may be many eigenvalues ω for which the flow disturbance satisfies prescribed homogeneous boundary conditions. The possible complex values of ω comprise the temporal eigenvalue spectrum.

On restricting attention to the transverse velocity component $u_3 = w$, one obtains the linearized equation

$$\left(\frac{\partial}{\partial t} + U\frac{\partial}{\partial x}\right)\nabla^2 w - U''\frac{\partial w}{\partial x} = R^{-1}\nabla^4 w. \tag{7.1}$$

When $w(x, y, z, t)$ consists of a single mode $w \propto \mathrm{Re}\,\{\phi(z)\exp[\mathrm{i}\alpha(x - ct)]\}$ with $\omega \equiv \alpha c$, this reduces to the Orr–Sommerfeld equation (3.1) for $\phi(z)$. On applying to (7.1) a double Fourier transform in x and y,

$$\tilde{f}(\alpha, \beta; z, t) = \frac{1}{2\pi}\iint_{-\infty}^{\infty}\exp[-\mathrm{i}(\alpha x + \beta y)]\,f(x, y, z, t)\,\mathrm{d}x\,\mathrm{d}y$$

and a Laplace transform in time,

$$\bar{g}(s; z) = \int_0^{\infty} \mathrm{e}^{-st}g(z, t)\,\mathrm{d}t,$$

the resultant transformed equation is (cf. Case 1961; Gustavsson 1979)

$$(U-c)(\Phi''-k^2\Phi)-U''\Phi-(i\alpha R)^{-1}(\Phi^{iv}-2k^2\Phi''+k^4\Phi) = (i\alpha)^{-1}(\nabla^2 w)_{t=0},$$

$$(7.2)$$

$$(\Phi = \bar{\bar{w}}, \quad c = -s/i\alpha, \quad k^2 = \alpha^2+\beta^2).$$

The left-hand side is just the Orr–Sommerfeld operator for oblique wave-modes and the right-hand side is determined by the initial disturbance. With boundary conditions $w = \partial w/\partial z = 0$ on plane boundaries $z = 0$, $z = H$, the corresponding conditions on Φ are $\Phi = \Phi' = 0$ on $z = 0$ and H. The homogeneous equation (7.2), with zero right-hand side, clearly has four linearly-independent solutions in $0 \leqslant z \leqslant H$; from these, the solution of the inhomogeneous equation and given boundary conditions for Φ may be constructed by the method of variation of parameters. Inversion of the Laplace transform yields

$$\tilde{w}(\alpha, \beta; z, t) = (2\pi i)^{-1} \int_{p-i\infty}^{p+i\infty} e^{st}\,\Phi\,\mathrm{d}s$$

where the constant p is chosen to give a path of integration to the right of all singularities of Φ in the complex s-plane. This integral may be evaluated by the method of residues on choosing a contour Γ which encloses all singularities. A line $\mathrm{Re}\{p\} = $ constant and a semi-circular arc at infinity would suffice provided there is no branch cut associated with a branch point of Φ. The poles of Φ then yield a discrete spectrum of eigenvalues. For viscous flows in finite domains $0 \leqslant z \leqslant H$ ($H = 1$ without loss of generality), the number of discrete eigenvalues is infinite and there is no continuous spectrum. The associated eigenfunctions then form a complete set (Lin 1961; DiPrima & Habetler 1969). On the other hand, inviscid flows ($R = \infty$) with finite H have a continuous spectrum (Dikii 1960a, b; Case 1961) and the relationship between the inviscid spectrum and the viscous spectrum as $R \to \infty$ is a matter of some subtlety.

For unbounded viscous flows, such as jets and shear layers, and for boundary-layer flows with just one wall, the discrete spectrum of the Orr–Sommerfeld operator contains only a finite number of modes. In such cases, the associated eigenfunctions cannot comprise a complete set and there must be a continuous spectrum. Gustavsson (1979) has identified the contribution of this continuous spectrum to the contour integral around Γ as due to the appearance of a branch point of Φ, the contour being deformed to pass around the branch cut. The continuous spectrum may also be found directly from the Orr–Sommerfeld operator by relaxing the boundary condition $\Phi \to 0$ as $z \to \infty$ to require only boundedness as $z \to \infty$.

The following simple example, given by Grosch & Salwen (1978), dispels any air of mystery surrounding the appearance of the continuous spectrum for $H = \infty$. The wave equation

$$u_{tt} = u_{xx}$$

has solutions of the form $u = f(x)\, e^{i\omega t}$ where

$$f_{xx} + \omega^2 f = 0.$$

With boundary conditions $u(0, t) = u(1, t) = 0$, there is an infinite set of discrete eigenvalues $\omega_n = n\pi$ with corresponding orthonormal eigenfunctions $f_n = 2^{-\frac{1}{2}} \sin(n\pi x)$ $(n = 1, 2, 3, \ldots)$. But for the infinite domain $0 \leqslant x < \infty$ with boundary conditions

$$u(0, t) = 0, \quad u(x, t) \to 0 \quad \text{as} \quad x \to \infty,$$

there is a finite number of discrete eigenvalues: in fact, none! With the relaxed boundary condition

$$u(x, t) \quad \text{bounded as} \quad x \to \infty,$$

the spectrum is continuous, with 'improper' eigenfunctions

$$f(x; \omega) = (2\pi)^{-\frac{1}{2}} \sin \omega x$$

for all real non-negative values of ω. In the finite case, an arbitrary disturbance may be represented by a Fourier sine series, summed over the discrete spectrum, while in the infinite case a Fourier integral over the continuous spectrum is required.

In the free stream, where $U(z) \to U_1(\text{constant})$ as $z \to \infty$, the Orr–Sommerfeld equation (3.1) reduces to

$$(D^2 - \lambda^2)(D^2 - k^2)\,\Phi = 0, \quad \lambda^2 \equiv i\alpha R(U_1 - c) + k^2, \quad D \equiv d/dz,$$

with four independent solutions

$$\phi_j \approx \exp(\lambda_j z)\,(j = 1, 2, 3, 4), \quad \lambda_1 = -\lambda, \quad \lambda_2 = \lambda, \quad \lambda_3 = -\alpha, \quad \lambda_4 = \alpha.$$

Here, ϕ_1, ϕ_2 are 'viscous' solutions and ϕ_3, ϕ_4 are 'inviscid' solutions. On choosing λ_1 and λ_2 to have negative and positive real parts respectively, it is seen that only ϕ_1 and ϕ_3 decay to zero as $z \to \infty$. The discrete eigenvalues c are those for which a linear combination of ϕ_1 and ϕ_3 can be found which satisfies the two boundary conditions at the wall $z = 0$.

Additionally, there are solutions which remain bounded as $z \to \infty$. These arise when λ^2 is real and negative; that is, when

$$c_r = U_1, \quad c_i = -(k^2 + l^2)/\alpha R$$

where l is an arbitrary real and non-zero number. There are then three permissible solutions $\phi_j \approx \exp(\lambda_j z)$, with $\lambda_1 = -il$, $\lambda_2 = +il$, $\lambda_3 = -\alpha$, bounded as $z \to \infty$. Clearly, it is always possible to find a linear combination of these which satisfies the two boundary conditions at $z = 0$. These solutions make up the continuous spectrum of the temporal stability problem. They all travel with the free-stream velocity U_1 and have damping rates no smaller than $k^2/\alpha R$. Their form is

$$\Phi \approx \exp\left[i(\alpha x + \beta y - \alpha ct)\right]\{e^{-ilz} + B\,e^{ilz} + C\,e^{-\alpha z}\}$$

as $z \to \infty$. The first two terms represent waves propagating towards and away from the wall respectively (though without energy flux in the z-direction); the third is a 'wall mode' which decays exponentially with z. Grosch & Salwen (1978) find 'reflection coefficients' $|B|$ which are less than unity for a Blasius boundary-layer flow, but which may be as great

Figure 2.11. Real and imaginary parts c_r, c_i of the complex phase velocity c *vs.* R at $\alpha = 0.179$ for the first 15 modes of Blasius flow. The continuous spectrum lies along $c_r = 1$ with $c_i < -\alpha/R$ (from Mack 1976).

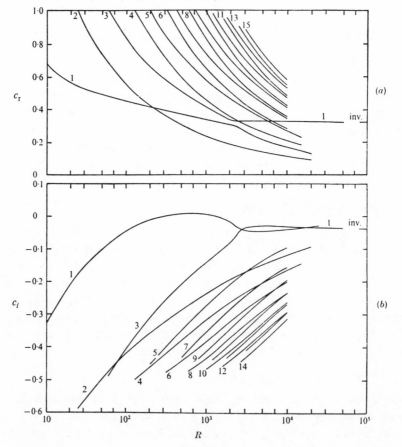

Figure 2.12. Eigenvalues of Blasius flow in the c_r–c_i plane at $\alpha = 0.179$ and four Reynolds numbers R: (a) 1000, (b) 2000, (c) 5000, (d) 10000 (from Mack 1976).

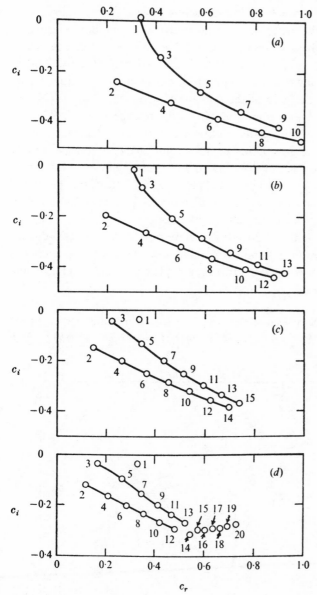

as $O(10^3)$ for a jet profile. The explanation of such large values of $|B|$ is unknown.

Jordinson (1971), Mack (1976), Grosch & Salwen (1978) and Antar & Benek (1978) have examined, numerically, the temporal eigenvalue spectrum of the Blasius boundary layer. The results of Antar & Benek nicely demonstrate how part of the discrete spectrum for a finite domain becomes increasingly dense and close to the continuous spectrum for an infinite domain as the size of the domain is increased. Mack (1976) has shown that there are just seven modes in the discrete spectrum for $R = 580$, $\alpha = 0.179$, $\beta = 0$; of these, only the first is unstable. As R increases, more modes emerge from the continuous spectrum: at the same wavenumber and $R = 5000$, there are at least fifteen and probably more, all damped. Such modes are shown in Figures 2.11 (a), (b) and 2.12 (a)–(d); apart from mode 1, they fall on two distinct curves in the c_r–c_i plane at each fixed R. The continuous spectrum lies along the line $c_r = 1$, $c_i < -\alpha/R$. Salwen & Grosch (1981) established that, for temporal evolution of two-dimensional disturbances in unbounded parallel flow, the discrete eigenmodes and continuous eigenfunctions form a complete set: this generalizes the

Figure 2.13. Distribution of eigenvalues of antisymmetric disturbances for plane Poiseuille flow at $\alpha = 1.0$, $R = 10^4$. ○ A-family, ▽ P-family, □ S-family (from Mack 1976).

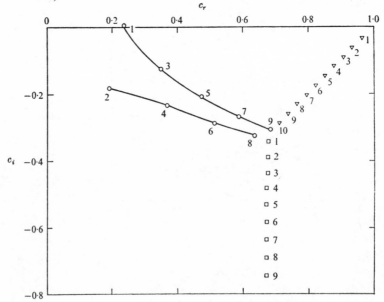

work of DiPrima & Habetler (1969), who proved the completeness of the discrete eigenfunctions in bounded flow. The nature of the continuous spectrum for non-parallel unbounded flows was briefly considered by Bouthier (1983).

Grosch & Salwen (1968), Orszag (1971) and Mack (1976) computed the temporal eigenvalue spectrum of plane Poiseuille flow. The eigenvalues are discrete and infinite in number, divided into three families: see Figure 2.13. The A-family is finite in number and qualitatively resembles the discrete spectrum of Blasius flow; the P- and S-families resemble results found by Antar & Benek (1978) for Blasius flow in *finite* domains, which correspond to the continuous spectrum of the unbounded case.

So far, we have considered only the transverse velocity component w, governed by (7.1). For two-dimensional disturbances independent of y, this is sufficient; but there also exist y-dependent modes for which w is identically zero. To see this, we need only consider the equation for the transverse vorticity component, $\omega_z = \partial v / \partial x - \partial u / \partial y$. This is

$$\left(\frac{\partial}{\partial t} + U \frac{\partial}{\partial x}\right) \omega_z - R^{-1} \nabla^2 \omega_z = -U' \frac{\partial w}{\partial y}, \qquad (7.3)$$

with w given by (7.1). A y-dependent w therefore acts as a source of vertical vorticity, by tilting the spanwise vortex lines associated with the primary flow $U(z)$. But the homogeneous equation for ω_z, with zero right-hand side, has its own eigenvalue spectrum, which must be taken into account in a complete description of three-dimensional disturbances. The spectrum is just that of the equation

$$D^2 \psi - i\alpha R(U - c')\psi = 0, \quad c' \equiv c + ik^2/\alpha R \qquad (7.4)$$

where c is the complex phase speed of wavelike disturbances of the form $\omega_z \propto \psi(z) \exp[i(\alpha x + \beta y - \alpha c t)]$. The appropriate boundary conditions are $\psi = 0$ on $z = 0, H$, or ψ finite as $|z| \to \infty$ in the unbounded case.

This eigenvalue problem is identical with that for two-dimensional temperature (or density) modes in plane stratified flows, examined by Davey & Reid (1977a, b) and already discussed in §6 above. Gustavsson & Hultgren (1980) and Gustavsson (1981) consider the interesting possibility of linear resonance, when an eigenvalue c of (7.2) coincides with one of (7.4). They find such resonant cases for plane Couette and plane Poiseuille flow and it is likely that they exist for most flows. The solution of (7.3) then contains a term in $t \exp(-i\alpha c t)$. Although all modes have $c_i < 0$ for plane Couette flow, this term grows to a maximum amplitude before decaying. The largest amplitudes are apparently attained for modes with $|\beta/\alpha| \gg 1$, which are highly-elongated in the streamwise direction.

The eigenvalue problem (7.4) has been considered analytically by Murdock & Stewartson (1977) for the discontinuous velocity profile $U = 0$ $(0 \leqslant z \leqslant 1)$, $U = 1$ $(1 < z < \infty)$. The structure of the spectrum is believed to resemble, qualitatively, that of the Orr–Sommerfeld equation. Murdock & Stewartson point out that the continuous temporal spectrum may be reinterpreted as a *discrete* spectrum of modes of heat conduction type, with z replaced, in $z > 1$, by the similarity variable $zR^{\frac{1}{2}} t^{-\frac{1}{2}}$.

For inviscid flows between plane parallel walls at $z = 0$ and $z = H$, exponentially-growing modes may occur only if $U'' = 0$ somewhere in $0 \leqslant z \leqslant H$. However, even when $U'' \neq 0$ everywhere, there exist inviscid disturbances which *grow algebraically* with time t (Shnol' 1974; Landahl 1980). These are essentially three-dimensional, being associated with longitudinal stretching and tilting of vortex lines. The total kinetic energy of an initially-localized disturbance may indeed increase at least as fast as linearly with time t, largely because the streamwise extent of the disturbance grows in proportion to t. The presence of viscosity would lead to the eventual decay of such linear disturbances (Gustavsson & Hultgren 1980; Hultgren & Gustavsson 1981). Corresponding two-dimensional disturbances always *decay* algebraically in time, in stratified as well as homogeneous flows (see, most recently, Brown & Stewartson 1980; also Blumen 1971 for a discussion of three-dimensional disturbances). All such disturbances may of course be represented in terms of the eigenfunctions of the complete three-dimensional temporal eigenvalue spectrum.

7.2   The spatial eigenvalue spectrum

As an alternative to specifying initial disturbances at $t = 0$, we now consider disturbances which are known, for all times t, at a particular location, say on $x = 0$. Such situations frequently arise in experimental configurations, where a known disturbance may be continuously driven by a wave-maker or vibrating ribbon. The spatial evolution of the disturbance, with x, is then required. For disturbances of fixed (real) frequency ω and (real) spanwise wavenumber β, it is necessary to find the complex eigenvalues of the downstream wavenumber, $\alpha(\beta, \omega, R) = \alpha_r + i\alpha_i$. The spectrum of α for the Orr–Sommerfeld equation has been investigated for Blasius flow by Jordinson (1971), Corner, Houston & Ross (1976), Grosch & Salwen (1978) and Salwen & Grosch (1981). As well as a finite discrete spectrum, there is a continuous spectrum with four branches. Two of these branches represent waves travelling upstream into the region $x < 0$; the other two describe downstream propagation into $x > 0$. All modes of the continuous spectrum decay exponentially with distance away from the source at $x = 0$.

Murdock & Stewartson (1977) also studied the structure of the spatial eigenvalue spectrum of equation (7.1) with real frequency $\omega = \alpha c$. They found that the continuous spectrum may be re-expressed as an infinite but discrete spectrum of modes of diffraction type.

For bounded viscous flows, there is an infinite discrete spectrum. The least stable spatial mode of plane Poiseuille flow has been calculated by Itoh (1974a, b) and agrees satisfactorily with the experiments of Nishioka *et al.* (1975).

An exact, but complicated, linear solution with specified localized periodic forcing is given by Jones & Morgan (1972): they consider an inviscid compressible fluid with Helmholtz vortex-sheet velocity profile subjected to acoustic radiation from a fixed, pulsating, line source. Such are the analytical complexities, even for this simple idealized flow, that modelling of more realistic flows demands approximation, via numerical and asymptotic methods. Mack & Kendall (1983) recently made such a study of the motion downstream of oscillatory point and line sources of disturbance in the Blasius boundary layer. Their theoretical results agree well with their own experiments and also those of Gilev, Kachanov & Kozlov (1982).

7.3 *Evolution of localized disturbances*

As discussed above, the evolution of a prescribed initial disturbance may be described by Fourier–Laplace transforms. In practice, a complete solution is rarely available in simple form; but good approximations may be obtained on making certain assumptions.

For each given real wavenumber pair (α, β), it is reasonable to retain only the contribution from the least-damped (most unstable) eigenvalue $\omega(\alpha, \beta)$, since this must eventually dominate the higher eigenmodes – and usually does so after just a few wave periods. Also, the contribution from the continuous spectrum may be neglected, when there exist unstable (or less-heavily damped) discrete modes for some range of (α, β).

In such cases, the disturbance is well-represented by a double Fourier integral over α and β,

$$u_j(\mathbf{x}, t) \approx \int\int_{-\infty}^{\infty} \hat{U}_j(\alpha, \beta; z) \exp\left[\mathrm{i}(\alpha x + \beta y - \omega t)\right] \mathrm{d}\alpha \, \mathrm{d}\beta, \qquad (7.5)$$

where $\omega = \alpha c(\alpha, \beta)$ is the complex dispersion relation for the least-damped discrete mode at each (α, β) and $\hat{U}_j$ is the contribution of this mode to u_j, determined by the initial conditions.

It often happens that the most unstable mode is two-dimensional, with wavenumber components $(\alpha, \beta) = (\alpha_0, 0)$ say. Then, a satisfactory approximation to $\omega(\alpha, \beta)$ in the vicinity of $(\alpha_0, 0)$ at fixed Reynolds number R is given by a truncated power series in $(\alpha - \alpha_0)$ and β^2; viz. $\omega = \omega_r + i\omega_i$ where

$$\left.\begin{aligned} \omega_r &\approx \omega_{r0} + \left(\frac{\partial \omega_r}{\partial \alpha}\right)_0 (\alpha - \alpha_0) + \frac{1}{2}\left(\frac{\partial^2 \omega_r}{\partial \alpha^2}\right)_0 (\alpha - \alpha_0)^2 + \frac{1}{2}\left(\frac{\partial^2 \omega_r}{\partial \beta^2}\right)_0 \beta^2, \\ \omega_i &\approx \omega_{i0} + \frac{1}{2}\left(\frac{\partial^2 \omega_i}{\partial \alpha^2}\right)_0 (\alpha - \alpha_0)^2 + \frac{1}{2}\left(\frac{\partial^2 \omega_i}{\partial \beta^2}\right)_0 \beta^2. \end{aligned}\right\} \quad (7.6)$$

Here, ω_{i0} is the maximum temporal growth rate and both $(\partial^2 \omega_i / \partial \alpha^2)_0$ and $(\partial^2 \omega_i / \partial \beta^2)_0$ are negative. Also $(\partial \omega_r / \partial \alpha)_0$ is the group velocity of the most unstable mode.

Employing such a representation, both Benjamin (1961) and Criminale & Kovasznay (1962) derived the asymptotic behaviour of the integral (7.5). For disturbances initially localized near the origin $x = y = 0$ (i.e. when the phases of $\hat{U}_j(\alpha, \beta; z)$ are strongly correlated) they found solutions of the form

$$u_j(\mathbf{x}, t) \approx \hat{U}_j(\alpha_0, 0; z)\, t^{-1} \exp(\omega_{i0} t + i\alpha_0 x - i\omega_{r0} t) \exp[-\mathcal{A}(X^2/t) - \mathcal{B}(y^2/t)] \quad (7.7)$$

$$X \equiv x - (\partial \omega_r / \partial \alpha)_0 t, \quad t \to \infty$$

where $\mathcal{A}$ and $\mathcal{B}$ are complex constants with $\mathcal{A}_r, \mathcal{B}_r > 0$. Contours of constant amplitude at each time t and on each plane of constant z are therefore ellipses. This result is apparently supported by Benjamin's experiment of an unstable film of water on an inclined plane, which shows a nearly circular wave-envelope. The centre of disturbance $X = y = 0$ propagates downstream with the group velocity $(\partial \omega_r / \partial \alpha)_0$ and no wave amplification occurs outside the ellipse $\mathcal{A}_r X^2 + \mathcal{B}_r y^2 = \omega_{i0} t^2$. Also, the wave amplitude within the packet satisfies a nearly Gaussian distribution, with characteristic widths $(t/\mathcal{A}_r)^{\frac{1}{2}}$, $(t/\mathcal{B}_r)^{\frac{1}{2}}$ in the respective X and y directions. Accordingly, the linear dimensions of the unstable wave-packet vary as t, but most of the disturbance is concentrated within a distance of order $O(t^{\frac{1}{2}})$ from the centre. Since X/t and y/t are $O(\omega_{i0}^{\frac{1}{2}})$ within the unstable region, the visible wave crests deviate only slightly from a plane two-dimensional wave-train whenever ω_{i0} is small.

Gaster (1968, 1979) and Gaster & Davey (1968) observed that a more appropriate expansion procedure than (7.6) is that about the saddle point

(α^*, β^*) of the integral (7.5), where α^* and β^* are complex quantities defined by

$$\frac{\partial \omega}{\partial \alpha} = \frac{x}{t}, \quad \frac{\partial \omega}{\partial \beta} = \frac{y}{t}.$$

The resultant solution differs from the above elliptical patch, but coincides with it near the centre of disturbance. Gaster's (1968) results incorporated an incorrect assumption which led to a spurious caustic, but this is corrected in Gaster (1979).

Gaster (1975) and Gaster & Grant (1975) undertook a detailed computational and experimental study of localized disturbances in the Blasius boundary layer. In their computations, the continuous Fourier representation (7.3) was replaced by a superposition of many discrete wave-modes. Their well-known results (see Figure 2.14) show rather crescent-shaped wave-packets with curved wave crests, which are very different from the elliptical patch predicted by Benjamin and Criminale & Kovasznay. For two-dimensional disturbances only, but including the effect of boundary-layer growth, Gaster's (1982a, b) critical comparison of approximate methods shows that direct computation and the saddle-point method are in good agreement.

An analytical approach by Craik (1981, 1982a) employed a model dispersion relation which coincides with (7.6) close to ($\alpha_0, 0$) but better represents the directional properties of three-dimensional wave-modes. Simple solutions for various limiting cases were obtained by the saddle-point method. These represent curved wave-packets and successfully reconcile (7.7) with Gaster's computed results. For much of the parameter range, the wave-envelopes and wave crests are curved concavely in the upstream or downstream direction respectively, according as the group velocity ($\partial \omega_r / \partial \alpha$)$_0$ is greater or less than the phase velocity ω_{r0}/α_0. But there are also solutions with roughly circular wave-envelopes, which typically occur when the group velocity and phase velocity are nearly equal: the latter situation may correspond to Benjamin's experiment.

A rather different method, based on ideas of ray theory, was used by Itoh (1980a) to study two-dimensional wave-packets in plane Poiseuille flow and is readily extended to three-dimensional packets (N. Itoh, private communication). His method correctly yields the neutral wave-envelope – beyond which no unstable waves are observed – but may not accurately describe growing wave-modes. Craik (1982a) obtains good agreement with Itoh's curved wave-envelope and also draws attention to cases of 'splitting' wave-packets with two maxima, one on either side of the centre line. The latter may arise when the most unstable waves are oblique modes with non-zero β, as often occurs at Reynolds numbers substantially above the critical one for onset of instability. Oblique modes are also prominent in

Figure 2.14. Comparison of experimental and computational results for a wave-packet 76 cm downstream of the excitation point (from Gaster 1975). Flow is from top right to bottom left.

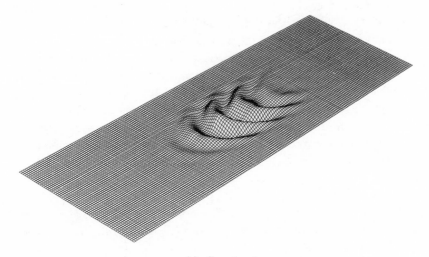

(*a*)   Experiment

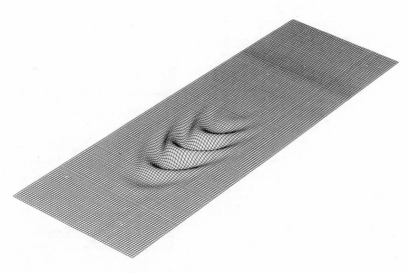

(*b*)   Theory

Spooner & Criminale's (1982) computational study of localized disturbances
in an Ekman boundary layer.

Yet another approach employs multiple scales, where the complex wave
amplitude $A(x, y, t)$ of a nearly-monochromatic wave-train with
wavenumbers $(\alpha_0, 0)$ is assumed to vary only slowly with respect to x, y
and t. One may then formally replace ω, α and β in the dispersion relation
$\omega = \omega(\alpha, \beta)$ by differential operators

$$\omega + i\,\partial/\partial t, \quad \alpha - i\,\partial/\partial x, \quad \beta - i\,\partial/\partial y$$

respectively, acting on the amplitude $A(x, y, t)$ (Newell & Whitehead 1969;
Hocking, Stewartson & Stuart 1972). The co-ordinate scaling

$$\tau = \epsilon t, \quad \xi = \epsilon^{\frac{1}{2}}(x - a_1 t), \quad \eta = \epsilon^{\frac{1}{2}} y,$$

where $a_1 = (\partial \omega/\partial \alpha)_0$ and ϵ is a small parameter which characterizes the
amplitude modulation, leads to a linear Schrödinger equation

$$\frac{\partial A}{\partial \tau} - \tfrac{1}{2}i\left(\frac{\partial^2 \omega}{\partial \alpha^2}\right)_0 \frac{\partial^2 A}{\partial \xi^2} - \tfrac{1}{2}i\left(\frac{\partial^2 \omega}{\partial \beta^2}\right)_0 \frac{\partial^2 A}{\partial \eta^2} = hA, \quad h \equiv \epsilon^{-1}\omega_{i0}, \tag{7.8}$$

at leading order in ϵ. Here the amplification rate ω_{i0} is assumed to be $O(\epsilon)$.
Among the solutions of this equation is the elliptical packet (7.7). Different
co-ordinate scalings are required to recover differential equations with
curved wave-packets as solutions (Craik 1981). Though the saddle-point
seems simplest and most natural for linear wave-packets, a multiple-scales
approach is obligatory for weakly *nonlinear* problems (see Chapter 6).

When the instability arises from the coalescence of neutral inviscid
modes, with resultant complex-conjugate frequencies, the form of localized
wave-packets is rather different from that just described. The partial
differential equation governing an almost two-dimensional wave-packet is
then no longer (7.8), but a linear *Klein–Gordon* or *telegraph equation*,

$$\partial^2 A/\partial t^2 - a\,\partial^2 A/\partial x^2 - b\,\partial^2 A/\partial y^2 - dA = 0, \tag{7.9}$$

in an appropriately-chosen reference frame (see Weissman 1979 and §18).
The approximate linear dispersion relation in this frame is just

$$(\omega - \omega_0)^2 = a(k - k_0)^2 + bl^2 - d,$$

where ω is frequency and $\mathbf{k} = (k, l)$ is the wavenumber vector. The term
d is small and denotes a small departure from the critical conditions $\omega = \omega_0$,
$(k, l) = (k_0, 0)$ for onset of instability. In the unstable case, a, b and d may
be normalized to unity without loss.

Then, $\omega = \omega_0 + \omega'$, $(k, l) = (k_0, 0) + (k', l)$ where $\omega'^2 = k'^2 + l^2 - 1$ and a
wave-packet evolves as

$$\text{Re}\left\{ e^{i(k_0 x - \omega_0 t)} \iint_D f(k', l)\, e^{i(k'x + ly - \omega' t)}\, dk'\, dl \right\}$$

where D denotes the unstable portion of wavenumber space, $k'^2 + l^2 \leq 1$.

The saddle-point method may be applied much as above to determine the asymptotic form of solutions as $t \to \infty$. Note, also, that the general solution of the telegraph equation is known (Copson 1975, p. 86). Weissman (1979) gave an asymptotic approximation independent of y and with $|x| \ll t$. A more general asymptotic representation is given by Craik & Adam (1978), who found that the saddle-point method fails at $x/t = \pm 1$, where the saddle-points go off to infinity. Even in the absence of instability, the large-time linear evolution of arbitrary localized disturbances may be dominated by the contribution from Fourier modes near points of direct (linear) resonance (Akylas & Benney 1982).

This chapter ends with a cautionary note on the use of discontinuous velocity profiles, when arbitrary initial disturbances are permitted. The eigenvalues (2.4) of Helmholtz instability imply that growth rates approach infinity as $k \to \infty$: Craik (1983) has observed that in such cases linear theory predicts that some, though not all, initial disturbances attain infinite amplitudes after a finite time. Moreover, some bounded initial disturbances exhibit this singularity at *all* times, however small, after initiation! Of course, the discontinuous profile ceases to be a realistic approximation for waves with length comparable to the shear-layer thickness: such waves cannot have indefinitely-large growth rates.

Chapter three

INTRODUCTION TO NONLINEAR THEORY

8 Introduction to nonlinear theory

8.1 *Introductory remarks*

Nonlinear theories are of three more or less distinct kinds. In one, properties of arbitrarily-large disturbances are deduced directly from the full Navier–Stokes equations. Consideration of integral inequalities yields bounds on flow quantities, such as the energy of disturbances, which give stability criteria in the form of necessary or sufficient conditions for growth or decay with time. An admirable account of such theories is given by Joseph (1976). They have the advantage of supplying mathematically rigorous results while incorporating very few assumptions regarding the size or nature of the disturbances. Sometimes, these criteria correspond quite closely to observed stability boundaries. The bounds for onset of thermal (Bénard) instability and centrifugal (Rayleigh–Taylor) instability in concentric rotating cylinders are particularly notable successes. Often, however, the bounds are rather weak: this is especially so for shear-flow instabilities, where local details of the flow typically play an important rôle which cannot be (or, at least, has not been) incorporated into the global theory.

The second class of theories relies on the idea that linearized equations provide a satisfactory first approximation for those finite-amplitude disturbances which are, in some sense, *sufficiently small*. Successive approximations may then be developed by expansion in ascending powers of a characteristic dimensionless wave amplitude. These are known as *weakly nonlinear* theories, and they have proved successful in revealing many important physical processes. Some of these theories have been developed with full regard to mathematical rigour, yielding firm results for particular limiting cases. Others rely on more heuristic methods in which questions of convergence of amplitude expansions, or the validity of their truncation at some chosen order, are left unresolved. The justification for adopting an unrigorous approach must rest on its success in modelling situations of real scientific interest which are not yet amenable to exact mathematical representation. In practice, many important theoretical ideas, eventually formulated with full rigour, have grown from heuristic models; and many apparently 'irrational' theories have stimulated valuable new experiments and provided new insights into old ones.

The third class may be called *numerical simulations*, which attempt to follow the development of some initial disturbance by direct computation. Recently, considerable successes have been achieved in this area and further rapid advances may confidently be predicted, despite the complexity and sensitivity of the numerical procedures. Since the present work has mode interactions as its theme, it is primarily concerned with theories of the second and, to lesser extent, the third class, which have been most successful in describing the evolution and interaction of wavelike disturbances.

8.2   *Description of a general disturbance*

For definiteness, we first consider finite-amplitude disturbances to a parallel shear flow $U(z)$ of constant-density fluid, between plane rigid boundaries at $z = 0, H$. The exact equations of motion are the Navier–Stokes equations $(1.1)'$. Since the boundary conditions are here linear and homogeneous, all nonlinearities are contained in the convective term $(\mathbf{u} \cdot \nabla)\mathbf{u}$. For other flow configurations, particularly those with fluid interfaces or free surfaces, nonlinearities may also appear in the boundary conditions. For instance, irrotational, inviscid water waves exactly satisfy Laplace's equation within the fluid, but the free-surface boundary conditions are nonlinear.

When the disturbance is independent of the spanwise co-ordinate y, the flow may be represented by

$$\mathbf{u}(x, z, t) = [U(z) + \partial\Psi/\partial z, 0, -\partial\Psi/\partial x]$$

where $\Psi(x, z, t)$ is a disturbance stream function. The nonlinear equation for Ψ, deduced from the Navier–Stokes equations, is (cf. Eckhaus 1965; Itoh 1977a)

$$(L - M\,\partial/\partial t)\,\Psi = \tfrac{1}{2}\hat{N}[\Psi, \Psi] \qquad (8.1)$$

where the operators L, M and $\hat{N}$ are defined as

$$L \equiv R^{-1}M^2 - (UM - U'')\partial/\partial x, \quad M \equiv \partial^2/\partial x^2 + \partial^2/\partial z^2,$$

$$\hat{N}[\Psi, \Phi] \equiv \left(\frac{\partial\Psi}{\partial z}\frac{\partial}{\partial x} - \frac{\partial\Psi}{\partial x}\frac{\partial}{\partial z}\right)M\Phi + \left(\frac{\partial\Phi}{\partial z}\frac{\partial}{\partial x} - \frac{\partial\Phi}{\partial x}\frac{\partial}{\partial z}\right)M\Psi$$

and R is the Reynolds number. The boundary conditions are $\partial\Psi/\partial x = \partial\Psi/\partial z = 0$ on $z = 0, H$. A treatment of three-dimensional disturbances would follow similar lines, but is omitted for simplicity.

The temporal eigenvalue spectrum of the linear operator $L - M\,\partial/\partial t$, which is just that of (7.1), was discussed in §7.1. For each Fourier component $\psi(\alpha; z, t)\exp(i\alpha x)$ of Ψ at fixed R and finite H, there exists a complete set of discrete eigenfunctions $\psi^{(n)}(\alpha, z)$ and eigenvalues $\omega_n(\alpha)$ ($n = 1, 2, \ldots, \infty$). Accordingly, any y-independent disturbance of finite amplitude may be expressed formally as an eigenfunction expansion

$$\Psi(x, z, t) = \int_{\alpha=-\infty}^{\infty} \sum_{n=1}^{\infty} A^{(n)}(\alpha, t)\,\psi^{(n)}(\alpha, z)\,e^{i\alpha x}\,d\alpha. \qquad (8.2)$$

For $H = \infty$, the corresponding representation must include the contribution of the continuous spectrum.

Adjoint linear eigenfunctions $\phi^{(n)}(\alpha, z)$ are found from

$(L' - M\,\partial/\partial t)\,\Phi = 0,$

$L' \equiv R^{-1}M^2 - (MU - U'')\partial/\partial x, \quad \partial\Phi/\partial x = \partial\Phi/\partial z = 0 \quad (z = 0, H).$

Each $\phi^{(n)}(\alpha, z)$ satisfies the adjoint Orr–Sommerfeld equation

$[(U - \omega_n/\alpha)(D^2 - \alpha^2) + 2U'D - (i\alpha R)^{-1}(D^4 - 2\alpha^2 D^2 + \alpha^4)]\,\phi^{(n)} = 0$

and boundary conditions

$$\phi^{(n)} = D\phi^{(n)} = 0 \quad (z = 0, H)$$

with $D \equiv d/dz$ and the same eigenvalues $\omega_n(\alpha)$ as above. The functions $\psi^{(n)}(z)$, $\phi^{(n)}(z)$ may be orthonormalized so that

$$\int_0^H \phi^{(m)}(D^2 - \alpha^2)\,\psi^{(n)}\,dz = \delta_{mn}$$

for each wavenumber α (cf. Eckhaus 1965, Chapter 6), together with any convenient normalization of the functions $\psi^{(n)}$.

At this stage, it is convenient to confine attention to flows which are periodic in x, thereby replacing the Fourier integral (8.2) by a Fourier series. Solutions of (8.1) which are strictly periodic in x at some instant t remain so at all times. Such solutions may, but need not, possess a single dominant Fourier mode. In general, all Fourier components of the form $\exp(ik\alpha x)$ are present, as

$$\Psi(x, z, t) = \sum_{k=-\infty}^{\infty} \Psi_k(z, t)\,e^{ik\alpha x} \quad (k = 0, \pm 1, \pm 2, \ldots) \qquad (8.3)$$

where α is a fixed wavenumber. Also, each component $\Psi_k(z, t)$ may be decomposed into a sum of linear eigenfunctions,

$$\Psi_k(z, t) = \sum_{n=1}^{\infty} A_k^{(n)}(t)\,\psi_k^{(n)}(z), \qquad (8.4)$$

where $\psi_k^{(n)}(z)$ is the nth Orr–Sommerfeld eigenfunction with wavenumber $k\alpha$. To ensure a real physical disturbance, it is necessary that $A_{-k}^{(m)} = [A_k^{(m)}]^*$ where * denotes complex conjugation.

Separation of the Fourier components of (8.1) gives

$$(L_k - M_k \, \partial/\partial t) \, \Psi_k = \frac{1}{2} \sum_{l=-\infty}^{\infty} N[\Psi_{k-l}, \Psi_l], \quad (k, l = 0, \pm 1, \pm 2, \dots)$$

(8.5)

where the operators L_k, M_k and N are obtained from L, M and $\hat{N}$ on replacing $\partial\Psi_q/\partial x$ by $(i\alpha q) \Psi_q$.

On substituting (8.4) in (8.5), multiplying by the corresponding adjoint eigenfunction $\phi_k^{(n)}(z)$ and integrating over z from 0 to H, an infinite set of coupled nonlinear equations is found for the unknown amplitude functions $A_k^{(n)}(t)$, namely

$$(d/dt + i\omega_k^{(m)}) A_k^{(m)} = -\frac{1}{2} \sum_{l=-\infty}^{\infty} \sum_{p,q=1}^{\infty} \sigma_{kl}^{(m, p, q)} A_{k-l}^{(p)} A_l^{(q)}$$

$$\left(\begin{matrix} k = 0, 1, 2, \dots \\ m, p, q = 1, 2, 3, \dots \end{matrix} \right), \quad (8.6a)$$

$$A_{-k}^{(m)} = (A_k^{(m)})^*, \quad \sigma_{kl}^{(m, p, q)} = \int_0^H \phi_k^{(m)} N[\psi_{k-l}^{(p)}, \psi_l^{(q)}] \, dz, \quad (8.6b)$$

where $\omega_k^{(m)} = \omega_m(k\alpha)$ is the mth eigenvalue at wavenumber $k\alpha$ and $i\omega_{-k}^{(m)} = [i\omega_k^{(m)}]^*$. Also, since $\psi_{-m}^{(r)} = \psi_m^{(r)*}$ and $\phi_{-m}^{(r)} = \phi_m^{(r)*}$, it follows that

$$\sigma_{-k-l}^{(m, p, q)} = [\sigma_{kl}^{(m, p, q)}]^*.$$

The nonlinear interaction between eigenmodes is expressed *exactly* on the right-hand side of (8.6a). The interaction of any two modes with wavenumbers $(k-l)\alpha$ and $l\alpha$ influences a third mode with wavenumber $k\alpha$. The strength of each interaction is determined by the interaction coefficients $\sigma_{kl}^{(m, p, q)}$ defined in (8.6b). Results similar to (8.5), (8.6a) and (8.6b) may be derived for a continuous wave spectrum, with summation over k and l replaced by integrations (cf. Ikeda 1977). Unfortunately, but predictably, little progress can be made in solving these equations without the introduction of further assumptions.

For other physical configurations, the theory may be similarly constructed. Ripa (1981) has formulated the problem of general nonlinear wave-interactions for two geophysical systems, inviscid barotropic Rossby waves and internal gravity waves without primary flow. The general interaction equations for a continuous spectrum of inviscid surface gravity waves are comprehensively discussed by Yuen & Lake (1982) and West (1981). For these, the eigenmodes are just solutions of Laplace's equation and the free-surface boundary conditions supply the nonlinearity. In this case, amplitude equations equivalent to (8.6a) may be deduced without integrating over the flow domain (see Chapters 6 and 7).

8.3 Review of special cases

(i) All modes damped

The simplest – and least interesting – case is that where *all* eigen-modes are damped. This is certainly so if each initial mode amplitude $A_j^{(n)}(0)$ is sufficiently small that

$$\sum_{l=-\infty}^{\infty} \sum_{p,q=1}^{\infty} |\sigma_{kl}^{(m,p,q)}| \, |A_{k-l}^{(p)}| \, |A_l^{(q)}| + \text{Im}\{-\omega_k^{(m)}\} |A_k^{(m)}| < 0 \quad (t = 0),$$

for every k and m. Then, every wave component decays to zero as $t \to \infty$, tending asymptotically to the linear approximation

$$A_k^{(m)}(t) = C_k^{(m)} \exp(-i\omega_k^{(m)} t), \quad C_k^{(m)} \text{ constant.}$$

(ii) Single dominant neutral mode

If all free linear modes, except one, are damped and the exceptional mode is maintained at a constant amplitude a with frequency Ω, either naturally or driven by suitable applied forces, there exists a non-zero time-periodic equilibrium state containing all harmonics. On choosing the forced mode to have $k = \pm 1$, $m = 1$ and setting $A_1^{(1)} = \frac{1}{2}a \exp(-i\Omega t)$, $A_{-1}^{(1)} = \frac{1}{2}a \exp(i\Omega t)$ where a is *small* and real, each $A_k^{(m)}$ may be ordered as

$$\left. \begin{aligned} |A_{\pm 1}^{(1)}| = O(a); \quad |A_0^{(m)}|, \ |A_{\pm 2}^{(m)}| &\leqslant O(a^2); \\ |A_{\pm 1}^{(s)}| \leqslant O(a^3) \quad (s > 1); \quad |A_{\pm r}^{(m)}| &\leqslant O(a^{|r|}) \quad (|r| > 2); \end{aligned} \right\} \tag{8.7}$$

provided each $|\sigma_{kl}^{(m,p,q)}(\omega_k^{(m)} - k\Omega)^{-1}|$ has magnitude of order $O(1)$ or less relative to a. Successive approximations may then be constructed, as ascending power series in a. The first few terms are

$$\Psi = \text{Re}\Big\{ a\psi_1^{(1)}(z) \exp[i(\alpha x - \Omega t)] + \tfrac{1}{4}a^2 \sum_{m=1}^{\infty} (i\sigma_{01}^{(m,1,1)}/\omega_0^{(m)}) \psi_0^{(m)}(z)$$

$$+ \tfrac{1}{4}a^2 \sum_{m=1}^{\infty} i\sigma_{21}^{(m,1,1)}(\omega_2^{(m)} - 2\Omega)^{-1} \psi_2^{(m)}(z) \exp[2i(\alpha x - \Omega t)] + O(a^3) \Big\}.$$

The first term represents the fundamental forced mode, the second group of terms an $O(a^2)$ mean flow induced by the self-interaction of the forced mode, and the third group the various components of a second harmonic. The higher harmonics are absent at $O(a^2)$, but the procedure may be continued to any desired order. The series may be expected to converge for sufficiently small values of a, but an appreciation of the actual radius of convergence requires detailed knowledge of all interaction coefficients.

If the higher eigenstates $\psi_k^{(m)}(z)$ $(m > 1)$ are heavily damped compared with the first eigenstate $\psi_k^{(1)}(z)$ at each k, satisfactory approximations may sometimes be obtained on retaining only the modes with $m = 1$. This will be so whenever

$$\left| \frac{\sigma_{kl}^{(m, p, q)} \psi_k^{(m)}(z)}{\sigma_{kl}^{(1, p, q)} \psi_k^{(1)}(z)} \right| \ll \left| \frac{\omega_k^{(m)} - k\Omega}{\omega_k^{(1)} - k\Omega} \right| \quad (\text{all } m > 1)$$

at the value of z considered.

In the same way, a single dominant mode which is *damped* according to linear theory as $\psi_1 = a\psi_1^{(1)}(z)\, e^{-\sigma t} \exp[i(\alpha x - \Omega t)]$, where $\Omega = \mathrm{Re}\,(\omega_1^{(1)})$ and $\sigma = \mathrm{Im}\,(-\omega_1^{(1)})$, will drive mean-flow and second-harmonic components proportional to $a^2\, e^{-2\sigma t}$ and $a^2\, e^{-2\sigma t} \exp[2i(\alpha x - \Omega t)]$ respectively. If only the $\psi_1^{(1)}$ mode is present initially, the requirement that there is no mean flow or second harmonic at $t = 0$ necessitates the presence, for all $t \geqslant 0$, of $O(a^2)$ free modes of the form $\psi_0^{(m)} \exp(-i\omega_0^{(m)} t)$, $\psi_2^{(m)} \exp[i(2\alpha x - \omega_2^{(m)} t)]$. Only if the respective damping rates of these modes exceed 2σ do the forced components dominate them after a sufficient time has elapsed.

When the fundamental $O(a)$ wave is amplified according to linear theory $(\sigma < 0)$ and $|\sigma|$ is $O(1)$ relative to a, the convergence of the amplitude expansion is certain to break down at sufficiently large times: this is clear since successive terms behave as $O(a^n\, e^{n|\sigma| t})$.

(iii) The Landau equation

If the dominant mode is not maintained at constant amplitude and its linear growth or damping rate σ is very small, of order a^2, the analysis requires modification to account for changes in amplitude. Then, $O(a^3)$ terms on the right-hand side of (8.6a), for $k = 1$, are of the same order as the linear term $\sigma A_1^{(1)}$. The linear approximation for the temporal evolution is inadequate and the appropriate evolution equation for the wave amplitude is found to have the form (see §18)

$$dB_1/dt + \sigma B_1 = \lambda\,|B_1|^2\,B_1 + O(a^5), \quad A_1^{(1)}(t) = B_1(t)\exp(i\Omega t).$$
$$(8.8)$$

This is generally known as the *Landau equation* (Landau 1944) and $\lambda = \lambda_r + i\lambda_i$ as the *Landau constant*, a complex number which is determined by the linear eigenfunction $\psi_1^{(1)}$ and its adjoint $\phi_1^{(1)}$. The cubic nonlinearity arises through interaction of the fundamental $O(a)$ mode with $O(a^2)$ mean-flow and second-harmonic components which it drives. This nonlinear term is stabilizing or destabilizing according as $\lambda_r < 0$ or $\lambda_r > 0$.

Provided higher-order terms may be ignored, a linearly-damped mode

($\sigma > 0$) will grow, if λ_r is positive, whenever its amplitude $|A_1^{(1)}|$ exceeds $(\sigma\lambda_r^{-1})^{\frac{1}{2}}$. Similarly, a linearly unstable mode ($\sigma < 0$) decays if $\lambda_r < 0$ and $|A_1^{(1)}|$ exceeds $(\sigma\lambda_r^{-1})^{\frac{1}{2}}$. In both cases, an equilibrium 'threshold' amplitude $|A_1^{(1)}| = (\sigma\lambda_r^{-1})^{\frac{1}{2}}$ separates growing and decaying solutions. When σ and λ_r have opposing signs, the nonlinear term reinforces the linear growth or decay rate and there is no equilibrium solution of the truncated equation.

As a flow parameter, typically the Reynolds number R, varies, so do the constants σ and λ. At Reynolds numbers very near the critical value R_c for onset of linear instability, we may write

$$\sigma \approx (d\sigma/dR)_c\,(R-R_c)+O[(R-R_c)^2], \quad \lambda_r = \lambda_r^{(c)}+O[(R-R_c)].$$

At $R = R_c$, there is a bifurcation of equilibrium solutions of (8.6). The trivial solution $|A_1^{(1)}|^2 = 0$ and the non-trivial solution

$$|A_1^{(1)}|^2 = (\lambda_r^{(c)})^{-1}\,(d\sigma/dR)_c\,(R-R_c)+O[(R-R_c)^2] \tag{8.9}$$

intersect at $R = R_c$. The latter solution curve lies in the region $R \geqslant R_c$ if $(d\sigma/dR)_c/\lambda_r^{(c)}$ is positive and in $R \leqslant R_c$ if it is negative. Since σ is normally negative (i.e. linear wave growth) for $R > R_c$ and positive (linear wave decay) for $R < R_c$, the bifurcating solution (8.9) is said to be *supercritical* when $\lambda_r^{(c)} < 0$ and *subcritical* when $\lambda_r^{(c)} > 0$. These cases are shown schematically in figures 3.1(a), (b), where the regions of growth and decay predicted by (8.8) are indicated by arrows (cf. Stuart 1963; Drazin & Reid 1981).

In the supercritical case, all waves decay when $R < R_c$ but initially-small wave amplitudes $A_1^{(1)}$ grow to the equilibrium amplitude of (8.9) when $R > R_c$. In the subcritical case, *all* amplitudes grow when $R > R_c$, and only those with amplitudes less than (8.9) decay to zero when $R \leqslant R_c$. Of course,

Figure 3.1. Schematic representation of bifurcating equilibrium solutions near R_c of Landau equation (8.8): (a) supercritical case $\lambda_r^{(c)} < 0$, (b) subcritical case $\lambda_r^{(c)} > 0$. Solid lines denote stable solutions and dashed lines unstable ones; arrows indicate regions of growth or decay.

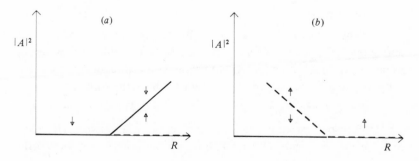

equation (8.8) is only a local approximation, formally justifiable in the limit $R \to R_c$.

Waves cannot grow to indefinitely large amplitudes. There may be equilibrium solutions at larger $| A_1^{(1)} |$ beyond the range of the approximation (8.8). More likely, other wave-modes attain magnitudes comparable with $A_1^{(1)}$ and a more complicated quasi-equilibrium state may be reached. When this state contains a *continuum* of strongly interacting modes in both wavenumber and frequency space, it is one of turbulence.

After Landau (1944), the nonlinear theory of hydrodynamic stability of shear flows was advanced by Stuart (1958, 1960) and Watson (1960). Stuart (1960) derived (8.8) together with an expression for λ by amplitude-expansion techniques. A subsequent rigorous derivation of the bifurcating periodic solutions also leads to result (8.9) (Joseph & Carmi 1969). More recently, higher-order amplitude expansions have been carried out in the quest for greater accuracy and wider range of validity (see §22).

(iv) Modulation in space and time

For nearly uniform wave-trains with amplitudes which vary slowly with both x and t, Stewartson & Stuart (1971) developed a rational nonlinear theory which is formally valid as $R \to R_c$. The complex amplitude $B_1(x, t) = \epsilon A$ then satisfies a nonlinear Schrödinger equation

$$\frac{\partial A}{\partial \tau} - a_2 \frac{\partial^2 A}{\partial \xi^2} = hA + \lambda | A |^2 A \tag{8.10}$$

(see Zakharov 1968; Taniuti & Washimi 1968; Watanabe 1969; DiPrima, Eckhaus & Segel 1971 for earlier related work). The scaled space and time variables ξ, τ are $\xi = \epsilon(x - c_g t)$, $\tau = \epsilon^2 t$ and the small parameter ϵ is chosen proportional to $| R - R_c |^{\frac{1}{2}}$; the $O(1)$ scaled growth rate h is just $-\epsilon^{-2}\sigma$ and $c_g = \partial\Omega/\partial\alpha$ is the dimensionless group velocity. An account of this theory and various extensions is given in Chapter 6. The case $\lambda = 0$ reduces to the equation for a linearized wave-packet noted in §7.3.

(v) Three- and four-wave resonance

When several dominant wave-modes are present, their mutual interaction is significant. This is especially so when some of these modes resonate. The simplest and most important case is three-wave resonance.

Suppose that three dominant linear modes have the form

$$\text{Re}\{a_j(t) \exp[i(\mathbf{k}_j \cdot \mathbf{x} - \text{Re}\,\omega_j t)]\} \quad (j = 1, 2, 3)$$

with small respective amplitudes a_j. Interaction of any two such modes, say $j = p$ and q, yields $O(a^2)$ quadratic terms – of (8.1) or any other nonlinear equation or boundary condition – with periodicities

$$\exp\{i[\pm(\mathbf{k}_p\cdot\mathbf{x}-\mathrm{Re}\,\omega_p\,t)\pm(\mathbf{k}_q\cdot\mathbf{x}-\mathrm{Re}\,\omega_q\,t)]\}.$$

Three-wave resonance occurs when any of these terms has the same $(\mathbf{x},t)$-periodicity as that of the third wave-mode: i.e. when

$$\mathbf{k}_1\pm\mathbf{k}_2\pm\mathbf{k}_3 = 0, \quad \mathrm{Re}\,\{\omega_1\pm\omega_2\pm\omega_3\} = 0 \tag{8.11a, b}$$

with corresponding signs being chosen.

A satisfactory $O(a^2)$ approximation to (8.6a) when $\mathbf{k}_1+\mathbf{k}_2 = \mathbf{k}_3$ and $\mathrm{Re}\,\{\omega_1+\omega_2\} = \mathrm{Re}\,\{\omega_3\}$ is

$$\left.\begin{array}{l}
da_1/dt+\sigma_1\,a_1 = \lambda_1\,a_2^*\,a_3+O(a^3) \\[4pt]
da_2/dt+\sigma_2\,a_2 = \lambda_2\,a_1^*\,a_3+O(a^3) \\[4pt]
da_3/dt+\sigma_3\,a_3 = \lambda_3\,a_1\,a_2+O(a^3)
\end{array}\right\} \tag{8.12}$$

where

$$\lambda_1 = -\tfrac{1}{2}(\sigma_{1-2}+\sigma_{13}), \quad \lambda_2 = -\tfrac{1}{2}(\sigma_{2-1}+\sigma_{23}), \quad \lambda_3 = -\tfrac{1}{2}(\sigma_{31}+\sigma_{32}).$$

These equations for three-wave resonance, and their extension to wave amplitudes with spatial as well as temporal variation, have many interesting properties, discussed in Chapter 5. Note that whenever the linear growth rates $-\sigma_j$ are $O(a)$ or less, the linearized approximation is unacceptable.

Not all systems exhibit three-wave resonance. Ironically, one of the first searches for such resonance (Phillips 1960, 1961), among inviscid surface gravity waves, yielded negative results and Phillips bravely continued his analysis to third order in a, to determine the cubic interaction coefficients of resonant quartets with

$$\mathbf{k}_1+\mathbf{k}_2-\mathbf{k}_3-\mathbf{k}_4 = 0, \quad \omega_1+\omega_2-\omega_3-\omega_4 = 0. \tag{8.13}$$

The resultant equations then have the form

$$\left.\begin{array}{l}
da_1/dt = ia_1\displaystyle\sum_{k=1}^{4} g_{1k}\,|a_k|^2+i\mathscr{K}\omega_1\,a_2^*\,a_3\,a_4+O(a^4), \\[10pt]
da_2/dt = ia_2\displaystyle\sum_{k=1}^{4} g_{2k}\,|a_k|^2+i\mathscr{K}\omega_2\,a_1^*\,a_3\,a_4+O(a^4), \\[10pt]
da_3/dt = ia_3\displaystyle\sum_{k=1}^{4} g_{3k}\,|a_k|^2+i\mathscr{K}\omega_3\,a_1\,a_2\,a_4^*+O(a^4), \\[10pt]
da_4/dt = ia_4\displaystyle\sum_{k=1}^{4} g_{4k}\,|a_k|^2+i\mathscr{K}\omega_4\,a_1\,a_2\,a_3^*+O(a^4),
\end{array}\right\} \tag{8.14}$$

where g_{rs} $(r,s = 1,2,3,4)$ and $\mathscr{K}$ are known real constants and the (real) linear wave freqencies ω_j equal $(g\,|\,\mathbf{k}_j\,|)^{\frac{1}{2}}$ where g is gravitational acceleration. The properties of this system of coupled equations, and similar systems which arise in other contexts, are discussed in Chapter 7.

When the amplitudes of resonant wave-modes vary slowly in space $\mathbf{x}$ as well as in time, the derivatives d/dt must be replaced by partial-differential operators $\partial/\partial t + \mathbf{c}_g^{(j)} \cdot \nabla$ where $\mathbf{c}_g^{(j)}$ denotes the (possibly complex) group velocity of each mode and ∇ is the gradient operator in the propagation space of one, two or three dimensions.

Three important subclasses of three- and four-wave interactions deserve special attention. These are (a) degenerate classes of three-wave resonance such as $\mathbf{k}_1 = \mathbf{k}_2 = \frac{1}{2}\mathbf{k}_3$, $\omega_1 = \omega_2 = \frac{1}{2}\omega_3$; (b) four-wave interactions among 'sidebands', $\mathbf{k}_1 = \mathbf{k}_2 = \mathbf{k}_3 - \boldsymbol{\delta} = \mathbf{k}_4 + \boldsymbol{\delta}$ where $\boldsymbol{\delta}$ is small; (c) three- and four-wave interactions between wave-modes of very different length scales, (e.g. $\mathbf{k}_2 = \mathbf{k}_1 - \boldsymbol{\delta}$ and $\mathbf{k}_3 = \boldsymbol{\delta}$ where $|\boldsymbol{\delta}| \ll |\mathbf{k}_1|$) or very different frequencies (e.g. $\omega_2 = \omega_1 - \Delta$, $\omega_3 = \Delta$ with $|\Delta| \ll |\omega_1|$). Such cases are discussed below, mainly in Chapters 5 and 7.

(vi) Wave-driven mean flows

Small-amplitude wave-trains usually have the capacity to induce mean flows and to modify pre-existing ones by their $O(a^2)$ self-interaction. Not only do such flows contribute at $O(a^3)$ to the evolution of the wave: they are frequently very important in their own right, as agents for the transport of mass and momentum. Such flows are subject to both inviscid and viscous effects and the outcome often depends rather sensitively on their relative magnitudes. For flows with a free surface, nonlinear boundary conditions are important as well as nonlinearities within the fluid. Amplitude modulations in space also induce mean flows. In the presence of a pre-existing primary flow or a free surface, wave-driven $O(a^2)$ mean flows make a significant contribution to the total energy of the disturbance. Such considerations are among those addressed in the following chapter.

Chapter four

WAVES AND MEAN FLOWS　　◦

9　Spatially-periodic waves in channel flows

9.1　The mean-flow equations

Incompressible flows comprising a primary unidirectional shear flow and a small-amplitude two-dimensional wave-train have the dimensionless form

$$\mathbf{u} = (u, v, w) = [U(z)+\partial\psi/\partial z, 0, -\partial\psi/\partial x],$$

$$\psi(x, z, t) = \epsilon \, \mathrm{Re}\,\{\phi(z)\exp[\mathrm{i}\alpha(x-ct)]\}+\psi_2(x, z, t). \qquad (9.1)$$

Here, ψ is the disturbance stream function, ϵ is a small real constant characteristic of the wave amplitude and ψ_2 denotes contributions to ψ which are $O(\epsilon^2)$ or smaller. The flow is confined between rigid walls at $z = 0$ and $z = H$, where H may be normalized to unity except for boundary-layer flows. For the latter, $H = \infty$ and the normalized boundary-layer thickness may be taken as unity.

The complex wave velocity $c = c_r+\mathrm{i}c_i$ is an eigenvalue and $\phi(z)$ is the corresponding eigenfunction of the dimensionless Orr–Sommerfeld equation (3.1b).

As indicated in the preceding chapter, the only $O(\epsilon^2)$ contributions to ψ_2 are a mean-flow component independent of x and a second-harmonic proportional to $\exp(\pm 2\mathrm{i}\alpha x)$. The equations for the $O(\epsilon^2)$ mean flow $[u^{(2)}, 0, w^{(2)}]$ and pressure $p^{(2)}$ are found from (1.1)' to be

$$\left(\frac{\partial}{\partial t} - R^{-1}\frac{\partial^2}{\partial z^2}\right) u^{(2)} = -\frac{\partial p^{(2)}}{\partial x}-\tfrac{1}{4}\epsilon^2\,\mathrm{e}^{2\alpha c_i t}\,\frac{\mathrm{d}}{\mathrm{d}z}\,[\mathrm{i}\alpha(\phi'\phi^*-\phi\phi'^*)],$$

$$\left(\frac{\partial}{\partial t} - R^{-1}\frac{\partial^2}{\partial z^2}\right) w^{(2)} = -\frac{\partial p^{(2)}}{\partial z}, \qquad\qquad (9.2\mathrm{a, b, c})$$

$$u^{(2)} = \overline{\partial\psi_2/\partial z}, \quad w^{(2)} = -\overline{\partial\psi_2/\partial x},$$

75

the overbar denoting an x-average. By continuity, the boundary conditions $w^{(2)} = 0$ at $z = 0, H$ require $w^{(2)}$ and so $\partial p^{(2)}/\partial z$ to be zero everywhere. Accordingly, $\partial p^{(2)}/\partial x$ must be a function of t only, since $u^{(2)}$ is independent of x. This pressure gradient may be externally and arbitrarily imposed. Two choices are commonly made: either $\partial p^{(2)}/\partial x = 0$, giving a constant imposed pressure gradient independent of ϵ, or the amplitude-dependent pressure gradient yielding constant mass flux at each instant t. The latter choice requires that

$$\int_0^H u^{(2)}(z, t)\,dz = 0. \tag{9.3}$$

The nonlinear term on the right of (9.2a) is just $-\partial(\overline{uw})/\partial z$, the $O(\epsilon^2)$ mean Reynolds stress gradient. For inviscid flows, and in regions where viscosity may be neglected when αR is large, this term becomes

$$-\frac{\partial(\overline{uw})}{\partial z} = \epsilon^2\,e^{2\alpha c_i t}\,\frac{\alpha c_i\,U''\,|\phi|^2}{2\,|U-c|^2} \tag{9.4}$$

where $\phi(z)$ satisfies Rayleigh's equation (3.3). As $c_i \to 0_+$, expression (9.4) approaches zero except at any critical levels $z = z_c$ where $U(z_c) = c$.

For neutral waves, the inviscid solution $\phi'(z)$ normally has a logarithmic singularity at such critical levels (see e.g. Drazin & Reid 1981) and the Reynolds stress $-\overline{uw}$ there has a discontinuity of magnitude $\frac{1}{2}\pi\alpha(U''_c/|\,U'_c\,|)|\phi_c|^2$, where the subscript 'c' denotes evaluation at z_c. A study of the asymptotic structure, near z_c, of solutions of the Orr–Sommerfeld equation reveals the same jump in Reynolds stress, but spread by viscosity over a layer of thickness $O[(\alpha R)^{-\frac{1}{3}}]$. The inviscid approximation for the Reynolds stress is valid, outside this thin layer, whenever $|c_i| \leqslant O[(\alpha R)^{-\frac{1}{3}}]$ and ϵ is sufficiently small to permit linearization. Also, it is valid *everywhere* along the real z-axis, except close to the walls, for amplified waves with $c_i \gg O[(\alpha R)^{-\frac{1}{3}}]$. In contrast, for damped waves, there is a finite interval around z_c within which the inviscid estimate is invalid (Lin 1955; Drazin & Reid 1981).

Near the walls, the linear eigenfunction $\phi(z)$ has a boundary-layer structure, with viscous effects confined to regions of thickness $O[(\alpha R)^{-\frac{1}{2}}]$. These viscous boundary layers remain distinct from any $O[(\alpha R)^{-\frac{1}{3}}]$ critical layer centred on z_c unless c_r is close to the flow velocity U at either wall. We assume in the following that this is so: when not, a different asymptotic theory as $\alpha R \to \infty$ or a direct numerical approach is required.

Though methods for numerical solution of the Orr–Sommerfeld equation are well established, the use of asymptotic approximations is justified on two grounds: they yield firm results in limiting cases, where numerical solutions are usually least reliable, and help provide valuable physical insight into the processes of instability. It is true, however, that asymptotic theories now play a less central rôle than previously. Very full discussions of the asymptotic solutions of the Orr–Sommerfeld equation are given by Reid (1965) and Drazin & Reid (1981).

9.2 *Particular solutions*

If the mean pressure gradient is held constant, $p^{(2)} = 0$ and solutions of (9.2a) which vary as $\exp(2\alpha c_i t)$ are

$$u^{(2)} = \tfrac{1}{4}\epsilon^2 \, \mathrm{e}^{2\alpha c_i t} \, R\mu^{-1}\left\{ \int_0^z \sinh\left[\mu(z-z_1)\right](\mathrm{d}f/\mathrm{d}z_1)\,\mathrm{d}z_1 \right.$$

$$\left. + A\cosh\mu z + B\sinh\mu z\right\} \quad (9.5)$$

where

$$\mu \equiv (2\alpha c_i R)^{\frac{1}{2}}, \quad f(z) \equiv \mathrm{i}\alpha(\phi'\phi^* - \phi\phi'^*).$$

The solution which satisfies $u^{(2)} = 0$ on $z = 0$ and $z = H$ has $A = 0$ and

$$B = -\operatorname{cosech}\mu H \int_0^H \sinh\left[\mu(H-z_1)\right](\mathrm{d}f/\mathrm{d}z_1)\,\mathrm{d}z_1. \quad (9.6)$$

For $\alpha c_i R \gg 1$, (9.5), (9.6) and Rayleigh's equation (3.3) together yield the inviscid approximation

$$u^{(2)} = \tfrac{1}{4}\epsilon^2 \, \mathrm{e}^{2\alpha c_i t} \frac{U''|\phi|^2}{|U-c|^2}, \quad (9.7)$$

a result most easily obtained from (9.2a) by ignoring the term in $R^{-1}\,\partial^2 u^{(2)}/\partial z^2$. As $c_i \to 0_+$, this solution develops a singularity at the critical layer z_c where $U(z_c) = c_r$ unless U'' or ϕ vanishes there. Result (9.5) shows how viscosity modifies this singular solution, the singularity as $c_i \to 0_+$ being replaced by a narrow jet-like profile of width $O[(\alpha c_i R)^{-\frac{1}{2}}]$. Note that the thickness of this jet greatly exceeds that of the oscillatory viscous critical layer for $\phi(z)$ when $O[(\alpha R)^{-1}] \ll c_i < O[(\alpha R)^{-\frac{1}{3}}]$.

When $\alpha c_i R \leqslant O(1)$, the inviscid expression (9.7) for the mean flow $u^{(2)}$ is *never* a good approximation, even though the linear eigenfunction $\phi(z)$ may be well-represented by inviscid theory over most of the flow domain. For the mean flow, viscous diffusion operates on a time-scale which is $O(R)$ whereas the nonlinear forcing terms evolve with time-scale $O[(\alpha c_i)^{-1}]$: the former must greatly exceed the latter for the inviscid approximation (9.7) to hold.

If $\alpha c_i R \ll 1$ but $\phi(z)$ remains well-approximated by the inviscid linear solution, $u^{(2)}$ approaches the 'quasi-steady' approximation

$$u^{(2)} = \begin{cases} [K_1 + K_2(z - z_c)] \, e^{2\alpha c_i t} & (z < z_c) \\ [K_1 + \bar{K}_2(z - z_c)] \, e^{2\alpha c_i t} & (z > z_c), \end{cases} \tag{9.8}$$

$$\bar{K}_2 = K_2 + \tfrac{1}{2}\epsilon^2 \pi \alpha R(U_c''/|U_c'|)|\phi_c|^2,$$

on either side of any critical layer. However, the constants K_1 and K_2 cannot immediately be evaluated by applying the no-slip boundary conditions at $z = 0$ and $z = H$, because the inviscid approximation to $\phi(z)$ is invalid in the viscous wall layers of thickness $O[(\alpha R)^{-\frac{1}{2}}]$. The effect of these boundary layers is considered below.

If one imposes a constant-mass-flux condition (9.3) in place of $p^{(2)} = 0$, there is normally an additional $O(\epsilon^2)$ pressure gradient P which induces a further contribution to $u^{(2)}$. The inviscid approximation (9.7) is an exception, for it already satisfies (9.3). To the quasi-steady approximation (9.8) a term in $\tfrac{1}{2}RPz(z - H)$ must be added, with appropriate P. For boundary-layer flows ($H = \infty$), there can be no such pressure gradient and (9.3) cannot normally be satisfied for x-periodic wave-trains. Furthermore, the quasi-steady solution can become established throughout the infinite flow domain only when $K_1 = \bar{K}_2 = 0$.

Since the complete solution $u^{(2)}(z, t)$ must correspond to some given initial state $u^{(2)}(z, 0)$, the above particular solutions which vary as $\exp(2\alpha c_i t)$ must be supplemented by a complementary function comprising the free modes of the diffusion equation. When H is finite, these are

$$\sum_{n=1}^{\infty} C_n \, e^{-\sigma_n t} \sin[(R\sigma_n)^{\frac{1}{2}} z] \tag{9.9}$$

with eigenvalues $\sigma_n = R^{-1}(\pi n/H)^2$ determined by the no-slip boundary conditions $u^{(2)}(0) = u^{(2)}(H) = 0$. The constants C_n may be found in terms of the initial distribution $u^{(2)}(z, 0)$ and particular solution (9.5). Since all these modes decay with time, (9.5) eventually dominates when the waves are neutrally-stable or amplified. But this is not necessarily so when the waves decay with time. Solution (9.5) then dominates, as $t \to \infty$, only if the slowest damping rate $\sigma_1 = R^{-1}(\pi/H)^2$ of the free modes exceeds $2\alpha|c_i|$. The corresponding diffusive solution for boundary-layer flows with $H = \infty$ is readily constructed.

9.3   *The viscous wall layer*

In a thin oscillating boundary-layer of thickness $O[(\alpha R)^{-\frac{1}{2}}]$ adjacent to each wall, the Orr–Sommerfeld equation (3.1) for the wave motion may be approximated by

$$i\alpha R(U_w - c)\phi'' - \phi^{iv} = 0, \tag{9.10}$$

where U_w is the value of $U(z)$ at the wall. This approximation is valid provided $\alpha R |U_w - c|$ is sufficiently large (Drazin & Reid 1981, Chapter 4). Its general solution is

$$\phi = A e^{-\lambda z} + B e^{\lambda z} + C + Dz, \quad \lambda \equiv [i\alpha R(U_w - c)]^{\frac{1}{2}}, \tag{9.11}$$

the terms in C and D being vestiges of the inviscid solution. Of the two independent viscous solutions, one grows and the other decays with distance from the wall. Only the latter may be retained, in order to match with the inviscid solution in the region beyond the wall layer.

For the wall layer adjacent to $z = 0$, the solution which satisfies the boundary conditions $\phi(0) = \phi'(0) = 0$ has $B = 0$, $C = -A$ and $D = \lambda A$. On substituting this approximation for $\phi(z)$ into (9.2a), the solution for $u^{(2)}$ which varies as $\exp(2\alpha c_i t)$ in the wall layer may be found. Provided $\alpha |c_i| \ll 1$,

$$|\partial u^{(2)}/\partial t| \ll R^{-1}|\partial^2 u^{(2)}/\partial z^2|$$

in this layer and so the time-derivative may be ignored. Also, any $O(\epsilon^2)$ pressure gradient has negligible effect. The appropriate quasi-steady approximation follows by direct integration, on using the boundary condition $u^{(2)}(0) = 0$. Just beyond the wall layer, at $z = \delta$ say, where $|\lambda|\delta \gg 1$ but $\delta \ll 1$ still, $u^{(2)}$ is found to equal

$$u^{(2)}(\delta, t) = \tfrac{3}{4}\epsilon^2 e^{2\alpha c_i t} \frac{|D|^2}{|U_w - c|} \tag{9.12}$$

at leading order. Moreover, D is just the value of $\phi'(0)$ given by the *inviscid* outer solution.

Result (9.12), together with a similar one at $z = H$, supply the missing conditions required to determine the arbitrary constants of the quasi-steady approximation (9.8). Note that (9.12) is independent of Reynolds number R (provided this is large). The induced velocity (9.12) just outside the wall layer therefore persists even as $R \to \infty$, provided (9.10) remains a valid approximation within the wall layer.

The above result is also applicable to the flow induced near the bottom of an open water channel along which propagates a train of constant-amplitude surface gravity waves (Longuet-Higgins 1953; Phillips 1977). Experiments by Collins (1963) and others show satisfactory agreement with (9.12), which then gives a corresponding dimensional velocity

$$\tfrac{3}{4}a^2 k\omega \operatorname{cosech}^2 kh \tag{9.13}$$

Figure 4.1. Wave-driven mean flow correction $u^{(2)}$ for critical plane Poiseuille flow: (*a*) with constant mass flux (from Reynolds & Potter 1967), (*b*) with constant pressure gradient. The normalized coordinate z equals 0 at the channel centre and 1 at the wall. Approximate widths of viscous critical layer and wall layer are denoted by ⊢⊣.

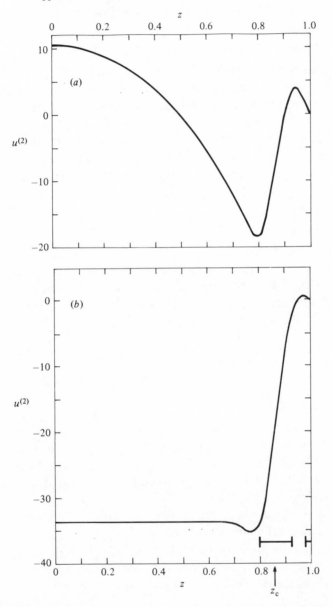

where a is the wave elevation at the free surface, k the dimensional wavenumber, h the mean water depth and ω the wave-frequency. A much earlier example, in acoustics, of a mean flow driven by an oscillatory boundary layer was given by Rayleigh (1896, p. 340). Further examples are discussed by Stuart (1966), Riley (1967) and Kelly (1970).

Wave-driven mean flows in closed channels have been calculated numerically, from (9.2) and the Orr–Sommerfeld equation, by Reynolds & Potter (1967), Pekeris & Shkoller (1967) and others. Figure 4.1(a) shows Reynolds & Potter's steady-state velocity profile $u^{(2)}(z)$ for plane Poiseuille flow at the critical point for neutral stability: $\alpha = 1.02$, $R = 5772$, $c = 0.264$ in the usual normalization.

Since Reynolds & Potter employed the constant mass-flux condition (9.3), their profile (a) is nearly parabolic in the region between the two critical layers. The lower curve (b) shows the corresponding mean flow with zero mean pressure gradient. The location and approximate width of the viscous critical layer and wall layer are indicated. This curve is in broad agreement with the asymptotic result (9.8), despite the fact that the critical layer is rather close to the wall in this example.

10   Spatially-periodic waves on deformable boundaries

10.1   *The Eulerian drift velocity of water waves*

When a bounding surface of the flow is deformable, the boundary conditions adopted above are inappropriate. For surface capillary–gravity waves with no primary flow U, the presence of viscosity, however small, modifies the linearized inviscid flow near the free surface. This modification takes the form of a thin oscillatory boundary layer of (dimensional) thickness $O[(\nu/\omega)^{\frac{1}{2}}]$, adjacent to the curved free surface, where ν is kinematic viscosity and ω is the wave frequency. In the presence of a primary shear flow, a similar boundary layer occurs, with equivalent dimensionless thickness $O\{[\alpha R(U_{\mathrm{s}} - c_{\mathrm{r}})]^{-\frac{1}{2}}\}$ where U_{s} is the (dimensionless) flow velocity at the mean free surface. This boundary layer is rather similar in structure to that at a plane rigid wall discussed above, but it is best described in curvilinear co-ordinates which follow the deformed surface (Longuet-Higgins 1953; Craik 1982b).

If the surface is truly free, supporting no tangential stress, this oscillatory boundary layer is rather weak, having vorticity which is $O(ak\omega)$, where ak is the maximum wave-slope at the surface. By proceeding to $O[(ak)^2]$, one may find the mean flow induced within this weak boundary layer. When the deformable surface is free of stress, there is found to be a non-zero velocity *gradient* just beyond the viscous layer. For free-surface waves in water of depth h which is otherwise at rest, this is

$$\mathrm{d}u^{(2)}/\mathrm{d}z_1 = -2a^2k^2\omega \coth kh \quad (z_1 = 0_+) \tag{10.1}$$

where $u^{(2)}$ is now the dimensional mean horizontal velocity just beyond the oscillatory boundary layer and z_1 is dimensional distance measured vertically downwards from the mean surface level.

For waves on an interface between viscous fluids of differing densities, the oscillatory boundary layers on either side of the interface are stronger, with vorticity which is $O[ak\omega R_w^{\frac{1}{2}}]$. Here, $R_w \equiv \omega/k^2\nu \gg 1$ is a 'wave Reynolds number'. Such boundary layers induce mean flow gradients stronger than (10.1), of magnitude $O[a^2k^2\omega R_w^{\frac{1}{2}}]$ when the viscosities of the two fluids are comparable. Details are given by Dore (1970, 1978a, b), who points out that, even at an air–water interface, result (10.1) ceases to be a good approximation for waves more than a few metres in length.

A similar increase in the strength of the oscillatory boundary layer, particularly important for short waves, may be brought about by contamination of an otherwise free surface by oil, detergents or other organic substances. Such contaminants impart to the surface an elasticity due to local changes in surface tension associated with their varying surface concentrations. Scott (1979) gives a comprehensive bibliography of work on surface-contamination effects. Such contamination greatly enhances the attenuation rate of short waves by viscosity, because of the increased vigour of the shearing oscillations within the boundary layer. It also gives rise to an enhanced mean flow gradient. In the limiting case of an inextensible surface which supports no mean concentration gradient of contaminant, Craik (1982b) has shown that (10.1) is replaced by

$$\mathrm{d}u^{(2)}/\mathrm{d}z_1 = -2^{-\frac{3}{2}}a^2k^2\omega \coth^2 kh R_w^{\frac{1}{2}}. \tag{10.2}$$

This is likely to be a satisfactory approximation for sufficiently short capillary–gravity waves with a contaminated surface. Indeed, it seems that (10.1) may *never* be a good approximation at a contaminated air–water interface!

For water waves with no primary mean flow, there is no critical layer. Accordingly, the predicted quasi-steady second-order Eulerian velocity outside the oscillatory boundary layers at $z_1 = 0$ and h is just

$$u^{(2)} = (\mathrm{d}u^{(2)}/\mathrm{d}z_1)_0 (z_1 - h) + (u^{(2)})_h \quad (0_+ < z_1 < h_-) \tag{10.3}$$

when $p^{(2)} = 0$. Here, $(u^{(2)})_h$ is given by (9.13) and $(\mathrm{d}u^{(2)}/\mathrm{d}z_1)_0$ by (10.1), (10.2) or similar result. For waves in channels closed at the downstream end, the zero-mass-flux condition must apply instead of $p^{(2)} = 0$. The mean surface would then adjust itself to incline at a slight angle to the horizontal, so that gravity can supply the necessary small pressure gradient $\partial p^{(2)}/\partial x = \rho \bar{P}$. The velocity profile (10.3) is then augmented by a term $\frac{1}{2}\nu^{-1}\bar{P}z_1 (z_1 - h)$, with $\bar{P}$ chosen to satisfy (9.3).

As the depth h increases, so does the time necessary to establish a quasi-steady mean flow. At shorter times, the solution $u^{(2)}(z, t)$ has a boundary-layer structure. The thickness of the boundary layers at the surface and the bottom increases with time and the quasi-steady flow is eventually established, unless other factors intervene. If the wave-train is not effectively infinite in length, but is generated by a wavemaker at some fixed location, the quasi-steady mean flow retains a boundary-layer structure near the wavemaker: the boundary layers then grow in thickness with downstream distance, rather than time. Such problems have been treated by Dore (1977) and Dore & Al-Zanaidi (1979). In infinitely deep water, an ever-deepening boundary layer would occur; but Madsen (1978) has shown that inclusion of the Coriolis force again permits a steady-state solution.

If a spatially-periodic wave-train decays with time under viscous action – as it must unless externally maintained – the quasi-steady solution (10.3) dominates the diffusive modes (cf. equation 9.9) only if the decay rate is sufficiently small. This was overlooked by Liu & Davis (1977) whose particular solution was generalized by Craik (1982b). Coriolis terms were included by Weber (1983).

Craik (1982b) also gives a physical explanation of the mean flow boundary condition for $du^{(2)}/dz_1$. In an Eulerian representation, all the mean momentum

$$\mathcal{M} = \tfrac{1}{2}\rho\omega a^2 \coth kh \tag{10.4}$$

per unit horizontal area within the fluctuating motion is concentrated between the horizontal planes containing the crests and troughs of the waves (Phillips 1977, p. 40). A decaying wave loses this momentum at a rate $2\sigma\mathcal{M}$ where σ is the exponential decay rate of the waves. This rate of momentum loss induces a stress just below the level of the troughs. Part of this is a constant Reynolds stress $-\rho\overline{uw}$ transmitted to the bottom boundary layer: this part equals $2\tilde{\sigma}\mathcal{M}$ where

$$\tilde{\sigma} = \tfrac{1}{2}k(2\omega\nu)^{\frac{1}{2}} \operatorname{cosech} 2kh$$

is the decay rate due to bottom friction alone. The remainder, $2(\sigma - \tilde{\sigma})\mathcal{M}$, equals the mean Eulerian stress $\rho\nu\,\partial u^{(2)}/\partial z_1$. Since the presence of surface contamination or a superposed viscous fluid increases the decay rate σ above that for a truly free surface, it naturally causes an increase in $\partial u^{(2)}/\partial z_1$. For waves maintained at constant amplitude by externally-imposed spatially-periodic normal stresses at the surface, similar mean velocity gradients are induced, the momentum of the fluctuating motion then being continuously replenished by the imposed stresses.

The second-order mean Eulerian velocity $u^{(2)}$ differs from the mean velocity of individual fluid particles. The latter is a Lagrangian mean, say $u_L^{(2)}$, and the difference $u_S^{(2)} = u_L^{(2)} - u^{(2)}$ is known as the *Stokes drift*. For a truly inviscid fluid, $u^{(2)}$ would be identically zero, since there would be no boundary layers to drive the flow; but a second-order Stokes drift $u_S^{(2)}$ remains, as Stokes (1847) first showed. However, for real fluids, results like (10.1) and (10.2) demonstrate that the induced Eulerian velocity $u^{(2)}$ can never be ignored, as it does not vanish when the viscosity approaches zero. The Lagrangian velocity gradient $\mathrm{d}u_L^{(2)}/\mathrm{d}z_1$ just below a clean free surface is precisely twice that given by (10.1), because of the contribution from the Stokes drift. For interfacial waves, and for waves at a contaminated surface, the Stokes-drift contribution to $\mathrm{d}u_L^{(2)}/\mathrm{d}z$ is normally a small part of the whole.

Various measurements of drift-current profiles $u_L^{(2)}$ have been made, most notably by Russell & Osorio (1958), but none shows satisfactory agreement with theory over the whole water depth. There are formidable experimental difficulties in controlling surface contamination and in eliminating end and side-wall effects. Also, the unidirectional drift-current profiles calculated above are almost certainly unstable to spanwise-varying disturbances for all but the smallest of wave amplitudes (see Craik 1982b and §13.2 below).

Mean flows generated by standing surface gravity waves and interfacial waves may be calculated rather similarly; see Longuet-Higgins (1953), Dore (1976), Crampin & Dore (1979).

10.2 'Swimming' of a wavy sheet

In a pioneering study of the propulsive mechanisms of micro-organisms, Taylor (1951) considered the idealized model of an almost plane infinite sheet which undergoes small wave-like distortions. If ω and k are the frequency and wavenumber of such undulations of periodicity $\exp[\mathrm{i}(kx - \omega t)]$, in a reference frame fixed relative to the unperturbed sheet, a characteristic Reynolds number is $R = \omega/\nu k^2$ where ν is the kinematic viscosity of the surrounding fluid. Taylor considered R to be sufficiently small to allow the Stokes approximation: in this case, the streamfunction ψ satisfies the biharmonic equation in x and z,

$$\nabla^2(\nabla^2\psi) = 0, \quad \nabla^2 = \partial^2/\partial x^2 + \partial^2/\partial z^2,$$

$$u = \partial\psi/\partial z, \quad w = -\partial\psi/\partial x.$$

Here, (u, w) are the Cartesian velocity components along and normal to the plane $z = 0$ denoting the unperturbed sheet position.

Since the biharmonic equation is linear, nonlinearity enters the problem only through the kinematic boundary conditions on the sheet, that

$$u(x_s, z_s, t) = \frac{dx_s}{dt}, \quad w(x_s, z_s, t) = \frac{dz_s}{dt}, \tag{10.5}$$

where $[x_s(t), z_s(t)]$ denote the position co-ordinates of material points of the sheet. For plane waves,

$$x_s = x + a\cos(kx - \omega t - \phi), \quad z_s = b\sin(kx - \omega t) \tag{10.6}$$

at leading order in wave amplitude, where the real constants a, b and ϕ depend on the nature of the distortion of the sheet. For instance, when $a \neq 0$ and $b = 0$, the sheet remains plane and performs purely tangential oscillations involving extension and compression, while, if $b \neq 0$ and $a = 0$, the leading-order displacement is purely perpendicular to the sheet and any extension or contraction enters only at $O(b^2)$.

A small-amplitude expansion of ψ may be developed in ascending powers of ak and bk, together with a Taylor expansion of the boundary conditions (10.5) about $(x, 0, t)$. Imposition of the remaining far-field boundary conditions $\partial u/\partial z$, $w \to 0$ as $z \to \pm\infty$ leads to a non-zero second-order mean velocity, as $z \to \pm\infty$, of

$$\bar{u} = U_s = \tfrac{1}{2}\omega k(b^2 + 2ab\cos\phi - a^2). \tag{10.7}$$

Clearly U_s is the (right-to-left) swimming speed of the sheet relative to the undisturbed fluid at infinity. The magnitude and sign of U_s depend on the values a, b and ϕ (see Childress 1981, for further details).

Taylor (1951) continued this analysis to higher order in wave amplitude for the case of a strictly inextensible sheet; for this, $a = 0$ and the kinematic conditions (10.5) yield

$$u = \tfrac{1}{4}\omega k b^2 \cos[2(kx - \omega t)] + O(b^4),$$

$$w = -\omega b\cos(kx - \omega t) + O(b^3)$$

on the surface $z = b\sin(kx - \omega t)$. The resultant second-order mean velocity as $z \to \pm\infty$ is found to be

$$\bar{u} = U_s = \tfrac{1}{2}\omega k b^2(1 - \tfrac{19}{16}b^2 k^2) + O[(bk)^6]. \tag{10.8}$$

Taylor's analysis was generalized to *all* finite Reynolds numbers R by Tuck (1968) and Brennen (1974) (see also Childress 1981). In this extension, nonlinear inertia terms contribute to the solution, the biharmonic equation being replaced by the full two-dimensional vorticity equation. Nevertheless, it must be assumed that the wave amplitude b is such that nonlinear convective terms remain small compared with viscous ones: that is, R is finite but $bkR \ll 1$ as well as $(bk)^2 \ll 1$. Tuck found that

$$U_s = \tfrac{1}{2}\omega k b^2 \left\{ \frac{1+F(R)}{2F(R)} + O[(kb)^4] \right\},$$ (10.9)

$$F(R) = 2^{-\frac{1}{2}}[1+(1+R^2)^{\frac{1}{2}}]^{\frac{1}{2}},$$

which is consistent with (10.7) as $R \to 0$ and yields precisely one half of Taylor's value as $R \to \infty$ and $kb \to 0$.

Quite clearly, the wavy sheet is not an acceptable model for flagellar propulsion of 'long-tailed' micro-organisms like spermatozoa. But, despite the obvious idealizations, this model is believed to be relevant to certain types of ciliary propulsion, with the wave-envelope describing the positions of the tips of closely-spaced beating cilia (Brennen 1974; Blake & Sleigh 1975).

The above solutions have the property that no net force is exerted on any portion of the sheet. Suppose, on the other hand, that a long but finite wavy sheet must develop a net thrust in order to overcome the viscous drag on a 'head' or other inactive portion of the body. This can only be accomplished by the exertion of an equal and opposite force on the surrounding fluid through a mean tangential stress. But, in this event, there is no quasi-steady x-independent solution when the surrounding fluid is unbounded. Instead, a boundary-layer structure must develop, growing either with x or t.

As a simple example, consider the configuration of Figure 4.2. In this, an infinitely long slender body has a plane lower boundary situated a

Figure 4.2. Slender-body configuration which cannot support constant progressive motion by means of undulations.

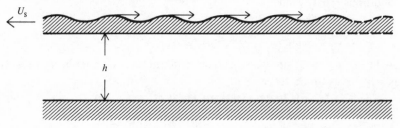

distance h from a plane stationary wall, the upper boundary undulates with constant amplitude and the surrounding fluid extends to $z = \infty$. The Stokes approximation is assumed to hold. To sustain a constant swimming speed $-U_s$, the upper boundary must provide a net thrust $-\rho\nu U_s/h$ per unit area to counteract the viscous drag on the inert lower boundary: but this is impossible! If released from rest, the body would at first accelerate; but as the viscous boundary layer above the upper surface deepens, the swimming speed must eventually decay to zero. The best swimming strategy in this instance is one of 'stop–go': thrust is provided with least diminution in speed when the upper boundary layer is thin. In contrast, if the lower boundary undulates and the upper is inert, a uniform swimming speed *can* be maintained.

11   Modulated wave-packets

11.1   *Waves in viscous channel flows*

Suppose that a primary unidirectional shear flow $U(z)$ is modified by a slowly-modulated, almost two-dimensional wave-train, with horizontal local (downstream) wavenumber everywhere close to α_0. The velocity field has the form

$$u = U(z) + \epsilon\, \partial\psi_1/\partial z + O(\epsilon^2), \quad v = O(\epsilon^2),$$
$$w = -\epsilon\, \partial\psi_1/\partial x + O(\epsilon^2)$$

where

$$\psi_1 = \mathrm{Re}\{A(x, y, t)\,\phi(z)\,\exp[i\alpha_0(x-ct)]\}; \tag{11.1}$$

ϵ is a small parameter which characterizes the wavemotion and $A(x, y, t)$ is a complex amplitude function which varies slowly in space and time. When the mode with wavenumber α_0 is neutrally stable ($c_i = 0$), the slow modulation is best described by new co-ordinates

$$\xi = \epsilon(x - c_g t), \quad \eta = \epsilon y, \quad \tau = \epsilon^2 t \tag{11.2}$$

where $c_g = \partial(\alpha c_r)/\partial\alpha$ is the real part of the wave's group velocity. Obviously, in the linear approximation, all weak amplitude modulations of a neutrally stable wave-train are carried along with the group velocity c_g. Evolution on the slow time-scale τ is associated with cubic nonlinearities studied in Chapter 6, and plays no part in the analysis at this stage.

As in Davey, Hocking & Stewartson (1974), the quasi-steady mean-flow and pressure modifications may be expressed as

$$\left.\begin{aligned}
\mathbf{u}^{(2)} &= \epsilon^2[u_2, v_2, w_2] + \epsilon^3[u_3, v_3, w_3] + \dots \\
p^{(2)} &= \epsilon P_1(\xi, \eta, \tau) + \epsilon^2 P_2(\xi, \eta, \tau) + \dots
\end{aligned}\right\} \tag{11.3}$$

where $u_2, v_2, w_2, u_3, v_3, w_3$ are functions of z, ξ, η and τ. The resultant $O(\epsilon^2)$ mean-flow equations are (cf. equations 9.2)

$$\left. \begin{aligned} D^2 u_2 - R \, \partial P_1/\partial \xi &= \tfrac{1}{4} i \alpha_0 R \, | A |^2 \, D[\phi^* \phi' - \phi \phi^{*\prime}], \\ D^2 v_2 - R \, \partial P_1/\partial \eta &= 0, \\ D w_2 &= 0, \quad (D \equiv d/dz). \end{aligned} \right\} \qquad (11.4a,b,c)$$

The first two are just the momentum equations in the x and y directions and the third is the equation of continuity. Also, at $O(\epsilon^3)$, the continuity equation is

$$\partial u_2/\partial \xi + \partial v_2/\partial \eta + D w_3 = 0. \qquad (11.4d)$$

For definiteness, we follow Davey *et al.* (1974) and confine attention to the critical neutral mode of plane Poiseuille flow between parallel walls at $z = \pm 1$. For this,

$$U(z) = 1 - z^2 \quad (-1 \leqslant z \leqslant 1), \quad R = 5772, \quad \alpha_0 = 1.02 \qquad (11.5)$$

and $\phi(z)$ is an even function of z. The boundary conditions are $u = v = w = 0$ on $z = \pm 1$ and a solution of (11.4a–c) is

$$\left. \begin{aligned} u_2 &= -\tfrac{1}{2}(1 - z^2) \, R \, \partial P_1/\partial \xi + | A |^2 \, S(z), \\ v_2 &= -\tfrac{1}{2}(1 - z^2) \, R \, \partial P_1/\partial \eta, \\ w_2 &= 0 \end{aligned} \right\} \qquad (11.6)$$

where $S(z)$ is an even function of z defined by

$$S(z) = -\tfrac{1}{4} i \alpha R \int_z^1 (\phi^* \phi' - \phi \phi^{*\prime}) \, dz \quad (0 \leqslant z \leqslant 1).$$

Also, from (11.4d),

$$\int_{-1}^1 (\partial u_2/\partial \xi + \partial v_2/\partial \eta) \, dz = 0$$

and so

$$\frac{\partial^2 P_1}{\partial \xi^2} + \frac{\partial^2 P_1}{\partial \eta^2} = \frac{3}{R} \frac{\partial | A |^2}{\partial \xi} \int_0^1 S(z) \, dz. \qquad (11.7)$$

For a given amplitude distribution $A(\xi, \eta, \tau)$, the $O(\epsilon)$ pressure field may be found from (11.7) and the $O(\epsilon^2)$ mean flow determined from (11.6).

For modulations independent of the spanwise co-ordinate η,

$$\partial P_1/\partial \xi = 3R^{-1} | A |^2 \int_0^1 S(z) \, dz \qquad (11.8)$$

there being no $O(\epsilon^2)$ pressure gradient when $| A | = 0$. Accordingly, $v_2 = 0$ and

$$u_2 = -\tfrac{3}{2}(1 - z^2) | A |^2 \int_0^1 S(z) \, dz + | A |^2 \, S(z) \qquad (11.9)$$

which satisfies the constant-mass-flux condition (9.3) at each station ξ. Numerical computation, employing the eigenfunction $\phi(z)$ of the critical neutral mode of plane Poiseuille flow normalized so that $\phi(0) = 2$,[†] yields

$$\int_0^1 S(z)\,\mathrm{d}z = -87.2. \tag{11.10}$$

Here, $R^{-1}\,\mathrm{d}S/\mathrm{d}z$ is just the wave Reynolds stress, which is virtually zero in the inviscid region $|z| < z_c$ – between the two critical layers. The term $|A|^2 S(z)$ of u_2 is therefore nearly constant in this region and it varies almost linearly with z between critical layer and wall layer in accord with the asymptotic result (9.8). Result (11.9) is precisely that of Reynolds & Potter (1967), shown in Figure 4.1(a).

For η-dependent modulations, the spanwise pressure gradient cannot be ignored and one first must solve the Poisson equation (11.7) before determining the velocity components (11.6).

The above quasi-steady theory also holds for small linear growth or decay rates, such that $|\alpha c_i|/R \ll 1$, on retaining exponential factors in $\exp(2\alpha c_i\,t)$. However, the quasi-steady state is attained only if sufficient time is available for the effect of viscous diffusion to penetrate throughout the mean flow. Since this viscous time scale is $O(R)$ and variations in $|A|^2$, convected past a fixed observer, occur on a time scale which is $O\{c_g^{-1}\,|A(\partial A/\partial x)^{-1}|\}$, the quasi-steady solution is attained only when the modulations are sufficiently weak that

$$|\alpha c_i|,\quad |A^{-1}\partial A/\partial x| \ll R^{-1}. \tag{11.11a, b}$$

Of course, this is tacitly assumed by the scaling (11.2), for ϵ is an over-riding small parameter which approaches zero and R is a fixed, though fairly large, finite constant. But (11.11b) is a very stringent condition requiring, when R is $O(10^3)$, that the wave amplitude changes little over as many as a thousand wavelengths.

An unsteady, inviscid alternative to the above quasi-steady approximation may be developed when the dimensionless $O(\epsilon^{-1})$ length scale $|A(\partial A/\partial x)^{-1}|$ of the modulation, or the amplification time scale $(\alpha c_i)^{-1}$, is *small* compared with R. Such an approximation led to result (9.7) in the case of spatially-uniform waves. When the modulated wave amplitude has the form $A = B(\xi, \eta)\,e^{\alpha c_i t}$, the viscous terms $D^2 u_2$ and $D^2 v_2$ of (11.4) are replaced by respective acceleration terms

$$-\epsilon R(2\alpha \hat{c}_i - c_g\,\partial/\partial\xi)\,u_2,\quad -\epsilon R(2\alpha \hat{c}_i - c_g\,\partial/\partial\xi)\,v_2,\quad \hat{c}_i \equiv \epsilon^{-1}c_i$$

† This normalization accords with Davey *et al.*'s $\phi(0) = 1$ as their definition of ψ_1 differs from (11.1) by a factor of 2.

for solutions which vary in time as $\exp(2\alpha c_i t)$. It readily follows that $P_1 = 0$ everywhere for localized wave-packets with $P_1 \to 0$ as $|\xi|$ and $|\eta| \to \infty$. Accordingly, the required particular solution is

$$u_2 = -\frac{i\alpha\, e^{2\alpha c_i t}}{4\epsilon c_g} D[\phi^*\phi' - \phi\phi^*] \int_\xi^\infty \exp[2\alpha \hat{c}_i(\xi - \xi_1)/c_g] \, |B(\xi_1, \eta)|^2 \, d\xi_1,$$

$$v_2 = 0, \tag{11.12}$$

which generalizes the inviscid approximation (9.7) to cover modulated waves.

For modulated wave-packets in boundary-layer flows of unbounded extent as $z \to \infty$, a quasi-steady approximation can never be valid throughout the flow domain, even when $c_i = 0$, unless u_2 is identically zero above the critical layer. If the x-dependence of the boundary-layer profile can be neglected – and this is *not* justified in general – an outer, inviscid solution with $P_1 = u_2 = 0$ may have to be matched to a time-dependent viscous solution with non-zero u_2, valid in a layer of ever-growing thickness. For the asymptotic suction boundary-layer, which *is* independent of x, the constant velocity component normal to the wall is responsible for extra convective effects, which limit the extent of the viscous region and so permit a quasi-steady flow. This case has been studied by Hocking (1975).

11.2   *Waves on a free surface*

Considerable modifications are necessary to deal with weakly nonlinear modulated waves on a free surface. This is because of the nonlinear kinematic and stress boundary conditions at the deformable boundary. The rôle of these boundary conditions is most simply demonstrated for a modulated train of gravity waves in strictly *inviscid* fluid and it is this which is first discussed. But it should not be forgotten that second-order flows driven by viscous, oscillatory boundary-layers do not generally vanish in real fluids as $R \to \infty$.

A rather general treatment of slowly-varying non-dissipative wave-trains is that pioneered by Whitham (1965 *et seq.*) and extended by Hayes (1973) (who lists relevent articles to that date). This employs a locally-averaged Lagrangian or Hamiltonian representation. In skilled hands, this powerful tool copes successfully not only with small local variations in wavenumber and frequency about fixed real values, but also with substantial overall changes, such as occur in fluid of varying depth, provided these changes take place sufficiently gradually. This theory is fully described in Whitham's own monograph (1974) and elsewhere (e.g. Leibovich & Seebass 1974) and so is not systematically expounded here; but several aspects are discussed later (§§ 11.4–11.5).

Unfortunately, Whitham's formulation lacks any definite measure of the 'slowness' of the modulation required for the theory to remain valid, for no small parameter is explicitly introduced (but see Whitham 1970). As Davey & Stewartson (1974) point out, 'since the rigorous theory of such equations is in its infancy many students feel a certain unease in using their solutions widely, particularly as the leading term of some asymptotic expansion': though further advances have been achieved since this statement was made, it remains broadly true.

For the moment, we continue to employ a multiple-scales approach, following that developed by Hasimoto & Ono (1972) and Davey & Stewartson (1974). This yields results in agreement with Whitham's while providing an explicit scheme of approximation. In one sense, a multiple-scales analysis is more restricted than an averaged-Lagrangian method in that it necessarily deals with weak modulations about a fixed frequency and wavenumber; in another, it is more general, for extension to dissipative flows is relatively straightforward in principle, whereas no Lagrangian then exists. Attempts to extend Whitham's variational method to dissipative flows (Usher & Craik 1974; Jimenez & Whitham 1976; Itoh 1981) have had some limited success.

Following Davey & Stewartson (1974), we consider an inviscid liquid of depth h with lower rigid boundary $z = -h$ and undisturbed free surface at $z = 0$. There is no $O(1)$ primary flow and the surface supports a modulated wave-train with elevation $z = \zeta(x, y, t)$ given by

$$\zeta = \mathrm{i}\epsilon\omega g^{-1}A(\xi, \eta, \tau)\exp[\mathrm{i}(kx-\omega t)]+\mathrm{c.c.}+O(\epsilon^2),$$

where 'c.c.' denotes complex conjugate. Here, k, ω are the dimensional wavenumber and frequency of the wave-train, ξ, η, τ are the dimensional counterparts of (11.2), ϵ is a small parameter which characterizes the wave-slope and g is gravitational acceleration. The dispersion relation for gravity waves is

$$\omega = (gk\sigma)^{\frac{1}{2}}, \quad \sigma = \tanh kh$$

and the group velocity c_g is $\mathrm{d}\omega/\mathrm{d}k$.

The liquid motion is described by a velocity potential $\phi(x, y, z, t)$, with $\mathbf{u} = \nabla\phi$, which satisfies Laplace's equation in $-h < z < \zeta$. The bottom boundary condition is just $\partial\phi/\partial z = 0$ on $z = -h$. The nonlinearities arise in the two surface boundary conditions, namely the kinematic and pressure conditions at $z = \zeta$:

$$\frac{\partial \phi}{\partial z} = \frac{\partial \zeta}{\partial t} + \frac{\partial \phi}{\partial x}\frac{\partial \zeta}{\partial x} + \frac{\partial \phi}{\partial y}\frac{\partial \zeta}{\partial y}, \tag{11.13}$$

$$\frac{\partial \phi}{\partial t} + \tfrac{1}{2}(\nabla \phi \cdot \nabla \phi) = -g\zeta. \tag{11.14}$$

The $O(\epsilon)$ wave motion is

$$\phi_{11} = \epsilon A \frac{\cosh k(z+h)}{\cosh kh}\exp[i(kx-\omega t)]+\text{c.c.} \tag{11.15}$$

Because of the modulations, one must include an $O(\epsilon)$ aperiodic function $\epsilon\phi_{01}(\xi,\eta,\tau)$ in the velocity potential and an $O(\epsilon^2)$ aperiodic surface displacement $\epsilon^2\zeta_{02}(\xi,\eta,\tau)$ in the expansions of ϕ and ζ in powers of ϵ. From (11.14),

$$g\zeta_{02} = c_g(\partial\phi_{01}/\partial\xi) - k^2(1-\sigma^2)|A|^2. \tag{11.16}$$

Also, on integrating Laplace's equation over the depth h and using (11.13) and (11.16) one obtains

$$(gh-c_g^2)\frac{\partial^2\phi_{01}}{\partial\xi^2}+gh\frac{\partial^2\phi_{01}}{\partial\eta^2} = -k^2\left\{\frac{2\omega}{k}+c_g(1-\sigma^2)\right\}\frac{\partial|A|^2}{\partial\xi}. \tag{11.17}$$

This coupled system may be integrated to yield ϕ_{01} and ζ_{02} subject to appropriate boundary conditions: e.g. the 'far-field' boundary conditions $\nabla\phi_{01}\to 0$ as $\xi^2+\eta^2\to\infty$, for a localized wave-packet with amplitude A which decays to zero sufficiently far from its centre.

The $O(\epsilon^2)$ horizontal mean flow $\epsilon^2[\partial\phi_{01}/\partial\xi,\partial\phi_{01}/\partial\eta]$ is *independent of depth* z and, to preserve continuity, there is a consequent $O(\epsilon^3)$ depth-dependent vertical velocity component

$$-\epsilon^3(z+h)(\partial^2/\partial\xi^2+\partial^2/\partial\eta^2)\phi_{01}. \tag{11.18}$$

When the modulation is independent of the spanwise variable η, the horizontal mean $O(\epsilon^2)$ velocity is solely in the x-direction, of magnitude

$$-\epsilon^2 k^2 |A|^2 \left\{\frac{2(\omega/k)+c_g(1-\sigma^2)}{gh-c_g^2}\right\} \tag{11.19}$$

and the $O(\epsilon^2)$ mean surface displacement $\epsilon^2\zeta_{02}$ is

$$-\epsilon^2 k^2 |A|^2 \left\{\frac{2c_g(\omega/k)+gh(1-\sigma^2)}{g(gh-c_g^2)}\right\}+\text{constant}. \tag{11.20}$$

The constant is chosen to conserve the total volume of fluid: for a localized wave-packet, with $|A|^2 \leqslant o(|x|^{-1})$ ad $|x|\to\infty$, it approaches zero.

For waves in deep water ($kh \gg 1$), ω and c_g are independent of h and $\sigma = 1$: accordingly both mean flow and surface displacement approach zero as $kh \to \infty$ (provided the wave-packet remains long compared with the depth: $\epsilon kh \ll 1$). However, the local x-momentum of the mean flow, integrated over the depth, remains finite, as

$$\mathcal{M}_1 \approx -2\rho\epsilon^2 k^2 \omega^{-1} |A|^2 = -\tfrac{1}{2}\rho\omega a^2 \quad (kh \to \infty) \tag{11.21}$$

per unit area of surface, where a is the modulated surface-wave amplitude. This exactly cancels the local mean momentum (10.4) associated with the fluctuating motion of deep-water waves. The cancelling of these contributions implies that, although work must be done, no net impulse need be exerted in order to generate a wave-packet in deep water. But, for finite depths h, the mean-flow momentum and that of (10.4) do not exactly cancel and the wave-packet possesses a finite net momentum. If, in creating a localized disturbance, an excess (or defect) of momentum is supplied over that required by the wave-packet, additional small changes in surface level are produced which eventually propagate far away from the main disturbance with the speed $\pm (gh)^{\frac{1}{2}}$ of long surface waves. A comprehensive discussion of questions relating to 'wave momentum' is given by McIntyre (1981 and forthcoming monograph).

Results (11.19) and (11.20) establish that when a localized inviscid, two-dimensional wave-packet propagates through otherwise undisturbed water, the induced mean flow which accompanies the packet is usually in the direction opposite to that of the packet's progress and the mean surface level is depressed below its equilibrium value. This is so provided $c_g < (gh)^{\frac{1}{2}}$, which is generally satisfied for gravity waves. Since $c_g \to (gh)^{\frac{1}{2}}$ as $kh \to 0$, the effect is most pronounced in relatively shallow water. Observational and experimental evidence support these conclusions for waves of small amplitude (Longuet-Higgins & Stewart 1962; Leblond & Mysak 1978) but one must guard against unwarranted extrapolation beyond the range of validity of the theory. In particular, breaking waves exhibit an opposite effect.

Results (11.19) and (11.20) may be derived rather more simply, if somewhat less rigorously, by introducing the concept of 'radiation stress' (Longuet-Higgins & Stewart 1960, 1962). In the present context, this is the depth-integrated quantity

$$S_{11}(\xi, \tau) = \int_{-h}^{\zeta} \overline{(p + \rho u^2)} \, dz - \int_{-h}^{\zeta} \rho g(\bar{\zeta} - z) \, dz \tag{11.22}$$

where $u = \partial\phi/\partial x$ and overbars denote local horizontal or time averages, taken over the fundamental wavelength or wave period. The first integral represents the total averaged x-momentum flux across each vertical plane $\xi = $ constant and the second denotes that part of the flux attributable to hydrostatic pressure alone. At leading order in ϵ, $\partial/\partial t = -c_g \partial/\partial x$ and the depth-integrated equations of continuity of mass and momentum reduce to

$$-\rho c_g \partial\bar{\zeta}/\partial x + \partial M/\partial x = 0,$$

$$\rho g h \partial\bar{\zeta}/\partial x - c_g \partial M/\partial x = -\partial S_{11}/\partial x$$

where

$$S_{11} = \tfrac{1}{2}\rho g a^2 \left(\frac{2kc_g}{\omega} - \frac{1}{2}\right)$$

(cf. Longuet-Higgins & Stewart 1962). Here, M is the total mean horizontal momentum, defined as

$$M = \overline{\int_{-h}^{\zeta} \rho u \, dz}$$

and $\bar{\zeta}$ is the $O(\epsilon^2)$ mean surface displacement. The term $\rho g h \partial\bar{\zeta}/\partial x$ derives from the horizontal hydrostatic pressure gradient associated with the mean surface slope. The solution corresponding to (11.19) and (11.20) is just

$$M = \frac{-c_g S_{11}}{gh - c_g^2}, \quad \rho\bar{\zeta} = \frac{-S_{11}}{gh - c_g^2},$$

which yields the $O(\epsilon^2)$ depth-averaged mean velocity

$$\bar{u}_2 = (\rho h)^{-1}(M - \mathcal{M})$$

where $\mathcal{M}$ is the momentum (10.4) within the fluctuating motion only. Since the mean flow is irrotational, $\bar{u}_2$ cannot vary with z and this expression reduces to (11.19).

Of course, the above inviscid results are modified by viscosity, because boundary conditions similar to (9.13) and (10.1) or (10.2) must be applied just outside the viscous boundary layers at the bottom and the free surface. If, in contrast to (11.11b), $|A^{-1} dA/dx| \gg R_w^{-1}$, the influence of viscosity is confined to thin, but growing, boundary layers as the wave-packet goes by. For example, a wave-packet of finite length L would leave behind it boundary layers of thickness $O[(\nu t_L)^{\frac{1}{2}}]$, where $t_L = L/c_g$ is the time taken for the wave-packet to pass. Subsequently, these boundary-layer flows further spread and decay by viscous diffusion. Inevitably, analysis of such problems is rather involved: the most comprehensive treatment so far is that of Grimshaw (1981b). Grimshaw (1977, 1979a, 1982) has also studied the mean flows generated by internal-wave-packets, both with and without viscous dissipation and primary shear flow. The internal-wave problem is further complicated by possible resonances involving the modulated mean flow and long internal-wave-modes (see also §19).

11.3   Wave propagation in inhomogeneous media

Longuet-Higgins & Stewart (1961, 1962) employed a 'radiation stress tensor' $S_{\alpha\beta}$ to determine the evolution of water waves in inhomogeneous environments with variable depth or horizontal current (see also Phillips 1977; Leblond & Mysak 1978). The local $O(\epsilon^2)$ depth-integrated mean energy density E associated with $O(\epsilon)$ fluctuations $\mathbf{u}'$ only is

$$E = \rho \int_{-h}^{0} \overline{\mathbf{u}'^2} \, dz = \tfrac{1}{2}\rho g a^2$$

for gravity waves with local amplitude a. (Recall that kinetic and potential energies are equal for such conservative oscillations.) Suppose that such waves propagate through a known slowly-varying, depth-independent, horizontal current $\mathbf{U}(x, y, t)$ and that the depth h also varies slowly.

When the imposed current variations are large compared with any $O(\epsilon^2)$ induced mean flow, the former may be retained while disregarding the latter and it may be shown that E evolves subject to

$$\left.\begin{aligned}
&\partial E/\partial t + \boldsymbol{\nabla}_1 \cdot [(\mathbf{U} + \mathbf{c}_g)\,E] + \tfrac{1}{2}S_{\alpha\beta}(\partial U_\alpha/\partial x_\beta + \partial U_\beta/\partial x_\alpha) = 0, \\
&S_{\alpha\beta} = E \left\{ \frac{|\mathbf{c}_g|\,k_\alpha k_\beta}{\omega\,|\mathbf{k}|} + \delta_{\alpha\beta}\left(\frac{\mathbf{c}_g \cdot \mathbf{k}}{\omega} - \frac{1}{2}\right)\right\} \quad (\alpha, \beta = 1, 2)
\end{aligned}\right\} \qquad \text{(11.23a, b)}$$

at $O(\epsilon^2)$, with summation over repeated indices. Here, $\boldsymbol{\nabla}_1 \equiv (\partial/\partial x, \partial/\partial y)$, $\mathbf{c}_g$ is the group velocity observed in a frame moving with the local mean flow $\mathbf{U}$, $\mathbf{k} = (k_1, k_2)$ denotes the horizontal wavenumber vector and ω the *local* frequency as seen by an observer moving with the current velocity $\mathbf{U}$. Accordingly,

$$\mathbf{c}_g = \partial\omega/\partial\mathbf{k}, \quad \omega^2 = g\,|\mathbf{k}|\,\tanh|\mathbf{k}h|.$$

Note that $S_{\alpha\beta}$ is determined, at $O(\epsilon^2)$, from the $O(\epsilon)$ wave field and that equation (11.23) is *linear* in E.

The wave-train is assumed to remain 'coherent': that is, it may always be described, locally, in terms of a *single* wave-mode of wavenumber $\mathbf{k}$. Its frequency relative to the *fixed* reference frame is $\omega' = \omega + \mathbf{U} \cdot \mathbf{k}$, where ω satisfies the usual linear dispersion relation given above. The periodicity may be written as $\exp(i\chi)$, with $\chi \equiv \mathbf{k} \cdot \mathbf{x} - \omega' t + \Delta$, where Δ is some phase shift. The quantities $\mathbf{k}$, ω, Δ may vary slowly in space and time and lines of constant phase carry constant values of χ. To define $\mathbf{k}$ and ω unambiguously, one chooses $\mathbf{k} = \boldsymbol{\nabla}\chi$, $\omega' = -\partial\chi/\partial t$. It follows that

$$\boldsymbol{\nabla} \times \mathbf{k} = 0, \quad \partial\mathbf{k}/\partial t + \boldsymbol{\nabla}(\omega + \mathbf{U} \cdot \mathbf{k}) = 0. \qquad \text{(11.24a, b)}$$

The latter is sometimes called the equation of 'conservation of wave crests'. For given $\mathbf{U}(x, y, t)$ and $h(x, y, t)$, the variations in $\mathbf{k}$, ω and E may be calculated from the above equations. The approximate equations outlined above constitute 'ray theory', so named by analogy with geometrical optics. They remain valid provided the wave-train is coherent, slowly-varying and of sufficiently small amplitude to permit linearization. We here consider just two illustrative examples, others being given by Longuet-Higgins & Stewart (1961, 1962); Peregrine (1976), Phillips (1977) and Leblond & Mysak (1978).

Case (a): $U = 0$. With variable depth $h(x, y)$ but no $O(1)$ mean flow, solutions with time-independent $\mathbf{k}$ must have constant frequency ω, from (11.24b). For simplicity, we restrict attention to the shallow-water case $|\mathbf{k}h| \ll 1$; then, the dispersion relation is $\omega = (gh)^{\frac{1}{2}}|\mathbf{k}|$, on choosing the positive root, and $\mathbf{c}_g = (gh)^{\frac{1}{2}}\mathbf{k}/|\mathbf{k}|$. Also from (11.24a), $\partial k_2/\partial x = \partial k_1/\partial y$, where $\mathbf{k} = (k_1, k_2)$.

A solution of (11.23) and (11.24) for which h, k_1 and E depend on x only is

$$k_1(x) = \left[\frac{(k_{10}^2 + k_{20}^2)\,h_0}{h(x)} - k_{20}^2\right]^{\frac{1}{2}}, \quad k_2 = k_{20}$$

$$E(x) = \frac{h_0 k_{10} E_0}{h(x)\,k_1(x)}, \tag{11.25}$$

where k_1, k_2, E and h take the constant values k_{10}, k_{20}, E_0 and h_0 respectively at $x = 0$. The wavenumber component k_1 remains real for all x if and only if

$$h(x)/h_0 \leqslant 1 + (k_{10}/k_{20})^2$$

everywhere. If $h(x)$ increases towards the value

$$h^* = h_0[1 + (k_{10}/k_{20})^2]$$

at some $x = x^*$, $k_1 \to 0$ and $E \to \infty$ there. Although the 'slowly-varying' approximation breaks down before this point is reached, relatively large amplitudes must nevertheless arise as waves approach this depth.

Wave refraction occurs because the phase speed increases with depth. The x-component of group velocity falls to zero at x^* and wave energy can penetrate no further in this direction. In this event, a caustic forms at x^* and elementary ray theory breaks down. A more complete analysis (Leblond & Mysak 1978, pp. 337–41; Lighthill 1978, §4.11), but still based on the linear approximation, resolves the singularity and finds the solution near x^* to have the form of an Airy function. The waves are turned back as a reflected wave-train with equal and opposite magnitude to the incoming one and the solution beyond x^* decays rapidly to zero. If the depth approaches a constant value $h_\infty > h^*$ beyond x^*, this decaying solution is an 'edge wave' with amplitude proportional to $\exp(-Kx)$ as $x \to \infty$, where $K = |k_1(\infty)|$.

Figure 4.3 shows ray trajectories for trapped waves which suffer total reflection at a vertical barrier $x = 0$ and which are 'turned back' by refraction at the caustic x^*. Trapped waves and edge waves are known to occur near coastlines.

Case (b): h constant. With constant depth but variable mean flow, $\mathbf{U} = [U_1(x, y), U_2(x, y), 0]$, it is necessary that

$$\partial U_1/\partial x + \partial U_2/\partial y = 0.$$

However, this condition may be relaxed if a small depth-dependent upwelling component U_3 is introduced (cf. Phillips 1977, Ch. 3). For simplicity, we consider a unidirectional shear flow $U_1 = U_1(y)$, $U_2 = 0$ and deep-water waves ($h \to \infty$) for which $\omega^2 = g |\mathbf{k}|$.

Time-independent solutions have

$$\omega' = \omega + \mathbf{U} \cdot \mathbf{k} = \omega_0 \text{ (constant)}, \quad \nabla \times \mathbf{k} = 0:$$

those also independent of x have constant $k_1 = k_{10}$ say, with $k_2(y)$ and $U_1(y)$ related by

$$g^2(k_{10}^2 + k_2^2) = (\omega_0 - U_1 k_{10})^4. \tag{11.26}$$

Also, from (11.23b),

$$S_{12} = S_{21} = \tfrac{1}{2} E k_{10} k_2 (k_{10}^2 + k_2^2)^{-1}$$

for deep-water waves and (11.23a) reduces to

$$\frac{E k_2}{k_{10}^2 + k_2^2} = \text{constant}. \tag{11.27}$$

If $U_1(y)$ is a shear-layer profile with $U_1 \to 0$ as $y \to -\infty$, and waves with $\mathbf{k} = (k_{10}, k_{20})$ as $y \to -\infty$ propagate towards it, the changes in k_2 and E within the shear layer are given by (11.26) and (11.27). However, (11.26) has real roots k_2 only when

$$|\omega_0 - U_1 k_{10}| > |g k_{10}|^{\frac{1}{2}}, \quad \omega_0 = g^{\frac{1}{2}}(k_{10}^2 + k_{20}^2)^{\frac{1}{4}}.$$

Figure 4.3. Ray trajectories showing wave trapping by reflection at a barrier $x = 0$ and 'turning' at a caustic $x = x^*$.

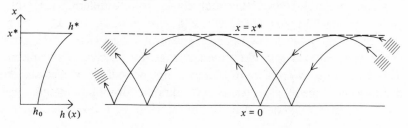

Consequently, when $k_{10} > 0$ and max $U_1 = V > 0$, the waves can propagate right through the shear layer only if

$$V < V^* \equiv (\omega_0/k_{10}) - (g/k_{10})^{\frac{1}{2}}.$$

Otherwise, they cannot pass the value y^* where $U(y^*) = V^*$, at which both k_2 and c_{g2} become zero and $E \to \infty$.

In the latter case, the wave is totally reflected and there is an evanescent wave beyond y^*. This reflection does *not* occur at the critical layer where $\omega'/k_{10} = U$, but before it is reached: this contrasts with the situation discussed in §5.3 for internal waves. The linear solution near the caustic may be constructed, by the WKB method, in terms of the Airy function as mentioned above. However, if the local wave amplitude near such a caustic becomes too large, nonlinear effects may interfere with the predicted total reflection and local dissipation of energy by wave breaking may occur.

With little modification, the above theory may be applied to study the effect of large-scale internal waves on relatively short ocean surface waves (see §14). Gargett & Hughes (1972) have drawn attention to short-wave modulation and apparent caustic formation, with local wave breaking, in this context. Internal waves are normally rendered visible at the ocean surface by such changes in the surface-wave field.

11.4 *Wave action and energy*

Because the radiation-stress tensor $S_{\alpha\beta}$ enters equation (11.23a), the energy E within the fluctuating motion is not conserved as the wave evolves, even in the absence of dissipation. This is because E is just a part of the total $O(\epsilon^2)$ energy associated with the disturbance; the remainder coming from any $O(\epsilon^2)$ mean flow. However, (11.23) may be recast as a conservation law, not of E, but of the *wave action* $\mathscr{A} \equiv E/\omega$, as

$$\partial \mathscr{A}/\partial t + \nabla_1 \cdot [(\mathbf{U} + \mathbf{c}_g) \mathscr{A}] = 0. \tag{11.28}$$

This was first introduced by Whitham (1965) and its derivation was further generalized by Bretherton & Garrett (1968) and Hayes (1970). In Cases (a) and (b) above, results (11.25) and (11.27) follow directly from (11.28).

A compact derivation of this result comes from considering a suitably-chosen averaged Lagrangian density $\mathscr{L}(\mathbf{k}, \omega', a; \mathbf{x}, t)$. The average is taken over the phase χ of the waves and the explicit dependence on the space variables $\mathbf{x}$ and t is weak. With 'modal waves' for which spatial variation in one or more co-ordinate directions is not periodic – for instance the depth-dependence of water waves – $\mathscr{L}$ takes the form of an integral over such co-ordinates and $\mathbf{x}$ denotes the remaining 'propagation space' co-ordinates (e.g. x, y above). The parameter a characterizes the local wave amplitude.

For coherent wave-trains, the function $\mathscr{L}$ satisfies a variational principle with respect to variations in amplitude a and phase $\chi = \mathbf{k} \cdot \mathbf{x} - \omega' t + \Delta$, namely

$$\frac{\partial \mathscr{L}}{\partial a} = 0, \quad \frac{\partial}{\partial t}\left(\frac{\partial \mathscr{L}}{\partial \chi_t}\right) + \nabla_1 \cdot \frac{\partial \mathscr{L}}{\partial(\nabla \chi)} - \frac{\partial \mathscr{L}}{\partial \chi} = 0. \quad (11.29\mathrm{a,b})$$

Because of the averaging, $\mathscr{L}$ is independent of χ. Also, by definition, $\chi_t = -\omega'$ and $\nabla \chi = \mathbf{k}$ and so the latter equation is

$$\frac{\partial A}{\partial t} + \nabla \cdot \mathbf{B} = 0, \quad A = \frac{\partial \mathscr{L}}{\partial \omega'}, \quad \mathbf{B} = -\frac{\partial \mathscr{L}}{\partial \mathbf{k}}.$$

The equation $\partial \mathscr{L}/\partial a = 0$ yields a nonlinear dispersion relation and the linear one is recovered on letting $a \to 0$. In the presence of a slowly-varying mean flow $\mathbf{U}(\mathbf{x}, t)$, the latter dispersion relation has the form $\omega' = \mathbf{U} \cdot \mathbf{k} + \Omega(\mathbf{k})$ relating $\mathbf{k}$ to ω'. Accordingly,

$$\mathbf{B} = -\frac{\partial \mathscr{L}}{\partial \mathbf{k}} = -\frac{\partial \mathscr{L}}{\partial \omega'}\frac{\partial \omega'}{\partial \mathbf{k}} = A(\mathbf{U} + \mathbf{c}_\mathrm{g}),$$

where $\mathbf{c}_\mathrm{g} = \partial \Omega/\partial \mathbf{k}$, and a result of the form (11.28) is established.

It remains to show that A may be identified with $\mathscr{A} = E/\omega$, with $\omega = \Omega(\mathbf{k})$. In a reference frame moving with velocity $\mathbf{V}$, A is conserved and $\mathbf{B}$ is altered by the addition of the quantity $A\mathbf{V}$ (Hayes 1970). Accordingly, A may be evaluated by constructing $\mathscr{L}$ in any reference frame and the most convenient choice has $\mathbf{V}$ exactly equal to the local mean flow $\mathbf{U}$. In this frame, $\partial \mathscr{L}/\partial \omega'$ equals $\partial T/\partial \omega$, where T is the kinetic energy. Also, T is proportional to $(\omega a)^2$ and so $\partial T/\partial \omega = 2T/\omega$. For linear waves, $2T = E$ in this frame, and the result follows. In any other frame, with observed mean flow $\mathbf{U}$, the frequency is $\omega' = \omega + \mathbf{U} \cdot \mathbf{k}$, and the $O(\epsilon^2)$ energy density E' associated with the fluctuations is

$$E' = E + \rho \int_{-h}^{\zeta} \mathbf{U} \cdot \mathbf{u}' \, \mathrm{d}z$$

$$= E + (\mathbf{U} \cdot \mathbf{k})(\rho a^2 \omega/2k \tanh kh)$$

$$= E\left(1 + \frac{\mathbf{U} \cdot \mathbf{k}}{\omega}\right) = E\left(\frac{\omega'}{\omega}\right).$$

Accordingly,

$$A = E/\omega = E'/\omega'. \quad (11.30)$$

Notice that E' is the energy density directly associated with the $O(\epsilon)$ fluctuations only. It does not take into account contributions to the total energy density associated with $O(\epsilon^2)$ mean changes $\overline{\zeta^{(2)}}$ in surface level or any wave-induced mean flows $\overline{\mathbf{u}^{(2)}}$ additional to the given $\mathbf{U}$, though these contributions are typically of comparable magnitude. It should also be noted that, although a conservation equation for A exists for fully nonlinear waves (see §12.2), the wave action A may be identified with E/ω only in the weakly-nonlinear limit.

For surface waves, the *total* $O(\epsilon^2)$ disturbance energy density, relative to that when no wave is present, is

$$\mathscr{E} = E' + \tfrac{1}{2}\rho U^2 \overline{\zeta^{(2)}} + \rho U h \overline{u^{(2)}} \tag{11.31}$$

when the fixed level of zero potential energy is chosen at the undisturbed free surface. If, say, the channel bottom were chosen instead, an additional term $\rho g h \overline{\zeta^{(2)}}$ must be added. In view of this arbitrariness of definition, no simple relationship between $\mathscr{E}$ and A should be expected here. In particular, $\mathscr{E}$ does not satisfy a conservation law of the form (11.28). In contrast, an example is given in the next section for which the total disturbance energy and wave action density *are* so connected.

11.5 Waves in inviscid stratified flow

Consider a train of small-amplitude internal gravity waves propagating through a stably stratified fluid which supports a primary shear flow $[U(z), 0, 0]$. For simplicity, the Boussinesq approximation is assumed to hold and wavelengths are supposed small compared with the length-scale of variations of $U(z)$ and the density $\rho(z)$. Consistent with (11.24a, b), we further suppose that the frequency ω' and horizontal wavenumber components k_1, k_2 are constant and we set $k_2 = 0$, though the latter is not an essential restriction. Also, there is now a vertical wavenumber component k_3 which is a function of z only, and the local wave amplitude is assumed to depend only on z and t.

The linear dispersion relation yields the 'ray-theory approximation' (cf. §4.1)

$$\omega' = \omega + k_1 U(z), \qquad \omega^2 = N^2(z) k_1^2 / (k_1^2 + k_3^2(z))$$

where $N^2(z)$ is the local Brunt–Väisälä frequency $(-g\rho_0^{-1}\,\mathrm{d}\rho/\mathrm{d}z)^{\frac{1}{2}}$. The requirement that ω' is constant yields

$$k_3(z) = \pm k_1 [N^2(\omega' - k_1 U)^{-2} - 1]^{\frac{1}{2}}. \tag{11.32}$$

The horizontally-averaged energy equation for such disturbances is, at $O(\epsilon^2)$,

$$\partial(E + \rho_0 U u^{(2)})/\partial t = -\partial(\overline{p'w'} + \rho_0 U\overline{u'w'})/\partial z \tag{11.33}$$

at each level z. Here, primes denote $O(\epsilon)$ fluctuations, $u^{(2)}$ is the wave-induced

$O(\epsilon^2)$ mean flow correction, ρ_0 is the constant reference density and E is the horizontally-averaged energy density at each z associated with the $O(\epsilon)$ fluctuations (cf. Acheson 1976).

Now, for plane waves,

$$u' = [-k_3(z)/k_1]\, w'$$

by continuity, and

$$E(z, t) = \tfrac{1}{2}\rho_0(\overline{u'^2 + w'^2} + N^2\overline{\zeta'^2}) = \rho_0\, \overline{w'^2}(k_1^2 + k_3^2)/k_1^2$$

where ζ' denotes the $O(\epsilon)$ vertical displacement of fluid particles. Clearly, E is positive definite: unlike the depth-integrated result (11.30), no term in U here occurs in E, whatever the reference frame.

Equation (11.33) may be rewritten as

$$\partial\mathscr{E}/\partial t + \partial\mathscr{F}/\partial z = 0, \qquad (11.34)$$

$$\mathscr{E} \equiv E + \rho_0\, U u^{(2)}, \qquad \mathscr{F} \equiv \overline{p'w'} + \rho_0\, U\overline{u'w'},$$

where $\mathscr{E}$ is the total $O(\epsilon^2)$ mean energy density of the disturbance. Clearly, $\mathscr{E}$ may take either positive or negative values. This expression is simpler than the depth-integrated gravity-wave result (11.31) owing to the absence of free-surface terms.

Integration of the linearized x-momentum equation yields

$$p' = \rho_0(c - U)u' - \rho_0\, dU/dz \int^x w'\, dx$$

and so

$$\overline{p'w'} = \rho_0(c - U)\overline{u'w'}$$

where $c = \omega'/k_1$ is the constant horizontal phase velocity of the waves. Accordingly,

$$\mathscr{F} = \rho_0\, c\overline{u'w'}.$$

Also, the mean $O(\epsilon^2)$ x-momentum equation is

$$\rho_0\, \partial u^{(2)}/\partial t = -\rho_0\, \partial(\overline{u'w'})/\partial z. \qquad (11.35)$$

Result (11.34) turns out to be the wave-action equation

$$\partial\mathscr{A}/\partial t + \partial(w_g\, \mathscr{A})/\partial z = 0 \qquad (11.36)$$

multiplied by the constant ω', where $w_g = \partial\omega/\partial k_3$ is the vertical component of group velocity and

$$\mathscr{A} = \mathscr{E}/\omega' = E/\omega. \qquad (11.37)$$

Here, wave action and total disturbance energy $\mathscr{E}$ *are* directly related. It follows that

$$\mathscr{F}/\omega' = \rho_0\, k_1^{-1}\overline{u'w'} = w_g\, \mathscr{A}$$

and, from (11.35) and (11.36), that

$$\rho_0 u^{(2)} = k_1 \mathscr{A} = c^{-1} \mathscr{E} \qquad (11.38)$$

since $u^{(2)} = 0$ when $\mathscr{A} = 0$.

This remarkably simple result for the induced mean flow $u^{(2)}$ has no counterpart in the superficially-similar case of surface gravity waves treated in §11.3(b): in the latter, the matter is complicated by variations of mean surface level.

The particular case of over-reflection in uniformly stratified Helmholtz flow (4.8) has already been discussed in §4.4. As a packet of incident x-periodic internal waves with local energy density E_I travels upwards through the lower fluid in which $U = 0$, it induces an $O(\epsilon^2)$ horizontal mean flow $u^{(2)} = E_I(\rho_0 c)^{-1}$, in accordance with (11.38). This disappears after the wave-packet has passed. A reflected wave induces a similar mean flow. In the upper fluid, where $U = U_1$, the mean flow modification $u^{(2)}$ is positive or negative in accordance with the sign of the transmitted disturbance energy $\mathscr{E} = \mathscr{E}_T$. When $\mathscr{E}_T$ is negative, the local reduction $-\rho u^{(2)} U_1$ of mean kinetic energy is

$$- U_1 \mathscr{E}_T / c = U_1 (U_1 - c)^{-1} E_T$$

which exceeds the local energy density E_T of the fluctuations. The upwards-transmitted disturbance then carries a net energy defect, while the reflected disturbance transmits positive net energy downwards. It is for this reason that the amplitude of the reflected wave must exceed that of the incident wave. The mean flows induced during over-reflection of such a wave-packet are illustrated schematically in Figure 4.4(a). In most respects, the above discussion follows Acheson (1976), who also incorporates hydromagnetic effects.

In *resonant* over-reflection, there is no incident wave but 'transmitted' and 'reflected' waves of opposite energies radiate away from the vortex sheet. In this case, the mean-flow modification is as indicated in Figure 4.4(b). It is of interest to note that the radiating waves do *not* continuously erode the velocity discontinuity: it is merely decreased from its original value by a fixed $O(\epsilon^2)$ amount. In contrast, a Kelvin–Helmholtz instability leads to the eventual (in fact, rapid) smoothing of the discontinuity, as in Figure 4.4(c). In the former case, the vortex sheet acts as a continuous transmitter of waves; in the latter, it produces an intense disturbance of finite duration.

Resonant over-reflection, without instability, provides a means of redistributing energy without greatly disrupting the primary shear flow which supports it. Lindzen (1974) has suggested this as a possible

Figure 4.4. Schematic representation of mean flow modifications of a vortex-sheet profile as time increases, for (*a*) a packet of over-reflecting waves meeting the vortex sheet; (*b*) resonant over-reflection; (*c*) Kelvin–Helmholtz instability (after Acheson 1976).

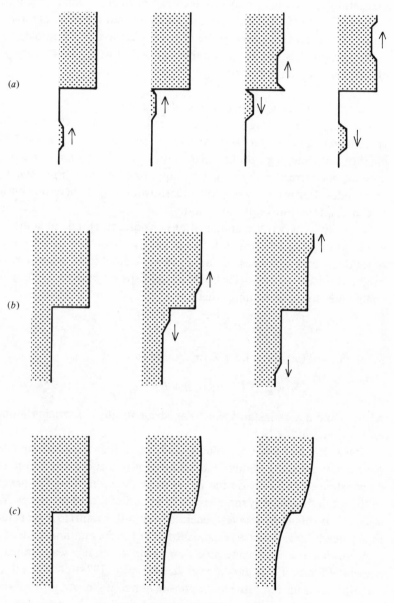

mechanism for generating clear-air turbulence at large altitudes. However, there are few known cases of resonant over-reflection for which the velocity profile is not unstable to other wavenumbers (see Acheson 1976, for a magnetohydrodynamic case) and, for most of these, instability reappears when more realistic velocity profiles are employed (Drazin, Zaturska & Banks 1979). The true importance of over-reflection in meteorology and oceanography has yet to be established: see Fritts' (1984) survey of atmospheric wave-phenomena.

11.6   *Mean flow oscillations due to dissipation*

In the preceding section, changes in mean flow are driven by a vertical gradient of Reynolds stress, which occurs only when the local wave amplitude is changing with time. However, when dissipation is taken into account, such stress gradients may be supported by *time-independent* wave amplitudes. For water waves, this was shown in § 10.1: here we consider a related internal-wave phenomenon.

As an internal wave, maintained at constant amplitude at some level $z = 0$, propagates upwards, it is progressively attenuated by viscous and other dissipative processes. Because the wave amplitude diminishes with height z, there is an associated Reynolds stress gradient $\partial \tau / \partial z$ which contributes to the governing equation

$$\frac{\partial \bar{u}}{\partial t} - \nu \frac{\partial^2 \bar{u}}{\partial z^2} = \frac{1}{\rho_0} \frac{\partial \tau}{\partial z}$$

for the mean horizontal flow $\bar{u}$. In equilibrium,

$$\bar{u}(z) = -(\nu \rho_0)^{-1} \int_0^z \tau \, dz + Az + B$$

where A and B are constants which vanish identically if $\bar{u}$ remains bounded as $z \to \infty$ and $\bar{u}(0) = 0$.

Holton & Lindzen (1972), Plumb (1977) and Plumb & McEwan (1978) have investigated the mutual interaction of waves and mean flow in such circumstances. Not only do the waves drive the mean flow: the presence of the mean flow slightly modifies the attenuation rate of the waves. With a single train of internal waves, generated at $z = 0$, Plumb (1977) concluded that a steady state is always established after a sufficient time has elapsed.

A much more interesting situation exists when *two* wave-trains are present. Plumb (1977) and Plumb & McEwan (1978) considered such wave-trains as having equal and opposite horizontal phase velocities: this corresponds to a standing wave maintained at $z = 0$. By symmetry, there is an equilibrium solution with zero mean flow, since mean-flow contributions from the respective wave-trains cancel. But this solution sometimes turns out to be *unstable*. The presence of a weak mean flow breaks the symmetry and alters the local attenuation rate of one wave component relative to the other. This, in turn, produces a Reynolds stress gradient which acts on the mean flow. Computations by Plumb (1977), both for a 'fully' nonlinear model and a weakly nonlinear approximation, revealed time-periodic oscillations of the mean flow. Experiments were conducted by Plumb & McEwan (1978), in an annular channel with flexible bottom made to perform standing-wave oscillations. Mean-flow reversals, of roughly one-hour period, were observed in the salt-stratified water within the channel. A similar mechanism is believed to be responsible for observed reversals, every 26 months or so, of the mean wind in the lower equatorial stratosphere.

Such phenomena deserve further theoretical and experimental investigation. The importance of dissipative processes, with sometimes unexpected manifestations, is all too easily overlooked.

12 Generalized Lagrangian mean (GLM) formulation
12.1 *The GLM equations*

It has long been recognized that the second-order Eulerian mean flow is not usually identical to the second-order averaged drift velocity of individual fluid particles. As mentioned in §10.1, the $O(\epsilon^2)$ Eulerian velocity is zero for constant amplitude *inviscid* water waves, but fluid particles travel with a non-zero Stokes drift velocity $u_s(z)$ which diminishes with depth. In a Lagrangian (particle) formulation, the Stokes drift contains all the mean momentum. In contrast, the Eulerian representation, with time-averages at fixed points in space, yields zero mean momentum except above the level of wave troughs. Only at those points, which are sometimes in the water and sometimes not, is there a non-zero average.

Rather different, but complementary, views of the flow are given by the Eulerian and Lagrangian representations. Traditionally, the former has predominated, but recent work has shown that a Lagrangian approach is often helpful. The stimulus for the latter work has come from the derivation of *exact* equations for Lagrangian-mean flows, by Andrews & McIntyre (1978a, b) following earlier attempts by Bretherton (1971) and others. The 'generalized Lagrangian mean' (or 'GLM') equations connect a Lagrangian velocity $\mathbf{u}^L(\mathbf{x}, t)$ and density $\tilde{\rho}(\mathbf{x}, t)$ to a 'pseudomomentum' vector $\mathbf{p}(\mathbf{x}, t)$, per unit mass. For weakly nonlinear waves, the pseudomomentum $\mathbf{p}$ may be evaluated, at order $O(\epsilon^2)$, directly from the linear wave solution and so the $O(\epsilon^2)$ Lagrangian velocity $\mathbf{u}^L$ may be found. The

$O(\epsilon^2)$ Eulerian velocity may be recovered from the Lagrangian velocity on subtracting the 'generalized Stokes drift' $\mathbf{u}^S(\mathbf{x}, t)$ which can also be evaluated at $O(\epsilon^2)$ from the linear solution. For many systems which support waves and weak mean flows, the GLM equations yield the mean flows more readily than would an Eulerian analysis, particularly where nonlinear free-surface boundary conditions are involved. We here give a brief outline of the GLM theory and some applications.

The GLM description is a hybrid Eulerian–Lagrangian one in which the GLM flow is described by equations in Eulerian form, with spatial position $\mathbf{x}$ and time t as independent variables rather than initial particle position and time. Let the particle displacement associated with the waves be $\xi(\mathbf{x}, t)$ where the (suitably-defined) Lagrangian mean of ξ is zero. The GLM operator $\overline{()}^L$ entails averaging over particles at the displaced positions $\mathbf{x}+\xi$: i.e.

$$\overline{\phi(\mathbf{x}, t)}^L = \overline{\phi^\xi(\mathbf{x}, t)}, \quad \phi^\xi(\mathbf{x}, t) \equiv \phi(\mathbf{x}+\xi, t), \tag{12.1a, b}$$

where $\overline{()}$ denotes a corresponding Eulerian average over positions in space. Averaging may be carried out over time t for time-periodic flows or over a single space co-ordinate x_i for flows which are spatially-periodic in x_i; alternatively, an ensemble average may be taken over some other suitable label, such as the phase of waves at some fixed position and time. The best choice of average depends on the problem to be examined.

Associated with any Eulerian velocity vector field $\mathbf{u}(\mathbf{x}, t)$, there is a uniquely defined 'related velocity field' $\mathbf{v}(\mathbf{x}, t)$ such that, when the actual fluid particle at $\mathbf{x}+\xi$ moves with its velocity $\mathbf{u}(\mathbf{x}+\xi, t)$; a notional particle at $\mathbf{x}$ may be regarded as moving with velocity $\mathbf{v}(\mathbf{x}, t)$. Here, $\mathbf{v}(\mathbf{x}, t)$ satisfies

$$(\partial/\partial t + \mathbf{v} \cdot \nabla)[\mathbf{x}+\xi(\mathbf{x}, t)] = \mathbf{u}(\mathbf{x}+\xi, t);$$

$\xi(\mathbf{x}, t)$ has zero Eulerian mean, $\overline{\xi(\mathbf{x}, t)} = 0$, and $\mathbf{v}(\mathbf{x}, t)$ is a mean quantity such that $\overline{\mathbf{v}(\mathbf{x}, t)} = \mathbf{v}(\mathbf{x}, t)$. So defined, $\mathbf{v}(\mathbf{x}, t)$ is the required Lagrangian-mean velocity, $\bar{\mathbf{u}}^L(\mathbf{x}, t)$. The Lagrangian means of all other flow quantities are similarly defined but it proves convenient to treat the density rather differently.

From the compressible Navier–Stokes equations, Andrews & McIntyre (1978a, b) derived the exact equations of GLM motion. For homentropic flows, the GLM momentum and continuity equations simplify to

$$\bar{D}^L(\bar{u}_i^L - p_i) + \bar{u}_{k,i}^L(\bar{u}_k^L - p_k) + 2(\mathbf{\Omega} \times \bar{\mathbf{u}}^L)_i + \bar{\Pi}_i^L = -\bar{X}_i^L - \overline{\xi_{k,i} X_k^l}, \tag{12.2}$$

$$\Pi_i \equiv F_i + \frac{\partial}{\partial x_i} \int \frac{dp}{\rho} \quad (i = 1, 2, 3),$$

$$\bar{D}^L \tilde{\rho} + \tilde{\rho} \nabla \cdot \bar{\mathbf{u}}^L = 0, \quad \bar{D}^L \equiv \partial/\partial t + \bar{\mathbf{u}}^L \cdot \nabla. \tag{12.3}$$

Here, F_i denotes all external body forces per unit mass except the GLM Coriolis force per unit mass $\boldsymbol{\Omega} \times \bar{\mathbf{u}}^{\mathrm{L}}$, where $\boldsymbol{\Omega}$ is the constant angular velocity of a rotating reference frame; the vector field $\mathbf{p} = p_i(x, t)$ is the *pseudomomentum per unit mass* defined as

$$p_i(\mathbf{x}, t) \equiv -\overline{\xi_{j,i}[u_j^l + (\boldsymbol{\Omega} \times \boldsymbol{\xi})_j]} \tag{12.4}$$

(not to be confused with pressure p) and $\mathbf{X} = \bar{\mathbf{X}}^{\mathrm{L}} + \mathbf{X}^l$ is a function which represents all dissipative terms. Symbols identified by the label $(\ )^l$ are 'wave quantities' with zero mean, such that

$$\phi^l \equiv \phi^\xi - \overline{\phi^\xi} = \phi^\xi - \bar{\phi}^{\mathrm{L}}.$$

Also, the 'wave fields' $\boldsymbol{\xi}(\mathbf{x}, t)$ and $\mathbf{u}^l(\mathbf{x}, t)$ are related by the kinematic condition

$$\mathbf{u}^l = \bar{\mathbf{D}}^{\mathrm{L}} \boldsymbol{\xi}. \tag{12.5}$$

Cartesian tensor notation is used throughout, with $(\ )_{,j} = \partial/\partial x_j$, $(\ )_{,t} = \partial/\partial t$ and summation of repeated indices over the values 1, 2, 3.

The density $\tilde{\rho}(\mathbf{x}, t)$ of the 'related flow' $\bar{\mathbf{u}}^{\mathrm{L}}(\mathbf{x}, t)$ is *defined* so as to satisfy (12.3), and is connected to the actual fluid density $\rho^\xi(\mathbf{x}, t) = \rho(\mathbf{x} + \boldsymbol{\xi}, t)$ by

$$\tilde{\rho} = \rho^\xi J, \quad J \equiv \det\{\delta_{ij} + \xi_{i,j}\} \tag{12.6}$$

where J is the Jacobian of the mapping $\mathbf{x} \to \mathbf{x} + \boldsymbol{\xi}$. We note that constant-density flows, $\rho^\xi = $ constant, do not usually give rise to a constant density $\tilde{\rho}$ of the related flow field. But this definition has the major advantage that $\tilde{\rho}$ is a mean quantity, $\overline{\tilde{\rho}(\mathbf{x}, t)} = \tilde{\rho}(\mathbf{x}, t)$.

In general, equations (12.2) and (12.3) must be supplemented by an equation of state linking pressure and density. The GLM equations are then *complete* if the wave field is known and may in principle be solved to find the GLM flow associated with any *specified* wave field $\boldsymbol{\xi}(\mathbf{x}, t)$. Of course, the wave field is itself usually dependent on the mean state: in this sense, the above equations are *not* complete. But, for weakly nonlinear systems, the $O(\epsilon)$ wave field is frequently known and equations (12.2) and (12.3) are then sufficient to yield the $O(\epsilon^2)$ mean flow supported by the waves.

For small-amplitude waves but $O(1)$ mean state, a Taylor expansion of (12.1b) gives

$$\phi^\xi = \bar{\phi} + \phi' + \xi_j \phi_{,j} + \tfrac{1}{2} \xi_j \xi_k \bar{\phi}_{,jk} + O(\epsilon^3)$$

where $\phi(\mathbf{x}, t) = \bar{\phi}(\mathbf{x}, t) + \phi'(\mathbf{x}, t)$ and $\phi'(\mathbf{x}, t)$ denotes the Eulerian wave field. Accordingly,

$$\phi^l = \phi' + \xi_j \bar{\phi}_{,j} + O(\epsilon^2)$$

and the mean fields $\bar{\phi}^{\mathrm{L}}(\mathbf{x}, t)$, $\bar{\phi}(\mathbf{x}, t)$ differ by the 'Stokes correction'

$$\bar{\phi}^{\mathrm{S}}(\mathbf{x}, t) = \bar{\phi}^{\mathrm{L}} - \bar{\phi} = \overline{\xi_j \phi'_{,j}} + \tfrac{1}{2} \overline{\xi_j \xi_k} \bar{\phi}_{,jk} + O(\epsilon^3). \tag{12.7}$$

In particular, result (12.7) yields the 'generalized Stokes drift' $\bar{u}^S = \bar{u}^L - \bar{u}$.

For incompressible (Boussinesq) flows, such that $\nabla \cdot u' = 0$ and the density ρ^ξ of each fluid particle is constant,

$$\tilde{\rho}/\rho^\xi = J = 1 - \tfrac{1}{2}(\overline{\xi_j \xi_k})_{,jk} + O(\epsilon^3).$$

Accordingly,

$$\nabla \cdot \bar{u}^L = \tfrac{1}{2}\bar{D}^L(\overline{\xi_j \xi_k})_{,jk} + O(\epsilon^3)$$

and the Lagrangian-mean velocity field is not usually divergence-free. Fortunately, it often happens that the $O(\epsilon^2)$ variations of $\tilde{\rho}$ drop out of the $O(\epsilon^2)$ equations for $\bar{u}^L$.

When the mean flow $\bar{u}$ is also weak, of order $O(\epsilon^2)$, $\phi^l = \phi' + O(\epsilon^2)$ and $\bar{\phi}^S = \overline{\xi_j \phi'_{,j}} + O(\epsilon^3)$. If the Coriolis force is absent, one then has, from (12.4), (12.5) and (12.7),

$$p_i = \bar{u}_i^S - \tfrac{1}{2}(\overline{\xi_j \xi_j})_{,it} + O(\epsilon^3). \tag{12.8}$$

When the wave field ξ_j is time-periodic, this gives $p_i = \bar{u}_i^S + O(\epsilon^3)$; a result which establishes the equality at $O(\epsilon^2)$ of the pseudomomentum $\tilde{\rho} p_i$ per unit volume and the $O(\epsilon^2)$ mean momentum per unit volume, $\tilde{\rho} \bar{u}_i^S$. In general, physical momentum and the pseudomomentum p_i are *not* equal but p_i and $\bar{u}_i^L$ are exactly related by

$$\nabla \times (\bar{u}^L - p) = 0$$

whenever the wave fluid is irrotational.

12.2   Pseudomomentum and pseudoenergy

The wave action $\mathscr{A}$ defined in §11.4 was originally introduced to describe an $O(\epsilon^2)$ property of weakly nonlinear waves. But the concept of wave action has been powerfully generalized by Andrews & McIntyre (1978b), who derived an *exact* equation of the form

$$\bar{D}^L A + \tilde{\rho}^{-1} \nabla \cdot B = \mathscr{H}. \tag{12.9}$$

The right-hand side $\mathscr{H}$ is zero for conservative motions and is here disregarded. The *generalized wave action density* A is defined as

$$A = \overline{\xi_{j,\beta}[u_j^l + (\Omega \times \xi)_j]} \tag{12.10}$$

where β is the label over which averages are taken. The non-advective flux of wave action B is

$$B_j = \overline{p^\xi \xi_{i,\beta} K_{ij}}$$

and K_{ij} is the (i, j)th cofactor of the Jacobian J defined in (12.6). For waves periodic in x_1, β may be chosen as $-x_1$; then A is identical to the x_1-component of pseudomomentum, $A = p_1$.

For plane-periodic, small-amplitude waves of frequency ω' in a (locally) constant mean flow $\mathbf{U}$, $\mathbf{u}^l = \mathbf{u}' = \overline{\mathbf{D}}^L \xi$ and $\xi \propto \exp i(\mathbf{k} \cdot \mathbf{x} - \omega' t - \delta)$ correct to $O(\epsilon)$. On choosing β as the phase δ varied over $(0, 2\pi)$, we have $\xi_{,\beta} = \omega^{-1} \overline{\mathbf{D}}^L \xi = \omega^{-1} \mathbf{u}'$ where $\omega = \omega' - \mathbf{k} \cdot \mathbf{U}$ is the intrinsic wave frequency observed in the local rest frame of the fluid. From (12.10),

$$A = \omega^{-1} \overline{\mathbf{u}' \cdot (\mathbf{u}' + \mathbf{\Omega} \times \xi)} + O(\epsilon^3),$$

which establishes that the integral of A over depth coincides with $\mathscr{A}$ of (11.28) at $O(\epsilon^2)$ when $\mathbf{\Omega} = 0$. General relationships between A and other GLM quantities, such as the flux of pseudomomentum and the 'pseudo-energy' are discussed by Andrews & McIntyre (1978b), within an energy–momentum-tensor formalism. Also, the $O(\epsilon^2)$ concept of 'radiation stresses' is connected to the exact GLM formulation by Andrews & McIntyre (1978a).

Briefly, for inviscid flows, a tensor $T_{\mu\nu}$, where μ, ν stand for x_i $(i = 1, 2, 3)$ or t, satisfies a set of conservation relations

$$T_{\mu\nu,\nu} = 0. \tag{12.11}$$

The component T_{tt} is the *pseudoenergy* per unit volume and the components T_{tj} $(j = 1, 2, 3)$ denote its flux. These are defined (see Andrews & McIntyre 1978b, §5) as

$$T_{tt} = \tilde{\rho} \mathrm{e} - \overline{L - L_0},$$
$$T_{tj} = \bar{u}_j^L \tilde{\rho} \mathrm{e} + \overline{p^\xi \xi_{m,t} K_{mj}} \tag{12.12}$$

where

$$\mathrm{e} = \overline{\xi_{,t} \cdot (\mathbf{u}^l + \mathbf{\Omega} \times \xi)};$$

L is the total Lagrangian density and L_0 the 'undisturbed' Lagrangian obtained from L by retaining all mean and ignoring all fluctuating quantities.

12.3   Surface gravity waves

As a first simple example, we consider the linear wave field $\mathbf{u}' = \nabla\phi$,

$$\phi(x, z, t) = \frac{\omega a \cosh\left[k(z+h)\right]}{k \sinh kh} \sin(kx - \omega t) + O(a^2) \quad (-h \leqslant z \leqslant 0)$$

$$\tag{12.13}$$

of inviscid surface capillary–gravity waves with frequency ω given by

$$\omega^2 = (gk + \rho^{-1}\gamma k^3) \tanh kh$$

(cf. §11.2 above). The associated periodic particle displacements ξ_i $(i = 1, 2, 3)$ satisfy $\xi_{i,t} = u_i' + O(\epsilon^2)$, where $\epsilon = ka$, and the Stokes drift velocity $\bar{u}_i^S = (u^S, 0, 0)$,

$$u^S = \frac{\omega k a^2 \cosh [2k(z+h)]}{2 \sinh^2 kh}$$

is immediately given by (12.7). The corresponding $O(\epsilon^2)$ solution of (12.2) is $\bar{u}_i^L = p_i = \bar{u}_i^S$. It immediately follows that the $O(\epsilon^2)$ Eulerian mean flow $\bar{u}_i$ is zero. Also, at $O(\epsilon^2)$, (12.6) gives

$$\tilde{\rho}/\rho - 1 = -\tfrac{1}{2}(ka)^2 \cosh^2 [k(z+h)]/\sinh^2 kh.$$

The reduction in mean density $\tilde{\rho}$ of the related velocity field may be explained, physically, in terms of a mean $O(\epsilon^2)$ change in level of an averaged set of fluid particles, as compared with that when no wave is present. This is most readily understood by considering a marked set of particles which, in the absence of waves, lie at equal intervals in x along a given line, say $z = z_0$. In a reference frame moving with the wave, these particles are observed to be more closely spaced at wave crests than at wave troughs. Their average height therefore exceeds z_0 by an $O(\epsilon^2)$ amount. A similar line of particles initially situated on $z = z_0 + \delta$ experiences a slightly greater or lesser $O(\epsilon^2)$ mean displacement according as $\delta > 0$ or < 0, because the local wave amplitude increases with height. As a result, the apparent mean density $\tilde{\rho}$ associated with the slab of fluid initially between $z = z_0$ and $z = z_0 + \delta$ decreases in order to conserve mass. It must be emphasized that this 'divergence effect' of the related velocity field $\bar{\mathbf{u}}^L$ is a property of the averaging procedure: the physical density of the fluid of course remains constant.

With averages taken over $-x$, the wave action density A and pseudo-momentum component p_1 are the same. Integrating over the liquid depth, we here have

$$\mathscr{A} = \int_{-h}^{\zeta} p_1 \, \mathrm{d}z = -\int_{-h}^{0} \overline{(\partial \xi_i/\partial x) u_i'} \, \mathrm{d}z$$

and

$$\int_{-h}^{\zeta} (\boldsymbol{\nabla} \cdot \mathbf{B}) \, \mathrm{d}z = \int_{-h}^{0} (\partial B_1/\partial x) \, \mathrm{d}z = -\frac{\partial}{\partial x} \int_{-h}^{0} \overline{p' \, \partial \xi_1/\partial x} \, \mathrm{d}z$$

at $O(\epsilon^2)$. Since the latter integral is just $\partial(c_g \mathscr{A})/\partial x$, where $c_g = \partial \omega/\partial k$, and $\bar{\mathrm{D}}^L \mathscr{A} = \partial \mathscr{A}/\partial t$ at $O(\epsilon^2)$, result (11.28) with $\mathbf{U} = 0$ is recovered from (12.9). A similar equivalence between (11.28) and (12.9) for small-amplitude waves is readily verified for other examples discussed above.

12.4 *Inviscid shear-flow instability*

The GLM formulation was used by Craik (1982c) to describe growing waves in unstable shear flows $[U(z), 0, 0]$ between plane boundaries at $z = z_1$ and $z = z_2$. In a reference frame moving with the wave velocity c_r, the disturbance stream function of a small-amplitude two-dimensional wave-train is (cf. §9.1)

$$\psi(x, z, t) = \epsilon\, e^{\alpha c_i t}\, \mathrm{Re}\{\phi(z)\, e^{i\alpha x_1}\} + O(\epsilon^2)$$

where $x_1 = x - c_r t$. We assume that $c_i \geqslant 0$. The corresponding linearized velocity components $(u_1', 0, u_3')$ and displacements $(\xi_1, 0, \xi_3)$ are†

$$
\left.
\begin{aligned}
u_1' &= \epsilon\, e^{\alpha c_i t}\, \mathrm{Re}\{\phi'(z)\, e^{i\alpha x_1}\}, \quad u_3' = \epsilon\, e^{\alpha c_i t}\, \mathrm{Re}\{-i\alpha\phi(z)\, e^{i\alpha x_1}\}, \\
\xi_1 &= \epsilon\, e^{\alpha c_i t}\, \mathrm{Re}\left\{\left(\phi' - \frac{\bar{u}'\phi}{\bar{u} - ic_i}\right)\frac{e^{i\alpha x_1}}{i\alpha(\bar{u} - ic_i)}\right\}, \\
\xi_3 &= \epsilon\, e^{\alpha c_i t}\, \mathrm{Re}\left\{\frac{-\phi\, e^{i\alpha x_1}}{\bar{u} - ic_i}\right\}, \quad \bar{u} \equiv U - c_r.
\end{aligned}
\right\}
\tag{12.14}
$$

From (12.7), the generalized Stokes drift $\bar{\mathbf{u}}^S = (u_1^S, 0, u_3^S)$ is found to be

$$
\left.
\begin{aligned}
u_1^S &= \tfrac{1}{4}\epsilon^2\, e^{2\alpha c_i t}\left\{-\left(\frac{\phi\phi^{*\prime}}{\bar{u} - ic_i} + \text{c.c.}\right)' + \frac{\bar{u}''|\phi|^2}{\bar{u}^2 + c_i^2}\right\}, \\
u_3^S &= \tfrac{1}{2}\epsilon^2\, e^{2\alpha c_i t}\left(\frac{\alpha c_i\, |\phi|^2}{\bar{u}^2 + c_i^2}\right)'.
\end{aligned}
\right\}
\tag{12.15a, b}
$$

As $\alpha c_i \to 0$, $u_3^S \to 0$ and

$$u_1^S = \tfrac{1}{4}\epsilon^2\left\{-\left[\frac{(|\phi|^2)'}{\bar{u}}\right]' + \frac{\bar{u}''|\phi|^2}{\bar{u}^2}\right\} + O(\epsilon^2\alpha c_i)$$

except near the critical layer where $\bar{u}$ vanishes. This singularity exists even for viscous flows, since it derives from the linearized kinematic condition $\bar{\mathrm{D}}^L\boldsymbol{\xi} = \mathbf{u}'(\mathbf{x} + \boldsymbol{\xi}, t)$. The linear approximation there breaks down and there is a region of closed streamlines – the Kelvin 'cats' eyes' – centred on the critical layer for which the chosen averaging procedure is invalid.

The pseudomomentum $\mathbf{p} = (p_1, 0, p_3)$ has

$$
\left.
\begin{aligned}
p_1 &= -\tfrac{1}{2}\epsilon^2\, e^{2\alpha c_i t}\left(\frac{\bar{u}}{\bar{u}^2 + c_i^2}\right)\left\{\left|\phi' - \frac{\bar{u}'\phi}{\bar{u} - ic_i}\right|^2 + \alpha^2|\phi|^2\right\} = -\alpha^2\bar{u}\,\overline{|\xi|^2}, \\
p_3 &= -\tfrac{1}{2}\epsilon^2\, e^{2\alpha c_i t}\, \mathrm{Re}\left\{\left(\frac{\phi^*}{\bar{u} + ic_i}\right)'\frac{1}{i\alpha}\left[(\bar{u} + ic_i)\left(\frac{\phi}{\bar{u} - ic_i}\right)'' - \alpha^2\phi\right]\right\}.
\end{aligned}
\right\}
\tag{12.16a, b}
$$

As $\alpha c_i \to 0$, $p_3 \to 0$ for *inviscid* disturbances but not for viscous ones.

The $O(\epsilon^2)$ Eulerian mean flow $u^{(2)}$ for *inviscid* disturbances was found to satisfy (9.7). This may also be derived from the GLM equations (12.2),

† Recall that primes denote d/dz only when applied to functions of z only.

(12.3) and (12.6), which reduce to

$$2\alpha c_i(\bar{u}_1^L - p_1) + \bar{u}_3^L \bar{u}' = 0,$$
$$2\alpha c_i(\bar{u}_3^L - p_3) = 0,$$
$$2\alpha c_i(J-1) + \partial \bar{u}_3^L / \partial z = 0,$$

when there is no $O(\epsilon^2)$ pressure gradient. The last of these equations is an identity and the other two yield

$$\bar{u}_1^L = p_1 - \frac{\bar{u}'p_3}{2\alpha c_i}, \quad \bar{u}_3^L = p_3.$$

Substitution from (12.14), (12.15) and (12.16) gives

$$u^{(2)} = \bar{u}_1^L - u_1^S - \bar{u} = \tfrac{1}{2}\bar{u}''\overline{\xi_3^2} \tag{12.17}$$

for the $O(\epsilon^2)$ Eulerian velocity. This is just result (9.7).

Since the net Eulerian mass flux must remain constant in the absence of any pressure gradient,

$$\int_{z_1}^{z_2} u^{(2)} \, dz = \int_{z_1}^{z_2} \bar{u}''\overline{\xi_3^2} \, dz = 0. \tag{12.18}$$

The inflexion-point criterion that $\bar{u}''$ must change sign in $[z_1, z_2]$ is an immediate consequence.

For inviscid flow with averages taken over $-x_1$, the conservation law (12.9) yields

$$\rho \, \partial p_1 / \partial t = \rho \, \partial A / \partial t = -\partial(\overline{p' \, \partial \xi_3 / \partial x}) / \partial z$$

at $O(\epsilon^2)$ where p' is the $O(\epsilon)$ Eulerian pressure fluctuation. Integration across the flow between rigid boundaries at $z = z_1$ and z_2 immediately yields

$$\int_{z_1}^{z_2} p_1 \, dz = \text{constant}:$$

net pseudomomentum in the x_1-direction is conserved. Moreover, the constant is zero since $p_1 = 0$ at $t = -\infty$. From (12.16a) this implies that

$$\int_{z_1}^{z_2} \bar{u} \, \overline{|\xi|^2} \, dz = 0, \tag{12.19}$$

from which follows the well-known result that the wave velocity c_r lies within the range of the flow velocity.

In the present case, the pseudoenergy density satisfies

$$\partial T_{tt} / \partial t + \partial T_{t3} / \partial z = 0$$

from (12.11), and net pseudoenergy is conserved,

$$\int_{z_1}^{z_2} T_{tt} \, dz = 0,$$

since T_{t3} vanishes on plane boundaries $z = z_1$ and z_2 and T_{tt} is zero at $t = -\infty$. But T_{tt} here reduces to

$$T_{tt} = \tfrac{1}{4}\rho(c_i^2 - \bar{u}^2)\,\alpha^2\,\overline{|\xi|^2}$$

at $O(\epsilon^2)$ (see Craik 1982c) and so

$$\int_{z_1}^{z_2} (c_i^2 - \bar{u}^2)\,\overline{|\xi|^2}\,dz = 0. \tag{12.20}$$

Equations (12.19) and (12.20) together imply that $\bar{u}^2 - c_i^2 + \lambda\bar{u}$ changes sign within $[z_1, z_2]$ for *all* constants λ: a result which leads immediately to Howard's (1961) semicircle theorem (cf. Eckart 1963).

The total energy is

$$\mathscr{E} = E + \rho \int_{z_1}^{z_2} u^{(2)}\bar{u}\,dz$$

where

$$E = \tfrac{1}{4}\epsilon^2\rho\,e^{2\alpha c_i t} \int_{z_1}^{z_2} (|\phi'|^2 + \alpha^2\,|\phi|^2)\,dz$$

is the energy within the fluctuations. Substitution for $u^{(2)}$ from (12.17) and use of (12.14) and Rayleigh's equation (3.3) yields $\mathscr{E} = 0$. Also, because of (12.18), $\mathscr{E}$ remains zero in any parallel reference frame with $\bar{u}$ replaced by $U - c_r$. Accordingly, spontaneously-growing disturbances in inviscid unstratified flows have zero total energy as well as zero net pseudo-momentum and pseudoenergy.

Applications of GLM theory to stratified shear flows are reviewed by Grimshaw (1984).

13 Spatially-periodic mean flows

13.1 *Forced motions*

The quadratic interaction of two $O(\epsilon)$ wave-modes with amplitudes a_j and periodicities $\exp[i(\mathbf{k}_j \cdot \mathbf{x} - \omega_j t)] + \text{c.c.}$ where $j = 1, 2$, $\mathbf{x} = (x, y)$ and $\mathbf{k} = (\alpha, \beta)$ yields $O(\epsilon^2)$ forcing terms of the form $b^\pm \exp[i(\mathbf{k}_1 \pm \mathbf{k}_2) \cdot \mathbf{x} - i(\omega_1 \pm \omega_2) t] + \text{c.c.}$ when $b^\pm$ are proportional to $a_1 a_2$ and $a_1 a_2^*$ respectively. When $\mathbf{k}_1 = \mathbf{k}_2$ and $\omega_1 = \omega_2$, the given wave-modes coincide and the forcing terms drive a spatially-uniform mean flow and a second harmonic.

More generally, on writing $\mathbf{k}_3^\pm = \mathbf{k}_1 \pm \mathbf{k}_2$, the respective Fourier components in $\mathbf{k}_3^\pm$ satisfy equations of the form

$$(L^\pm\,\partial/\partial t + M^\pm)\phi_3^\pm = b^\pm\Psi_{1,\,\pm 2}(z)\,e^{(\omega_{1i} + \omega_{2i})t}\,e^{-i(\omega_{1r} \pm \omega_{2r})t}, \tag{13.1}$$

where $L^\pm$, $M^\pm$ are known linear operators in $\partial/\partial z$, and $\Psi_{1,\,\pm 2}$ may be found in terms of the linear eigenfunctions of the given wave-modes. The corresponding homogeneous equation, with zero right-hand side, together with appropriate homogeneous boundary conditions, defines an eigenvalue problem for the frequency. This yields a set of discrete eigenvalues $\omega_{\bar{n}3}^{\pm}$ ($n = 1, 2, \ldots$) and sometimes a continuous spectrum. In particular, for two-dimensional modes with $\mathbf{k}_j = (\alpha_j, 0)$, the appropriate homogeneous equation is just the Orr–Sommerfeld equation for wavenumbers $\alpha_1 \pm \alpha_2$.

Any solution of (13.1) may be expressed as a sum (or integral) over a complete set of the linear eigenfunctions of the homogeneous problem with time-dependent coefficients. If $\omega_1 \pm \omega_2$ coincides, or nearly coincides, with an eigenvalue $\omega_{\bar{n}3}^{\pm}$, a *three-wave resonance* occurs. In this case, the third mode need not remain small compared with the two original modes. Consideration of such resonance is deferred to the next chapter. Here, we assume that $\omega_1 \pm \omega_2$ are *not* close to any linear eigenvalue $\omega_{\bar{n}3}^{\pm}$ and that the solution $\phi_3^{\pm}$ remains of order $O(\epsilon^2)$ at all times. We also assume that the growth rates $\mathrm{Im}\{\omega_{\bar{n}3}^{\pm}\}$ of all linear eigenmodes of (13.1) are less than $\omega_{1\mathrm{i}} \pm \omega_{2\mathrm{i}}$, so that $\phi_3^{\pm}$ is eventually dominated by a particular integral of the form $\Phi_3^{\pm} \exp[-\mathrm{i}(\omega_{1\mathrm{r}} \pm \omega_{2\mathrm{r}}) t] \exp[(\omega_{1\mathrm{i}} \pm \omega_{2\mathrm{i}}) t]$ where

$$[-\mathrm{i}(\omega_1 \pm \omega_2) \mathbf{L}^{\pm} + \mathbf{M}^{\pm}] \Phi_3^{\pm} = b^{\pm} \Psi_{1, \pm 2}(z).$$

When two wavenumbers $\mathbf{k}_1$, $\mathbf{k}_2$ have identical real frequencies $\omega_{1\mathrm{r}} = \omega_{2\mathrm{r}}$, and $\omega_{1\mathrm{i}} = \omega_{2\mathrm{i}} = 0$, the forcing terms in $\exp[\pm\mathrm{i}(\mathbf{k}_1 - \mathbf{k}_2)\cdot\mathbf{x}]$ and the corresponding particular integral are independent of time t. This gives rise to a spatially-periodic mean flow. Such flows may result from symmetry of the wave fields, as with two equal oblique waves $\exp[\mathrm{i}(\alpha x \pm \beta y - \omega t)]$ with real frequency ω in a parallel flow $[U(z), 0, 0]$. More generally, if two wavenumbers $\mathbf{k}_1$, $\mathbf{k}_2$ have nearly equal frequencies, $|(\omega_1 - \omega_2)/\omega_1| \ll 1$, the forced solution ϕ_3^{-} varies slowly in time.

The best-known observations of wave-induced spatially-periodic mean flows are those in Klebanoff, Tidstrom & Sargent's (1962) study of three-dimensional disturbances in the unstable Blasius boundary layer. A small-amplitude, nearly two-dimensional, Tollmien–Schlichting wave was generated by a ribbon which vibrated with fixed frequency. As the wave progressed downstream, growing in amplitude, marked spanwise variations in amplitude spontaneously appeared. To control the spanwise wavelength and phase of these variations, small strips of adhesive tape were placed at equal intervals on the plate, beneath the vibrating ribbon. The local wave amplitude at fixed x and z then varied almost periodically with spanwise distance y, about a non-zero mean. Also, a spanwise-periodic mean flow developed, in the form of 'longitudinal vortices' which varied only slowly with x. A somewhat similar development of three-dimensionality was found by Nishioka, Iida & Ichikawa (1975), in plane Poiseuille flow. The longitudinal-vortex structure is generally believed to play an important part in inducing transition to turbulence (e.g. Tani 1969). Certainly, turbulent 'spots' are observed to develop first at spanwise ('peak') locations where the local wave amplitude is greatest (see §27.2).

Benney & Lin (1960) first proposed a theoretical model of the development of such longitudinal vortices, which was subsequently extended by Benney (1961, 1964). They considered a wave field comprising $O(\epsilon)$ two-dimensional and spanwise-periodic modes – say, $a\exp(i\alpha x - i\omega t)$ and $b\exp(i\alpha x \pm i\beta y - i\omega' t)$ plus complex conjugates – with $\omega_r = \omega_r'$ and small amplification rates ω_i, ω_i'. They then calculated the resultant mean flow components in $ab^*\exp(\pm i\beta y)$ and $|b|^2\exp(\pm 2i\beta y)$ plus conjugates. The former derive from the interaction of the two-dimensional mode with an oblique mode; the latter from the interaction of the two oblique modes. In addition, there are y-independent mean flow components in $|a|^2$ and $|b|^2$ from the self-interaction of each mode. The resultant mean flows were found to have a longitudinal-vortex structure qualitatively similar to that observed by Klebanoff *et al.* However, this theoretical model postulates, rather than explains, the existence of a strongly y-periodic wave field downstream of a nearly uniformly-vibrating ribbon. In addition, the model has been criticised by Stuart (1962a) and others on the grounds that the linear frequencies ω_r, ω_r' in fact differ by around 15% at wavenumbers corresponding to the experimental data.

Antar & Collins (1975) improved upon Benney's model analyses both by relaxing the assumption $\omega_r = \omega_r'$ to account for the actual frequency mismatch $\Delta\omega_r = \omega_r - \omega_r'$ and by computing second-order flows for actual Blasius and Falkner–Skan velocity profiles. The component of mean-flow distortion with spanwise wavenumber β then oscillates in time with frequency $\Delta\omega$ while that with wavenumber 2β does not. An alternative model, proposed by Nelson & Craik (1977), concerns waves of equal frequency, but with a mismatch $\Delta\alpha$ in downstream wavenumber. Mean flow components then have wavenumbers $(\Delta\alpha, \beta)$ and $(0, 2\beta)$. The latter model seems closer to Klebanoff *et al.*'s experimental situation and deserves a more detailed study. It is also possible that higher-order nonlinearities, especially those at $O(\epsilon^3)$, may lead to synchronization of wave-modes with the same downstream wavenumber (see §26). Examples of Antar & Collins' calculated mean-flow structure are shown in Figure 4.5, while Figure 4.6 reproduces some of Klebanoff *et al.*'s experimental data. Fuller discussions of mean-flow distortion in unstable boundary layers are given by Tani (1980) and Craik (1980).

Figure 4.5. Theoretical results of Antar & Collins (1975) for streamline pattern of wave-driven longitudinal vortices in Blasius flow at $R_\delta = 1630$, $\alpha\delta = 0.1336$, $\beta\delta = 0.0614$, where δ is displacement thickness. Cases (a), (b), (c) correspond to $\lambda_1/\lambda_2 = 25$, 1, 0.025 respectively, where λ_1/λ_2 measures the ratio of two-dimensional to oblique-wave amplitudes.

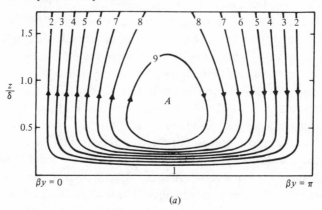

(a)

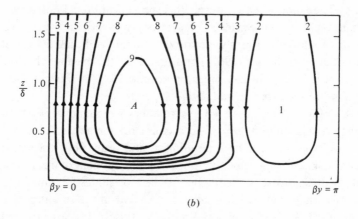

(b)

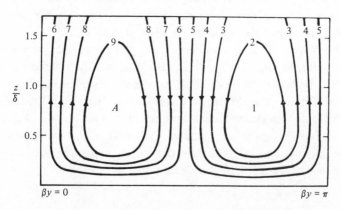

Figure 4.6. Experimental results of Klebanoff, Tidstrom & Sargent (1962) showing 'peak-valley' and longitudinal-vortex structure in the Blasius boundary layer. (*a*) shows spanwise variations of mean (U, V) and r.m.s. fluctuating (u', v') velocity components in the downstream and spanwise directions, at a fixed downstream location and fixed distance z from the wall: circles denote measurements at $z = 0.31 \, \delta$, crosses at $0.11 \, \delta$. (*b*), overleaf, shows z-distributions of mean velocity V at two spanwise locations separated by 0.4 in. and two downstream locations: those marked by $\bigcirc$, $\triangledown$ correspond to the position further downstream.

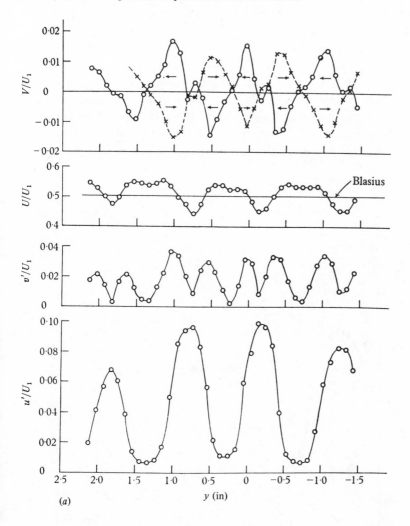

(*a*)

A rather similar model was proposed by Craik (1970) in an early attempt to describe the phenomenon of 'Langmuir circulations'. These are structures of longitudinal-vortex type which commonly develop in bodies of water subjected to moderate winds. Craik (1970) calculated the $O(\epsilon^2)$ spanwise-periodic flows associated with a pair of equal surface gravity waves with periodicities $\epsilon \exp(i\alpha x \pm i\beta y - i\omega t)$ propagating through an inviscid $O(1)$ primary shear flow $\mathbf{u} = [Kz, 0, 0]$. Without shear, there is no Eulerian mean flow; but when K is non-zero there is a spanwise-periodic distortion of the mean vorticity field. Leibovich & Ulrich (1972) gave a neat physical interpretation of this process in cases of weak $O(\epsilon^2)$ mean shear. Then the linear wave field is irrotational and has a Stokes drift which contains a spanwise-periodic component. Since vortex lines and fluid particles coincide in inviscid flows, the Stokes drift must induce spanwise-periodic distortions of an initially-uniform vorticity field. The spanwise-periodic x-component of vorticity at first grows linearly with time t, and this induces a further distortion of the mean flow to give a spanwise-periodic

Fig. 4.6b. For caption see previous page.

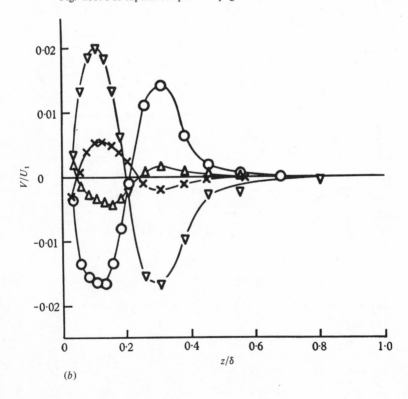

(b)

x-velocity component which grows initially as t^2. Viscous diffusion must inhibit the growth of the vortices and lead to an ultimate steady state.

Craik & Leibovich (1976) and Leibovich (1977a) subsequently derived the general equations governing weak $O(\epsilon^2)$ mean flows in the presence of arbitrary $O(\epsilon)$ fluctuations. The mean vorticity equation was found to be

$$(\partial/\partial\tau - \nu_1 \nabla^2)\,\overline{\omega} = (\overline{\omega}\cdot\nabla)\,(\overline{\mathbf{u}}+\mathbf{u}^S) - (\overline{\mathbf{u}}+\mathbf{u}^S)\cdot\nabla\overline{\omega} \qquad (13.2)$$

at leading order. Here, the mean Eulerian velocity, vorticity and generalized Stokes drift are $\epsilon^2\overline{\mathbf{u}}$, $\epsilon^2\overline{\omega}$ and $\epsilon^2\mathbf{u}^S$ respectively, while $\tau = \epsilon^2 t$ and $\nu_1 = \epsilon^{-2}\nu$. The viscosity ν is supposed small, of order $O(\epsilon^2)$, where ϵ characterizes the wave-slope. The average $(\,\overline{}\,)$ is taken over a time-scale large compared with the periods of the $O(\epsilon)$ wave-components but small compared with the slow $O(\epsilon^{-2})$ time-scale of the mean flow evolution. Note that the influence of the waves in (13.2) is entirely represented by the Stokes drift $\mathbf{u}^S$, in line with the convective processes described above. This, and the associated momentum equations, were initially derived using an Eulerian formulation, and entailed considerable effort, since the Stokes drift $\mathbf{u}^S$ does not arise naturally in such a representation. Subsequently, Leibovich (1980) derived these equations much more simply from the GLM equations (12.2) and (12.3). Craik (1982d) similarly showed that stronger mean flows, with an $O(\epsilon)$ downstream component, satisfy nearly identical equations which differ only by scaling factors in the definitions of $\overline{\mathbf{u}}$, $\overline{\omega}$, ν_1 and τ. For instance, if the mean flow is initially given as $\epsilon[\overline{u}(z), 0, 0]$ and the known $O(\epsilon)$ wave-field provides a spanwise-periodic Stokes drift $\epsilon^2[u^S(y, z), 0, 0]$, the inviscid GLM equations (12.2) and (12.3) yield

$$\left.\begin{array}{l} \partial\hat{u}_1/\partial\tau = -(\mathrm{d}\overline{u}/\mathrm{d}z)\,\hat{u}_3, \\[4pt] \partial(\partial\hat{u}_2/\partial z - \partial\hat{u}_3/\partial y)/\partial\tau = -(\mathrm{d}\overline{u}/\mathrm{d}z)\,(\partial u^S/\partial y), \\[4pt] \partial\hat{u}_2/\partial y + \partial\hat{u}_3/\partial z = 0, \quad \tau \equiv \epsilon t, \end{array}\right\} \qquad (13.3)$$

at leading order for the $O(\epsilon^2)$ Eulerian mean flow modification $\epsilon^2(\hat{u}_1, \hat{u}_2, \hat{u}_3)$. (Note that the pseudomomentum here equals the Stokes drift at $O(\epsilon^2)$.) At first, the spanwise-periodic longitudinal vorticity, and hence $\hat{u}_2$ and $\hat{u}_3$, grow linearly with time τ while the downstream component $\hat{u}_1$ develops as τ^2.

For weak mean flows, Leibovich (1977a) and Leibovich & Radhakrishnan (1977) have computed the developing longitudinal vortex system driven by a pair of oblique waves, with viscous terms included. Laboratory experiments of this configuration by Faller & Caponi (1978) and Faller & Cartwright (1982) confirm the appearance of such vortices.

For strong $O(1)$ primary shear flows, the GLM equations are normally insufficient to determine the developing vortices, because distortions of the initial Stokes drift and pseudomomentum cannot be neglected as in (13.3) (the case studied by Craik (1970) is an exception). The GLM equations must then be complemented by those governing the evolution of the wave field.

13.2   *Wave-driven longitudinal-vortex instability*

Craik (1977, 1982d) and, for stratified flows, Leibovich (1977b) and Leibovich & Paolucci (1981) have considered the stability of unidirectional shear flows in the presence of $O(\epsilon)$ two-dimensional surface gravity waves independent of the spanwise direction. The disturbances are regarded as x-independent, of longitudinal-vortex form.

When the primary flow is weak, of magnitude $O(\epsilon^2)$, let

$$\mathbf{u} = [O(\epsilon)\,\text{wave field}] + \epsilon^2[u^0(z) + \delta\hat{u}_1,\ \delta\,\partial\hat{\psi}/\partial z,\ -\delta\,\partial\hat{\psi}/\partial y] + O(\epsilon^3, \epsilon^2\delta^2)$$

where $\hat{u}_1$ and $\hat{\psi}$ depend on y, z and $\tau = \epsilon^2 t$, and δ is sufficiently small to permit linearization. The mean-flow equations then reduce to

$$(\partial/\partial\tau - \nu_1\,\nabla^2)\,\hat{u}_1 = (\mathrm{d}u^0/\mathrm{d}z)\,(\partial\hat{\psi}/\partial y),$$

$$(\partial/\partial\tau - \nu_1\,\nabla^2)\,\nabla^2\hat{\psi} = (\mathrm{d}u^S/\mathrm{d}z)\,(\partial\hat{u}_1/\partial y),$$

$$\nabla^2 \equiv \partial^2/\partial y^2 + \partial^2/\partial z^2,$$

where the primary flow $\epsilon^2 u^0(z)$ and Stokes drift $\epsilon^2 u^S(z)$ are known (cf. Craik 1977).

These equations bear a close similarity to those governing thermal (Bénard) and centrifugal (Rayleigh–Taylor) instability. Boundary conditions $\hat{u}_1 = 0$, $\nabla\hat{\psi} = 0$ hold at a rigid bottom $z = -H$ (or $\hat{u}_1$, $\nabla\hat{\psi} \to 0$ as $z \to -\infty$). At the mean free surface, $z = 0$, the kinematic and tangential-stress boundary conditions give

$$\partial\hat{u}_1/\partial z = 0, \quad \partial\hat{\psi}/\partial y = (\partial^2/\partial y^2 - \partial^2/\partial z^2)\,\hat{\psi} = 0.$$

It is readily shown that instability occurs, unless inhibited by viscosity, whenever the gradients of u^0 and u^S have the *same sign* in some part of the flow domain. Leibovich & Paolucci (1981) comprehensively discuss such instability of *time-dependent* mean flows induced by a constant surface stress: such flows are unstable for all but exceedingly-small wave amplitudes. For stronger, but still weak, primary flows of magnitude $O(\epsilon)$, the stability analysis is virtually identical. Treatment of strong $O(1)$ shear flows with $O(\epsilon)$ plane waves is necessarily more complex, but a rather similar instability mechanism is still likely to operate in such flows (Craik 1982d), with pseudomomentum rather than Stokes drift playing a central rôle.

Langmuir circulations are certainly responsible for the parallel 'windrows' often observed in lakes (see Figure 4.7) and similar longitudinal-vortex structures are an important ingredient of all turbulent shear flows. Such longitudinal vortices provide an efficient mixing mechanism which is believed to play a part in determining the location and structure of the thermocline in lakes and oceans (see Leibovich's (1983) review); but other factors such as temperature 'fronts' or 'ramps' (Thorpe & Hall 1982) and intermittent Kelvin–Helmholtz instability of internal-wave modes (Thorpe 1971, 1973) also contribute.

Even without the wind to supply a constant surface stress, surface gravity waves induce an $O(\epsilon^2)$ Eulerian drift velocity by viscous action (see §10.1). It would appear that unidirectional drift-velocity profiles like (10.3) are inherently unstable to spanwise-periodic perturbations.

The physical mechanism for instability of weak primary flows is readily understood in terms of vortex-line deformation. An initially-unidirectional shear flow u^0 has vortex lines extending uniformly in the y-direction while the Stokes drift u^S of a plane-wave field depends only on depth z. A small spanwise-periodic x-velocity perturbation $\hat{u}_1$ provides a small periodic

Figure 4.7. Photograph of windrows (Banana River). Those normally observed are rather less regular than this. (Photograph by A. Woodcock, published in Stommel 1951.)

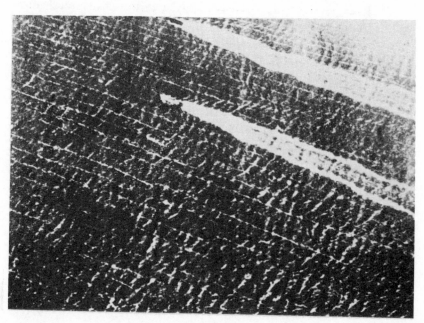

component of vertical vorticity: the resultant vortex lines are 'wavy' and lie in vertical planes. In the absence of viscosity, vortex lines travel with the fluid particles: accordingly, they are tilted by the Stokes drift gradient du^S/dz to yield a periodic component of longitudinal (x-) vorticity. This induces periodic upwelling and downwelling which convects the x-momentum of the primary flow $u^0(z)$. When du^S/dz and du^0/dz have the same sign, this momentum supplements the perturbation $\hat{u}_1$ and leads to instability: when the signs are opposite, $\hat{u}_1$ is diminished and the flow is stable.

An alternative, complementary, physical interpretation is given by Craik (1982d). In this, the forcing term is identified with that of centrifugal Taylor–Görtler instability on curved walls, a mean $O(\epsilon^2)$ curvature deriving from the x-averaged curvature of streamlines in the wavy flow. These equivalent kinematical and dynamical explanations typify the different insights provided by Lagrangian and Eulerian viewpoints.

A rather different physical model for generation of Langmuir circulations was proposed by Garrett (1976). The amplitudes of short gravity waves are modified by their interaction with variable horizontal currents associated with large-scale longitudinal vortices. Garrett envisages enhanced viscous or turbulent dissipation where the wave amplitudes are greatest and a resultant radiation stress acting to maintain the spanwise-periodic mean flow. Though Garrett's physical hypotheses are less secure than those of the Craik–Leibovich theory, it is reassuring that this alternative mechanism acts to reinforce the circulations.

The longitudinal-vortex instability is quite distinct from the Lin–Benney type of mechanism, in which vortices are forced by a pre-existing spanwise structure of the wave field. In contrast, the instability mechanism operates with two-dimensional waves independent of the spanwise co-ordinate. Further work is necessary to assess the importance of such instability in strong shear flows such as boundary layers. A recent theory developed in the latter context by Benney (1984) has similarities with that of Craik (1982d): see §20.3.

Chapter five

THREE-WAVE RESONANCE

A derivation of the three-wave resonance equations (8.12) was outlined in §8, for wave amplitudes a_j ($j = 1, 2, 3$) depending only on time t. Here, a fuller account is given of such resonance, including cases where the wave amplitudes a_j vary slowly in both time and space.

Many investigations of resonance followed Phillips' (1960, 1961) pioneering studies: Phillips (1981a) himself gives an interesting overview. Inviscid surface gravity waves in still water do not exhibit three-wave resonance with quadratic interactions. The lowest-order resonance then involves quartets of waves, with cubic nonlinearities. Discussion of such higher-order terms is mostly deferred to Chapters 6 and 7. Three-wave resonance in conservative systems is examined in § 14 and § 15; systems with linear growth or damping are treated in § 16; non-conservative interactions, such as arise in shear flows, are considered in § 17.

14 Conservative wave interactions

14.1 *Conditions for resonance*

McGoldrick (1965) showed that three capillary–gravity waves may interact resonantly, although gravity waves cannot. The frequencies of three two-dimensional waves with periodicities $a_j \exp(ik_j x - i\omega_j t)$ ($j = 1, 2, 3$) are

$$\omega_j = (gk_j + \gamma\rho^{-1}k_j^3)^{\frac{1}{2}}$$

for positive wavenumbers k_j. Here, γ denotes the coefficient of surface tension. If k_3 is chosen to equal $k_1 + k_2$, the resonance condition $\omega_1 + \omega_2 = \omega_3$ is met when k_1 and k_2 satisfy

$$\Gamma = \frac{(1+r^2)+(1+r)(1+7r+r^2)^{\frac{1}{2}}}{r(9+14r+9r^2)}, \tag{14.1}$$

$$\Gamma \equiv \gamma k_1^2/2\rho g, \quad r \equiv k_2/k_1.$$

This may be rearranged as a cubic equation in r, with coefficients depending on Γ.

For waves propagating in different horizontal directions,

$$a_j \exp(i\mathbf{k}_j \cdot \mathbf{x} - i\omega_j t), \quad \mathbf{k}_3 = \mathbf{k}_1 + \mathbf{k}_2, \quad \mathbf{x} = (x, y),$$

the frequencies ω_j are as above, with k_j replaced by $|\mathbf{k}_j|$. Roots of $\omega_3 = \omega_1 + \omega_2$ are then best located by computation or graphical construction: see Simmons (1969) for the latter.

In stratified fluid, three-wave resonance exists among internal gravity waves and between two surface gravity waves and one internal wave (Ball 1964; Thorpe 1966). In the latter case, the wavenumber $\mathbf{k}_1$ and frequency ω_1 of the internal wave are typically much smaller than those of the surface waves. Accordingly, let

$$\omega_1 = \omega_3 - \omega_2 = \Delta\omega, \quad \mathbf{k}_1 = \mathbf{k}_3 - \mathbf{k}_2 = \Delta\mathbf{k}$$

at resonance, where $\Delta\omega$ and $|\Delta\mathbf{k}|$ are small. Since the frequencies and wavenumbers of the surface waves are nearly equal,

$$\omega_3 \approx \omega_2 + (\partial\omega/\partial\mathbf{k})_2 \cdot \Delta\mathbf{k}.$$

But $\partial\omega/\partial\mathbf{k}$ is the group velocity $\mathbf{c}_g$ of the surface waves and so the resonance condition becomes

$$\Delta\omega \approx \mathbf{c}_g \cdot \Delta\mathbf{k}.$$

This means that the horizontal phase velocity of the internal wave,

$$\mathbf{c}_{in} = \frac{\Delta\omega\,\Delta\mathbf{k}}{|\Delta\mathbf{k}|^2}$$

equals the component of $\mathbf{c}_g$ in the direction of $\Delta\mathbf{k}$. When all three waves travel in the same direction, resonance occurs when $\mathbf{c}_{in} = \mathbf{c}_g$.

In most regions of the ocean, lowest-mode internal waves on the thermocline cannot have $|\mathbf{c}_{in}|$ greater than about 50 cm s^{-1}, and $|\mathbf{c}_g|$ is as small as this only for rather short surface waves, of under 60 cm wavelength. But when the directions of $\mathbf{k}_2$ and $\Delta\mathbf{k}$ differ, the internal wave resonates with longer surface waves. One should therefore expect a periodic surface-wave field to generate internal waves which propagate at a considerable angle to that of $\mathbf{k}_2$.

Internal waves on the thermocline are frequently visible in aerial photographs as periodic light and dark bands. These bands indicate regions of differing 'surface roughness', brought about by modification of the surface waves by the long internal waves. Such interactions of long and short waves may conveniently be treated by regarding the short waves as propagating in a slowly-varying current: variations of the short-wave field may then be found as described in §11.

Internal waves in stratified fluid of finite depth have a denumerable set of frequencies $\omega^{(n)}$ ($n = 1, 2, \ldots$) for every horizontal wavenumber $\mathbf{k}$. Each yields a separate surface in $\omega-\mathbf{k}$ space and these surfaces characterize families of modes with differing vertical mode-structures. For three-wave resonance to occur among these, Thorpe (1966) showed that the waves cannot belong to the same family. Longuet-Higgins & Gill (1967) and Ripa (1981) consider resonance in various other systems of geophysical interest.

14.2 *Resonance of capillary–gravity waves*

McGoldrick's (1965) derivation of the three-wave resonance equations was generalized by Simmons (1969) and Case and Chiu (1977) to waves with amplitudes varying slowly in both space and time. Before describing Simmons' variational method, we outline Case & Chiu's perturbation analysis.

The inviscid equations and boundary conditions for deep-water capillary–gravity waves are

$$
\left.
\begin{aligned}
\nabla^2\phi = 0, \quad \mathbf{u} = \nabla\phi, & \\
p/\rho + gz = -\partial\phi/\partial t - \tfrac{1}{2}|\nabla\phi|^2 &
\end{aligned}
\right\} \quad -\infty < z \leqslant \zeta(x, y, t);
$$

$$
\left.
\begin{aligned}
\partial\zeta/\partial t + \nabla\phi \cdot \nabla\zeta = \partial\phi/\partial z & \\
p = -\gamma\kappa(x, y, t) &
\end{aligned}
\right\} \quad z = \zeta(x, y, t);
$$

$$
\nabla\phi \to 0, \quad z \to -\infty.
$$

$$(14.2a\text{–}e)$$

Here, ϕ is the velocity potential, $\zeta(x, y, t)$ the free-surface elevation, γ the coefficient of surface tension and $\kappa(x, y, t)$ the surface curvature. Since $\kappa = \nabla \cdot \hat{\mathbf{n}}$ where $\hat{\mathbf{n}}$ is the unit outwards normal to the free surface $z = \zeta(x, y, t)$,

$$
\left.
\begin{aligned}
\kappa &= \nabla \cdot \{\nabla(\zeta - z)/|\nabla(\zeta - z)|\} \\
&= \nabla^2\zeta(1 + |\nabla\zeta|^2)^{-\frac{1}{2}} + \nabla\zeta \cdot \nabla[(1 + |\nabla\zeta|^2)^{-\frac{1}{2}}].
\end{aligned}
\right\}
\tag{14.3}
$$

For small-amplitude waves, the kinematic and pressure boundary conditions at the deformed free surface may be approximated by conditions at the mean level $z = 0$, through Taylor expansions about this value. At $O(a^2)$, where a is a dimensionless measure of wave-slope, these are

$$
\left.
\begin{aligned}
\phi_t - (\gamma/\rho)\nabla^2\zeta + g\zeta &= -\tfrac{1}{2}|\nabla\phi|^2 - \phi_{zt}\zeta, \\
\zeta_t - \phi_z &= \phi_{zz}\zeta - \nabla\phi \cdot \nabla\zeta,
\end{aligned}
\right\} \quad (z = 0)
\tag{14.4}
$$

the terms on the left-hand sides being $O(a)$ and those on the right $O(a^2)$. Subscripts t and z denote partial differentiation.

At leading order, $O(a)$, let

$$
\left.
\begin{aligned}
\phi &= \sum_{j=1}^{3} \phi_j(\mathbf{x}, z, t)\, e^{k_j z}, \quad \zeta = \sum_{j=1}^{3} h_j(\mathbf{x}, t), \\
\phi_j &= \mathrm{i}\omega_j k_j^{-1}\{P_j(\mathbf{x}, z, t)\exp[\mathrm{i}(\mathbf{k}_j\cdot\mathbf{x}+\omega_j t)] \\
&\qquad\qquad - Q_j(\mathbf{x}, z, t)\exp[\mathrm{i}(\mathbf{k}_j\cdot\mathbf{x}-\omega_j t)]\}+\text{c.c.}, \\
h_j &= P_j(\mathbf{x}, 0, t)\exp[\mathrm{i}(\mathbf{k}_j\cdot\mathbf{x}+\omega_j t)] \\
&\qquad\qquad + Q_j(\mathbf{x}, z, t)\exp[\mathrm{i}(\mathbf{k}_j\cdot\mathbf{x}-\omega_j t)] +\text{c.c.}
\end{aligned}
\right\} \quad (14.5)
$$

with $\mathbf{x} = (x, y)$, $k_j = |\mathbf{k}_j|$ and $\mathbf{k}_j$, ω_j chosen so that

$$
\mathbf{k}_2 = \mathbf{k}_1 + \mathbf{k}_3, \quad \omega_2 = \omega_1 + \omega_3. \tag{14.6}
$$

The functions P_j, Q_j are $O(a)$ and *slowly-varying* in $\mathbf{x}$, z and t, with derivatives which are $O(a^2)$. From (14.2a), P_j and Q_j approximately satisfy

$$
\partial P_j/\partial z = -\mathrm{i}k_j^{-1}(\mathbf{k}_j\cdot\nabla P_j), \quad \partial Q_j/\partial z = -\mathrm{i}k_j^{-1}(\mathbf{k}_j\cdot\nabla Q_j). \tag{14.7}
$$

Substitution of (14.5)–(14.7) in the boundary conditions (14.4) and isolation of Fourier components eventually yields

$$
\left.
\begin{aligned}
(\partial/\partial t + \mathbf{v}_1\cdot\nabla)Q_1 &= \mathrm{i}\gamma_1 Q_2 Q_3^*, \\
(\partial/\partial t + \mathbf{v}_2\cdot\nabla)Q_2 &= \mathrm{i}\gamma_2 Q_1 Q_3, \\
(\partial/\partial t + \mathbf{v}_3\cdot\nabla)Q_3 &= \mathrm{i}\gamma_3 Q_1^* Q_2,
\end{aligned}
\right\} \tag{14.8}
$$

along with similar equations for the P_j with $\mathbf{v}_j$ and γ_j replaced by $-\mathbf{v}_j$ and $-\gamma_j$. In these, the $\mathbf{v}_j = (\partial\omega/\partial\mathbf{k})_j$ denote the linear group velocities of the waves. But the $\mathbf{v}_j$ do not arise 'naturally' in this derivation and must be reconstituted from the linear dispersion relation. The coupling coefficients γ_j are found to be

$$
\begin{aligned}
\gamma_1 &= \tfrac{1}{2}\{k_1\omega_1^{-1}(\omega_2^2+\omega_3^2)+(\omega_3 k_3-\omega_2 k_2-k_1\omega_1^{-1}\omega_2\omega_3) \\
&\quad + \mathbf{k}_2\cdot\mathbf{k}_3[\omega_2 k_2^{-1}-\omega_3 k_3^{-1}-\omega_2\omega_3 k_1(k_2 k_3\omega_1)^{-1}]\}, \\
\gamma_2 &= \tfrac{1}{2}\{k_2\omega_2^{-1}(\omega_1^2+\omega_3^2)-(\omega_1 k_1+\omega_3 k_3-k_2\omega_2^{-1}\omega_1\omega_3) \\
&\quad - \mathbf{k}_1\cdot\mathbf{k}_3[\omega_1 k_1^{-1}+\omega_3 k_3^{-1}+\omega_1\omega_3 k_2(k_1 k_3\omega_2)^{-1}]\}, \\
\gamma_3 &= \tfrac{1}{2}\{k_3\omega_3^{-1}(\omega_1^2+\omega_2^2)+(\omega_1 k_1-\omega_2 k_2-k_3\omega_3^{-1}\omega_1\omega_2) \\
&\quad + \mathbf{k}_1\cdot\mathbf{k}_2[\omega_2 k_2^{-1}-\omega_1 k_1^{-1}-\omega_1\omega_2 k_3(k_1 k_2\omega_3)^{-1}]\}.
\end{aligned}
$$

Here, *two* independent resonant wave triads are present, since the Q_j and P_j modes have equal and opposite frequencies $\pm\omega_j$. When $|P_j| = |Q_j|$, these may be viewed as a resonant triad of standing waves.

The variational approach of Simmons (1969) gives the same results rather more quickly. This was inspired by Whitham's (1965, 1967) variational method, with local averaging of a suitable Lagrangian $\mathscr{L}$ over the 'short' length scales and 'fast' time-scales of waves. But Whitham's method was not directly applicable to resonant triads, for it allowed slow variations in frequency and wavenumber incompatible with (14.6). Modifying a previous Lagrangian formulation by Luke (1967), Simmons chose

$$\mathscr{L} = \int_{t_0}^{t} \int \int_{\text{all } x,\, y} L(\phi, \zeta)\, dx\, dy\, dt,$$

$$L = \tfrac{1}{2} g \zeta^2 + \gamma \rho^{-1}[(1 + |\nabla \zeta|^2)^{\frac{1}{2}} - 1] + \int_{-\infty}^{\zeta} (\phi_t + \tfrac{1}{2} |\nabla \phi|^2)\, dz.$$

Variation of $\mathscr{L}$ is subject to constraints $\delta\phi = 0$ at times t_0 and t_1 for all x, y, z. Variation of ϕ alone, by $\delta\phi(x, y, z, t)$ yields

$$\int_{t_0}^{t_1} \int \int \int_V \delta\phi [\nabla^2 \phi]\, dV\, dt + \int_{t_0}^{t_1} \int \int_S \{\delta\phi[\phi_z]\}_{z \to -\infty}\, dS\, dt$$

$$+ \int_{t_0}^{t_1} \int \int_S \{\delta\phi[\phi_z - \nabla\zeta \cdot \nabla\phi - \zeta_t]\}_{z=\zeta}\, dS\, dt$$

along with two other integrals which vanish if $\delta\phi$ is restricted to be zero at $x = \pm\infty$ and $y = \pm\infty$. The volume integral is taken over the domain V comprising $-\infty < z \leqslant \zeta(x, y, t)$ and the horizontal plane $S = \{-\infty < x < \infty, -\infty < y < \infty\}$. Since the variation $\delta\phi$ is arbitrary, each term in square brackets must vanish. The volume integral yields the field equation (14.2a) and the two surface integrals the boundary conditions (14.2c) and (14.2e). Similarly, variation $\delta\zeta$ of $\zeta(x, y, t)$, subject to the restriction that $\delta\zeta = 0$ at $x = \pm\infty$ and $y = \pm\infty$, gives the dynamical boundary condition (14.2d) with p eliminated through (14.2b).

The next step is to substitute into $\mathscr{L}$ expressions equivalent to (14.5), (14.6) for resonant waves, and then to construct the *averaged* Lagrangian $\bar{\mathscr{L}}$ taken over horizontal distances and times large compared with the wavelengths and wave periods but short compared with the modulations in wave amplitude. The algebra at this stage is rather tedious, but less so than in Case & Chiu's method. One then postulates arbitrary variations δQ_j of each complex wave amplitude Q_j and constructs the associated Euler–Lagrange equations

$$\frac{\partial}{\partial t}\left(\frac{\partial \bar{\mathscr{L}}}{\partial(\partial q/\partial t)}\right) + \frac{\partial}{\partial x}\left(\frac{\partial \bar{\mathscr{L}}}{\partial(\partial q/\partial x)}\right) + \frac{\partial}{\partial y}\left(\frac{\partial \bar{\mathscr{L}}}{\partial(\partial q/\partial y)}\right) = \frac{\partial \bar{\mathscr{L}}}{\partial q},$$

$$q = Q_j \quad (j = 1, 2, 3).$$

At $O(a^2)$ these simply yield the linear dispersion relation for each wave-mode; but at next order they give the interaction equations (14.8). A similar Lagrangian formulation of three- and four-wave resonance in plasmas is given by Turner & Boyd (1978).

Simmons actually considered independent variations of *real* amplitudes b_j and phases η_j of wave components of the form

$$h_j = b_j \cos(\mathbf{k}_j \cdot \mathbf{x} - \omega_j t + \eta_j) \quad (j = 1, 2, 3)$$

with $\mathbf{k}_1 + \mathbf{k}_2 + \mathbf{k}_3 = \omega_1 + \omega_2 + \omega_3 = 0$. These lead to the equivalent interaction equations

$$\left.
\begin{aligned}
&\tfrac{1}{2}\{\partial(b_i^2)/\partial t + \boldsymbol{\nabla} \cdot (\mathbf{v}_i\, b_i^2)\} = b_1 b_2 b_3 (Jk_i/\omega_i) \sin\eta, \\
&b_i^2\{\partial\eta_i/\partial t + \boldsymbol{\nabla} \cdot (\mathbf{v}_i\, \eta_i)\} = b_1 b_2 b_3 (Jk_i/\omega_i) \cos\eta \quad (i = 1, 2, 3) \\
&4J = -\sum_{j=1}^{3} \omega_j \omega_{j+1}(1 + \hat{\mathbf{k}}_j \cdot \hat{\mathbf{k}}_{j+1}) \quad (\hat{\mathbf{k}}_4 \equiv \hat{\mathbf{k}}_1), \\
&\eta = \sum_{j=1}^{3} \eta_j,
\end{aligned}
\right\} \quad (14.9)$$

where $\hat{\mathbf{k}}_j$ are unit vectors along the directions of wave propagation. The variational approach leads naturally to the simple form Jk_i/ω_i for the interaction coefficients, which was obscured in the 'direct' perturbation analysis. This form was confirmed by Hasselmann (1967a) for more general conservative systems.

Equations (14.9) hold for all resonant triads with non-equal wave numbers $\mathbf{k}_j$. If, say, $\mathbf{k}_1 = \mathbf{k}_2 = -\tfrac{1}{2}\mathbf{k}_3$, further terms which normally have zero spatial averages must be retained, for the wavenumber $\mathbf{k}_1$ is then in resonance with its own second harmonic. This case was separately dealt with by Simmons and has also been considered, using perturbation theory, by Nayfeh (1970). The phenomenon of second-harmonic resonance of capillary–gravity waves in deep water was first encountered by Wilton (1915), who found that a Stokes expansion in powers of wave amplitude became singular at certain wavenumbers (see §22.1). The greatest of these is $k_1 = 2^{-\frac{1}{2}}(\rho g/\gamma)^{\frac{1}{2}}$, which is just result (14.1) with $r = 1$. This corresponds to ripples of wavelength 2.44 cm in clean water. Wilton's other singularities relate to higher-order resonances; the next, at $k = 3^{-\frac{1}{2}}(\rho g/\gamma)^{\frac{1}{2}}$, is a 'four-wave' resonance between k and its third harmonic.

The form of (14.8) may be inferred from a simple, but general, heuristic analysis. If the linear dispersion relation is $D(\omega, \mathbf{k}) = 0$, slow modulations of each complex wave amplitude Q_j satisfy

$$D\left(\omega_j + \mathrm{i}\frac{\partial}{\partial t}, \mathbf{k}_j - \mathrm{i}\boldsymbol{\nabla}\right) Q_j \approx \mathrm{i}\left(\frac{\partial D}{\partial \omega_j}\frac{\partial}{\partial t} - \frac{\partial D}{\partial \mathbf{k}_j}\cdot\boldsymbol{\nabla}\right) Q_j = \text{n.l.t.}$$

where the nonlinear terms 'n.l.t.' denote quadratic interactions. Also, small perturbations $\delta\mathbf{k}$, $\delta\omega$ of wavenumber and frequency satisfy

$$\delta\omega(\partial D/\partial\omega)+\delta\mathbf{k}\cdot(\partial D/\partial\mathbf{k}) = 0$$

at leading order, and so

$$-\frac{\partial D}{\partial\mathbf{k}_j}\bigg/\frac{\partial D}{\partial\omega_j} = \frac{\partial\omega_j}{\partial\mathbf{k}_j} = \mathbf{v}_j \quad (j = 1, 2, 3),$$

the group velocity of each wave-mode. Accordingly,

$$i\,\partial D/\partial\omega_j(\partial/\partial t+\mathbf{v}_j\cdot\boldsymbol{\nabla})\,Q_j = \mu_j\,Q_k^*\,Q_l^* \quad (j \neq k \neq l)$$

for resonant triads with $\omega_1+\omega_2+\omega_3 = 0$ and $\mathbf{k}_1+\mathbf{k}_2+\mathbf{k}_3 = 0$.

If, further, the Q_j and $D(\omega, \mathbf{k})$ are defined as in §2.3, each mode has energy density

$$E_j = \tfrac{1}{4}\omega_j\,\partial D/\partial\omega_j\,|\,Q_j\,|^2$$

where $\partial D/\partial\omega = 2\rho\omega\,|\,\mathbf{k}\,|^{-1}$ for capillary–gravity waves. Since total energy density $E_1 + E_2 + E_3$ is conserved for spatially-uniform resonant wave-trains, each interaction coefficient μ_j must have the *same* constant value λ, in line with (14.9); but there is no obvious shortcut for evaluating this constant, which is $-\rho J$ for capillary–gravity waves.

It follows that the general conservative interaction equations for resonance with $\mathbf{k}_1+\mathbf{k}_2+\mathbf{k}_3 = \omega_1+\omega_2+\omega_3 = 0$ are

$$\left.\begin{aligned}
(\partial/\partial t+\mathbf{v}_1\cdot\boldsymbol{\nabla})\,A_1 &= s_1\,A_2^*\,A_3^*, \\
(\partial/\partial t+\mathbf{v}_2\cdot\boldsymbol{\nabla})\,A_2 &= s_2\,A_3^*\,A_1^*, \\
(\partial/\partial t+\mathbf{v}_3\cdot\boldsymbol{\nabla})\,A_3 &= s_3\,A_1^*\,A_2^*, \\
s_j &\equiv \mathrm{sgn}\,[(\partial D/\partial\omega)_j]
\end{aligned}\right\} \tag{14.10}$$

with renormalized amplitudes

$$A_j = K\,|\,(\partial D/\partial\omega)_j\,|^{\frac{1}{2}}Q_j, \quad K \equiv -i\lambda\left|\left(\frac{\partial D}{\partial\omega}\right)_1\left(\frac{\partial D}{\partial\omega}\right)_2\left(\frac{\partial D}{\partial\omega}\right)_3\right|^{-\frac{1}{2}}$$

such that

$$E_j = \tfrac{1}{4}\omega_j s_j\,|\,A_j/K\,|^2.$$

14.3   Some properties of the interaction equations

When the A_j depend on t alone, exact solutions of (14.10) are known in terms of elliptic functions. When the three signs s_j differ, these solutions are mostly periodic, but there are non-periodic limiting cases. These solutions were found independently by Jurkus & Robson (1960), Armstrong *et al.* (1962) and Bretherton (1964) – the universality of three-wave resonance in optics, electronics, plasma physics and fluid mechanics not being matched, then as now, by rapid communication! When the three signs s_j are the same, a singularity may, or may not, develop at a particular value of t: such a singularity signifies an 'explosive' breakdown of the equations after a finite time. Such solutions were given by Coppi, Rosenbluth & Sudan (1969). All these are described below in § 15.1. Long before, the case with $A_j(t)$ real and differing signs s_j was solved by Euler (1765): the equations are then Euler's equations for free rotation of a rigid body about a fixed point.

An 'explosion', in which all three wave amplitudes simultaneously become infinitely large, does not violate energy conservation: $E_1 + E_2 + E_3$ remains constant, with the participating waves having energies E_j of differing sign. For most waves in fluid at rest, including surface, interfacial and internal waves, $\partial D / \partial \omega$ is necessarily proportional to $\omega \, | \, \mathbf{k} \, |^{-1}$: therefore, such resonant waves with $\omega_1 + \omega_2 + \omega_3 = 0$ must have $(\partial D / \partial \omega)_j$ of differing sign, positive definite energies E_j, and bounded periodic modulations.

Interactions among waves of differing energy sign need not lead to breakdown: this only occurs when the wave of greatest absolute frequency $| \omega |$ has energy of different sign from the other two. In other words, breakdown occurs *only if the wave actions E_j / ω_j have the same sign*. Examples where this occurs are frequent in plasma physics (see, for example, Weiland & Wilhelmsson 1977). Craik & Adam (1979) have shown that the three-layer Kelvin–Helmholtz flow (2.5) supports resonant triads of both the explosive and periodic types, even when the flow is linearly stable. In contrast, the resonant triads identified by Ma (1984a) in simple Kelvin–Helmholtz flow (cf. §2.1) exist only when the flow is linearly unstable.

In addition to energy conservation, there are three other conserved quantities of (14.10), expressed by the *Manley–Rowe relations*

$$\frac{\mathrm{d}}{\mathrm{d}t} [s_1 \, | \, A_1 \, |^2 - s_2 \, | \, A_2 \, |^2] = \frac{\mathrm{d}}{\mathrm{d}t} [s_2 \, | \, A_2 \, |^2 - s_3 \, | \, A_3 \, |^2]$$

$$= \frac{\mathrm{d}}{\mathrm{d}t} [s_3 \, | \, A_3 \, |^2 - s_1 \, | \, A_1 \, |^2] = 0. \quad (14.11)$$

From these, it is obvious that the A_j remain bounded whenever the signs s_j differ. Equivalently, each $s_j \, | \, A_j \, |^2$ may be replaced by the wave actions E_j / ω_j.

When the A_j of (14.10) depend on a single variable, $\xi = \mathbf{a} \cdot \mathbf{x} + \beta t$ ($\mathbf{a}, \beta$ constant), similar solutions are easily constructed (Armstrong, Sudhanshu & Shiren 1970). For near-resonance, with $\mathbf{k}_1 + \mathbf{k}_2 + \mathbf{k}_3 = 0$ but small frequency mismatch $\Delta \omega = \omega_1 + \omega_2 + \omega_3$, equations analogous to (14.10) have additional exponential factors $\exp(-\mathrm{i} \, \Delta \omega t)$ on the right-hand side.

These, too, yield Jacobi elliptic function solutions (Armstrong *et al.* 1962), though $E_1 + E_2 + E_3$ is no longer constant.

When two of the waves, say A_2 and A_3, are much smaller than the third, A_1, and each depends on t alone, (14.10) yield the linearized approximation

$$dA_1/dt = 0, \quad dA_2/dt = s_2 A_1^* A_3^*, \quad dA_3/dt = s_3 A_1^* A_2^*$$

or

$$A_1 = A_1^0 \text{ (constant)}, \quad d^2 A_k/dt^2 = s_2 s_3 |A_1^0|^2 A_k \quad (k = 2, 3).$$

Obviously, A_2 and A_3 are periodic, with frequency $|A_1^0|$ if the signs s_2, s_3 are opposite; and exponential, with growth and decay rates $\pm |A_1^0|$ if $s_2 s_3 = 1$. For waves in fluid at rest, $s_j = \text{sgn } \omega_j$ and the waves A_2, A_3 remain small compared with A_1 except when A_1 corresponds to the wave of greatest absolute frequency. When $|\omega_1|$ is largest, the wave A_1 is unstable to wave-modes A_2, A_3 (Hasselmann 1967b). In plasma physics, A_1 is called a 'pump wave' when it can pump up A_2 and A_3 from infinitesimal levels; and the instability of the latter is usually described as 'parametric-resonance' instability. When all three signs s_j are the same, growth of A_2 and A_3 is limited by the depletion of A_1 at larger amplitudes; periodic modulations occur as described by the elliptic-function solutions.

A 'pump-wave approximation' may also be applied to (14.10) to describe modulations in both time and space, when one wave is dominant (Craik & Adam 1978), but this inevitably breaks down if the magnitudes of the A_j become comparable, unless the 'pump wave' is artificially maintained at constant amplitude. The governing equations may then be combined as a linear Klein–Gordon (or telegraph) equation (7.9) which is easily solved.

General conservation laws of (14.9), derived by Simmons (1969) for capillary–gravity waves, are

$$\sum_{i=1}^{3} (\partial/\partial t + \mathbf{v}_i \cdot \nabla) E_i = 0, \quad \sum_{i=1}^{3} (\partial/\partial t + \mathbf{v}_i \cdot \nabla) \mathbf{M}_i = \mathbf{0},$$

$$\mathbf{M}_i \equiv \tfrac{1}{2}\rho \hat{\mathbf{k}}_i \omega_i b_i^2.$$

Here, $\mathbf{M}_i$ denotes the momentum (per unit horizontal area) of the ith mode. Corresponding conservation laws of (14.10) are

$$\left. \begin{aligned} \sum_{i=1}^{3} (\partial/\partial t + \mathbf{v}_i \cdot \nabla)(s_i \omega_i |A_i|^2) = 0, \\ \sum_{i=1}^{3} (\partial/\partial t + \mathbf{v}_i \cdot \nabla)(s_i \mathbf{k}_i |A_i|^2) = 0. \end{aligned} \right\} \tag{14.12}$$

These yield energy conservation and the Manley–Rowe relations (14.11) when the A_i depend on t only. When the A_i depend on the two variables

x, t and each $A_i \to 0$ as $x \to \pm\infty$, (14.12) readily reveals that the quantities

$$\int_{-\infty}^{\infty} (s_1 |A_1|^2 - s_2 |A_2|^2)\,dx, \quad \int_{-\infty}^{\infty} (s_2 |A_2|^2 - s_3 |A_3|^2)\,dx,$$

$$\int_{-\infty}^{\infty} (s_3 |A_3|^2 - s_1 |A_1|^2)\,dx$$

are constants. Similar constants of motion exist when the A_i depend on more than one space variable, provided $A_i \to 0$ sufficiently rapidly as $|\mathbf{x}| \to \infty$. The solution of (14.10) for waves modulated in both space and time is accomplished by the method of inverse scattering, described in §15.2.

14.4   *Wave-interaction experiments*
(i)   *Capillary–gravity waves*

The resonance of three capillary–gravity waves, analysed in §14.1, was first investigated experimentally by McGoldrick (1970a). He demonstrated the second-harmonic resonance of two-dimensional waves with $k_1 = k_2 = -\frac{1}{2}k_3 = (\rho g/2\gamma)^{\frac{1}{2}}$ and confirmed that viscous damping is well-represented by linear theory. Figure 5.1, taken from his results, shows typical waveforms at resonance and not, when a single wave is excited mechanically.

Recent experiments of Bannerjee & Korpel (1982) exhibit resonance

Figure 5.1. Capillary–gravity waveforms downstream of a sinusoidally vibrating wavemaker (from McGoldrick 1970a). Cases (a) and (b) are off resonance and show typical 'gravity type' and 'capillary type' waves (frequencies 8.82 c s^{-1} and 11.0 c s^{-1} respectively). Cases (c) and (d) show second-harmonic resonance (frequency about 10 c s^{-1}); (d), 25 cm further from wavemaker than (c), has amplitude reduced by viscosity. Wave amplitude in (a) is about 0.11 mm.

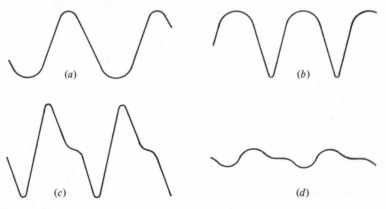

among waves with differing directions of propagation. A mechanically generated wave $\mathbf{k}_3 = (k, 0)$ gave rise to obliquely-propagating subharmonics $\mathbf{k}_{1,2} = (\frac{1}{2}k, \pm l)$ with half the fundamental frequency. As reinterpreted by Hogan (1984b), their results show satisfactory agreement with theory. One difficulty, which must be accounted for, is that viscous damping of the waves induces a significant second-order mean flow and this, in turn, causes a frequency shift of the waves.

Kim & Hanratty (1971) also demonstrated second-harmonic resonance of capillary–gravity waves in shallow liquid layers and found evidence of higher-harmonic generation. These higher harmonics were further investigated by McGoldrick (1972), who attributed them to higher-order near-resonances of the form $k_2 = Nk_1$, $\omega_2 \approx N\omega_1$ (integer $N \geq 2$): cf. §22.2. Clearly, the Nth harmonic must have phase speed close to that of the fundamental: it is then generated as a parametric instability by Nth order nonlinearities and the resultant wave profiles typically have N crests per period. In contrast, Kim & Hanratty attempted to explain their higher-harmonic generation by *quadratic* interactions among four waves with wavenumbers nk ($n = 1, 2, 3, 4$): this would be effective only if the waves were weakly dispersive, with nearly equal phase speeds.

The onset of three-dimensionality in wind-generated ripples, reported by Craik (1966) – see Figure 2.5 – may also be due to three-wave resonance.

(ii) Internal gravity waves

Davis & Acrivos (1967) showed that an internal wave, propagating along a diffuse stratified layer between fluids of different densities, may become distorted from its original sinusoidal shape by the growth of a three-wave resonance. This was revealed by use of neutrally-buoyant dyed droplets which showed surfaces of constant density: their photographs are reproduced in Figure 5.2(a), (b), (c). The 'lumps' which develop in the stratified layer give rise to local turbulent mixing.

McEwan (1971), McEwan, Mander & Smith (1972) and McEwan & Robinson (1975) investigated standing internal gravity waves in fluid with linear density variation with depth. The first and second of these studies concern the degeneration of a single forced mode into more complex structures through three-wave resonances. The third deals with parametric instability of waves due to periodic oscillations of the container; this may be viewed as a limiting case of three-wave resonance where one wave, represented by the moving container, is much longer than the other two. When, because of geometrical constraints of the apparatus, the given

Figure 5.2. Progressive 'disintegration' of an internal wave due to three-wave resonance (from Davis & Acrivos 1967).

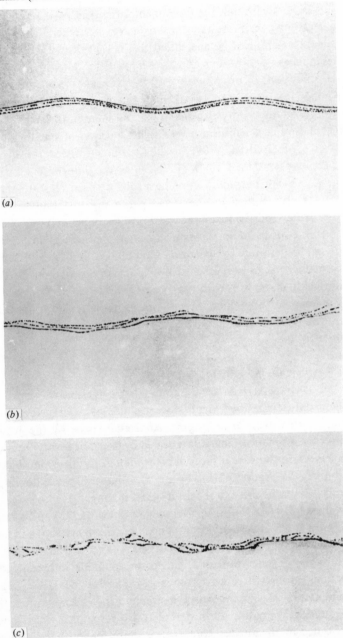

fundamental wave can participate in only one resonant triad, the resultant motion is easily understood in terms of the growth of two other modes; but when several participating triads co-exist the behaviour is much more complex. Similar effects were demonstrated, for progressive internal waves, by Martin, Simmons & Wunsch (1972), who considered up to five triads with a member in common. Such cases are analysed in § 16.2 (v) and (vi).

(iii) Surface and internal waves

With two interfaces, or with a free surface above continuously stratified fluid, resonant interactions occur between surface and internal waves. There are many oceanographic observations of such interacting modes – see, for example, Gargett & Hughes (1972) – but few quantitative investigations in the ocean. Experiments in a large tank were conducted by Lewis, Lake & Ko (1974) and Koop & Redekopp (1981). In the former, a layer of water was bounded above by a free surface and below by a denser freon–kerosene mixture. Since the density difference across the lower interface was quite small, resonant triads comprised relatively short surface gravity waves and a long wave at the lower interface. Both surface and internal waves were generated mechanically with the same direction of propagation. It was confirmed that the strongest modulations of the former occurred when their group velocity was close to the phase velocity of the interfacial wave, in agreement with theory (see § 14.1). The study of Koop & Redekopp concerned similar interaction of long and short waves on the two interfaces of a three-layer configuration (see § 19.1).

A novel application of three-wave interaction theory is that examined by Davies (1982), Davies & Heathershaw (1984) and Mei (1984): they consider propagation of a wave-train over fixed periodic sandbars on the bottom. When the incident waves have wavenumber and frequency (k, ω) and the zero-frequency bottom undulations have wavenumber $2k$, their interaction produces resonant *reflected* waves with wavenumber and frequency $(-k, \omega)$. Because the undulations have fixed amplitude, the 'pump-wave approximation' then applies and solutions are readily constructed. Predicted reflection coefficients in resonant and non-resonant cases agree quite well with the experimental results of Davies & Heathershaw (1984). Such partial reflection can induce second-order drift currents which act to maintain or enhance the sandbars by sediment transport. This resonant excitation of reflected waves is similar to Bragg reflection of X-rays by crystal lattices (see e.g. Pinsker 1978).

As yet, no experiment has been undertaken, in fluid mechanics, to demonstrate the potentially explosive conservative interactions among waves with differing energy signs. The situation envisaged by Craik & Adam (1979) is certainly realizable, at least approximately, in the laboratory. Corresponding, and more difficult, experiments in plasmas were undertaken by Hopman (1971), though the expected 'explosion' turned out to be rather undramatic. Experiments on the *non*conservative interaction of waves in shear flows are discussed in §17.1.

15   Solutions of the conservative interaction equations

15.1   *The one-dimensional solutions*

Here we give the solutions of the three-wave interaction equations when the amplitudes A_j depend on the single variable t. In doing so, a small frequency mismatch $\Delta\omega = \omega_1 + \omega_2 + \omega_3$ is incorporated, so that the governing equations are

$$\left.\begin{array}{l} \mathrm{d}A_1/\mathrm{d}t = s_1 A_2^* A_3^* \, \mathrm{e}^{-\mathrm{i}\,\Delta\omega t}, \quad \mathrm{d}A_2/\mathrm{d}t = s_2 A_3^* A_1^* \, \mathrm{e}^{-\mathrm{i}\,\Delta\omega t}, \\[2mm] \mathrm{d}A_3/\mathrm{d}t = s_3 A_1^* A_2^* \, \mathrm{e}^{-\mathrm{i}\,\Delta\omega t}. \end{array}\right\} \quad (15.1)$$

These equations yield constant total energy $E_1 + E_2 + E_3$ only at exact resonance, $\Delta\omega = 0$. Otherwise, $E_1 + E_2 + E_3$ oscillates periodically with frequency $\Delta\omega$. This may seem paradoxical, for the original system is non-dissipative and energy conserving; but the effect is a simple consequence of treating combinations of modes with slightly differing frequencies as single modes. For instance, two linear wave-trains

$$p \cos(kx - \Omega t), \quad q \cos[kx - (\Omega + \delta)\,t]$$

yield net energy $\tfrac{1}{2}(p^2 + q^2)$ when summed separately; but $\tfrac{1}{2}(p^2 + q^2 + 2pq \cos \delta t)$ when regarded as a single, slowly-modulated wave-mode. Accordingly, we regard (15.1) as the equations of *conservative* three-wave resonance, even when $\Delta\omega \neq 0$. An account similar to that following is given by Weiland & Wilhelmsson (1977).

On writing $b_j = |A_j|$, $\eta_j = phA_j$, (15.1) may be rewritten as four real equations

$$\left.\begin{array}{ll} \mathrm{d}b_1/\mathrm{d}t = s_1 b_2 b_3 \cos\eta, & \mathrm{d}b_2/\mathrm{d}t = s_2 b_1 b_3 \cos\eta \\[2mm] \mathrm{d}b_3/\mathrm{d}t = s_3 b_1 b_2 \cos\eta, & \mathrm{d}\eta/\mathrm{d}t = \Delta\omega - b_1 b_2 b_3 \left(\dfrac{s_1}{b_1^2} + \dfrac{s_2}{b_2^2} + \dfrac{s_3}{b_3^2}\right) \sin\eta, \end{array}\right\} \quad (15.2)$$

where $\eta \equiv \eta_1 + \eta_2 + \eta_3 + \Delta\omega t$. A similar formulation applies for amplitudes depending on x only, with small wavenumber mismatch $\Delta k = k_1 + k_2 + k_3$ and matched frequencies $\omega_1 + \omega_2 + \omega_3 = 0$.

The Manley–Rowe relations (14.11) are satisfied by (15.1): these may be written as

$$s_1[b_1^2(t) - b_1^2(0)] = s_2[b_2^2(t) - b_2^2(0)] = s_3[b_3^2(t) - b_3^2(0)] \equiv x(t).$$

A further constant of motion is

$$\Gamma \equiv b_1 b_2 b_3 \sin \eta - \tfrac{1}{2} \Delta \omega s_j \, b_j^2,$$

for any choice of $j = 1, 2, 3$.

On using these, (15.2) reduces to a single equation for $x(t)$,

$$\mathrm{d}x/\mathrm{d}t = \pm 2\{[s_1 x + b_1^2(0)] \, [s_2 x + b_2^2(0)] \, [s_3 x + b_3^2(0)]$$
$$- \langle \Gamma + \tfrac{1}{2}\Delta\omega[x + s_j \, b_j^2(0)]\rangle^2\}^{\frac{1}{2}}$$
$$\equiv \pm\{-2\pi(x)\}^{\frac{1}{2}}, \tag{15.3}$$

where the $\pm$ signs indicate that of $\cos\eta$. Since at least two of the s_j

Figure 5.3. Typical potential function $\pi(x)$ when $\Delta\omega = \Gamma = 0$: (a) $s_1 = -1$, x oscillates between $-b_2^2(0)$ and $b_1^2(0)$; (b) $s_1 = 1$, x grows without bound.

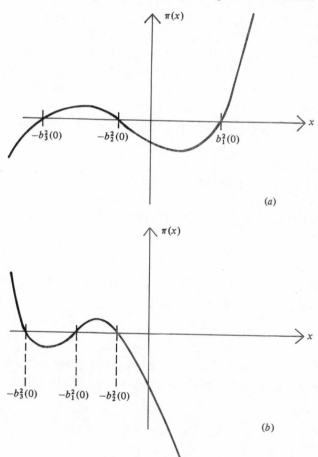

(a)

(b)

($j = 1, 2, 3$) are equal, there is no loss in setting $s_2 = s_3 = 1$ and in taking the term $s_j b_j^2(0)$ as $b_2^2(0)$.

Equation (15.3) has the form

$$\tfrac{1}{2}(dx/dt)^2 + \pi(x) = 0 \qquad (15.3)'$$

where $\pi(x)$ is a cubic function, negative in cases of interest. Clearly, $\pi(x)$ may be thought of as a potential. If $\Delta\omega = \Gamma = 0$, $\pi(x)$ is typically as in Figure 5.3(*a*) when $s_1 = -1$ and as in Figure 5.3(*b*) when $s_1 = 1$. In the former, the wave actions E_j/ω_j have differing signs; in the latter they have the same sign and the signs of the energies E_j differ. For $s_1 = -1$, the solution $x(t)$ must oscillate in the potential well. For $s_1 = 1$, $x(t)$ must always become infinite: in fact, it does so 'explosively', in a finite time.

This explosive growth is inhibited, but not eliminated, by non-zero frequency mismatch $\Delta\omega$ which changes the location of the roots of $\pi(x)$. Explosive growth then occurs only with sufficiently large initial amplitudes. The constant of motion Γ is related to the initial phases of the waves and the coupling is strongest when $\Delta\omega = \Gamma = 0$. When $\Gamma \neq 0$, the roots of $\pi(x)$ remain real when $s_1 = -1$ but two may be complex conjugates when $s_1 = 1$. Solutions for $\Gamma \neq 0$, $s_1 = -1$ remain periodic and those with $\Gamma \neq 0$, $s_1 = 1$, $\Delta\omega = 0$ remain explosive for all initial amplitudes.

The solution of (15.3) may be given explicitly as an elliptic integral,

$$t = \pm \int_0^{x(t)} [-2\pi(\xi)]^{-\frac{1}{2}} d\xi$$

which gives $x(t)$ in terms of 'sn' and 'cn' Jacobi elliptic functions (Abramowitz & Stegun 1964). Corresponding solutions for the amplitudes b_j are as follows (cf. Weiland & Wilhelmsson 1977; Coppi, Rosenbluth & Sudan 1969).

(*i*) $s_1 = -1$.

$$b_1 = [b_1^2(0) - x(t)]^{\frac{1}{2}}, \quad b_k = [b_k^2(0) + x(t)]^{\frac{1}{2}} \quad (k = 2, 3),$$

$$x(t) = (\alpha_2 - \alpha_1) \operatorname{sn}^2 [(\alpha_1 - \alpha_3)^{\frac{1}{2}} t + \theta, k] + \alpha_1,$$

$$\theta = \operatorname{sn}^{-1} \left[\left(\frac{\alpha_1}{\alpha_1 - \alpha_2} \right)^{\frac{1}{2}}, k \right], \quad k = \left(\frac{\alpha_1 - \alpha_2}{\alpha_1 - \alpha_3} \right)^{\frac{1}{2}},$$

where $\alpha_1 > \alpha_2 > \alpha_3$ are the three real roots of the cubic $\pi(x) = 0$. Cases of equal roots are simpler but require separate treatment. Solutions have period

$$\tau = \int_0^{2\pi} (1 - k \sin^2 u)^{-\frac{1}{2}} du.$$

(ii) $s_1 = 1$; *real roots* $0 > \alpha_1 > \alpha_2 > \alpha_3$.

$$b_j = [b_j^2(0) + x(t)]^{\frac{1}{2}} \quad (j = 1, 2, 3),$$

$$x(t) = (\alpha_1 - \alpha_3)\{\text{sn}^2[(t_\infty - t)(\alpha_1 - \alpha_3)^{\frac{1}{2}}, k']\}^{-1} + \alpha_3,$$

$$t_\infty = (\alpha_1 - \alpha_3)^{-\frac{1}{2}} \text{sn}^{-1}\left[\left(\frac{\alpha_1 - \alpha_3}{-\alpha_3}\right)^{\frac{1}{2}}, k'\right], \quad k' = \left(\frac{\alpha_2 - \alpha_3}{\alpha_1 - \alpha_3}\right)^{\frac{1}{2}}.$$

Here, t_∞ denotes the time of explosion.

(iii) $s_1 = 1$; *real non-positive* α_1, *complex conjugate* α_2, α_3.

$$b_j = [b_j^2(0) + x(t)]^{\frac{1}{2}} \quad (j = 1, 2, 3),$$

$$x(t) = 2|\alpha_1 - \alpha_3|\{1 + \text{cn}[2t|\alpha_1 - \alpha_3|^{\frac{1}{2}} + \phi, k'']\}^{-1} + \alpha_1 - |\alpha_1 - \alpha_3|,$$

$$\phi = \text{cn}^{-1}\left[\frac{|\alpha_1 - \alpha_3| + \alpha_1}{|\alpha_1 - \alpha_3| - \alpha_1}, k''\right], \quad k'' = \left(\frac{|\alpha_1 - \alpha_3| + \alpha_2 + \alpha_3 - \alpha_1}{2|\alpha_1 - \alpha_3|}\right)^{\frac{1}{2}}.$$

These also explode at finite time.

(iv) $s_1 = 1$; *real roots with* $\alpha_1 < 0 < \alpha_2 < \alpha_3$.

Solutions oscillate in the potential well $\alpha_1 \leqslant x \leqslant \alpha_2$. These periodic solutions resemble (i) and arise when the initial amplitudes $b_j(0)$ lie below the threshold for explosion imposed by frequency mismatch $\Delta\omega$.

When $\Gamma = \Delta\omega = 0$, there are particularly simple solutions of (15.2) when two or more wave amplitudes are initially equal. These are

(a) $b_1 = b \tanh(bt)$, $b_2 = b_3 = b \operatorname{sech}(bt)$, $\cos\eta = -1$,

$\quad [s_1 = -1, \quad b_1(0) = 0, \quad b_2(0) = b_3(0) = b]$;

(b) $b_1 = b \tan(bt)$, $b_2 = b_3 = b \sec(bt)$, $\cos\eta = 1$,

$\quad [s_1 = 1, \quad b_1(0) = 0, \quad b_2(0) = b_3(0) = b]$;

(c, d) $b_j = (b^{-1} - t)^{-1}$ $(j = 1, 2, 3)$, $\cos\eta = 1$

$\quad$ and

$\quad b_j = (b^{-1} + t)^{-1}$ $(j = 1, 2, 3)$, $\cos\eta = -1$,

$\quad [s_1 = 1, \quad b_1(0) = b_2(0) = b_3(0) = b]$.

Note that (b) and (c) explode at $t = \pi/2b$ and $t = 1/b$ respectively.

15.2 *Inverse-scattering solution in two dimensions*

For modulations in one space dimension and time, or in two space dimensions only, Zakharov & Manakov (1973, 1975) and Kaup (1976) have obtained solutions of (14.10) by the method of inverse scattering. Subsequently, solutions for modulation in two space dimensions and time (or three space dimensions alone) were given by Zakharov (1976), Cornille (1979) and Kaup (1980). This work is admirably reviewed by Kaup,

Reiman & Bers (1979) and Kaup (1981b) (frequent misprints in the former are mercifully corrected in the reprints). Here, a brief account is given of the two-dimensional case, with A_j depending on x and t.

First, let the indices 1, 2, 3 be ordered such that $c_1 < c_2 < c_3$ where c_j denotes the x-component of $\mathbf{v}_j$. The strategy, as in all inverse-scattering problems, is to find an associated *linear* eigenvalue problem which helps towards constructing the solutions of the nonlinear equations (14.10).

The eigenvalue problem of Zakharov & Manakov is

$$
\left.
\begin{aligned}
(\mathrm{i}\,\mathrm{d}/\mathrm{d}x - c_1\,\zeta)\,u_1 &= V_{12}\,u_2 + V_{13}\,u_3, \\
(\mathrm{i}\,\mathrm{d}/\mathrm{d}x - c_2\,\zeta)\,u_2 &= V_{21}\,u_1 + V_{23}\,u_3, \\
(\mathrm{i}\,\mathrm{d}/\mathrm{d}x - c_3\,\zeta)\,u_3 &= V_{31}\,u_1 + V_{32}\,u_2,
\end{aligned}
\right\}
\tag{15.4}
$$

where $\mathbf{u} = (u_1, u_2, u_3)$ is an eigevector, ζ the eigenvalue and V_{ij} are six unknown 'potentials' related to the unknown amplitudes $A_j(x, t)$ by

$$
\left.
\begin{aligned}
V_{23} &= \frac{-\mathrm{i}A_1}{[(c_2-c_1)(c_3-c_1)]^{\frac{1}{2}}}, & V_{32} &= -s_2 s_3 V_{23}^*, \\
V_{31} &= \frac{-\mathrm{i}A_2}{[(c_2-c_1)(c_3-c_2)]^{\frac{1}{2}}}, & V_{13} &= s_1 s_3 V_{31}^*, \\
V_{12} &= \frac{-\mathrm{i}A_3}{[(c_3-c_1)(c_3-c_2)]^{\frac{1}{2}}}, & V_{21} &= -s_1 s_2 V_{12}^*.
\end{aligned}
\right\}
\tag{15.5}
$$

The amplitudes A_j are assumed to be localized in x so that $V_{ij} \to 0$ as $|x| \to \infty$.

Suppose that $\mathbf{u}^{(n)} = u_j^{(n)}$ ($n = 1, 2, 3$) are three linearly-independent solutions, each associated with an eigenvalue ζ, which satisfy respective boundary conditions

$$
u_j^{(n)} \approx \delta_{jn} \exp(-\mathrm{i}c_j\,\zeta x), \quad x \to -\infty,
$$

$$
\delta_{jn} = \begin{cases} 1 & (n = j) \\ 0 & (n \neq j). \end{cases}
$$

Note that these are consistent with (15.4). Each solution may be thought of (loosely) as a different wave-mode incoming from $x = -\infty$. As $x \to +\infty$, $V_{ij} \to 0$ and each solution must behave as

$$
u_j^{(n)} \approx a_{nj}(\zeta, t) \exp(-\mathrm{i}c_j\,\zeta x), \quad x \to +\infty.
$$

The a_{nj} may be viewed as a 'scattering matrix' which relates the outgoing to the incoming wave field. On using (14.10) and, in particular, the invariant properties (14.12), it may be shown – with some effort – that *the eigenvalue ζ is independent of t and that the time-dependence of $a_{ij}(\zeta, t)$ is exponential, as*

$$
a_{ij}(\zeta, t) = a_{ij}(\zeta, 0) \exp[\mathrm{i}\zeta t c_1 c_2 c_3 (c_j^{-1} - c_i^{-1})].
\tag{15.6}
$$

Accordingly, if $a_{ij}(\zeta, 0)$ can be found for the *initial* wave-envelopes $A_j(x, 0)$ it is known at all later times. But the spectrum of eigenvalues ζ and the associated $a_{ij}(\zeta, 0)$ may be found by solving the linear eigenvalue problem at $t = 0$, since the V_{ij} are then known. The problem then reduces to reconstructing the potentials V_{ij} at all later times t from a knowledge of the scattering data ζ and $a_{ij}(\zeta, t)$. The formal solution of the latter 'inverse-scattering' problem may be expressed as that of a linear integral equation.

The disadvantage of having to solve the third-order eigenvalue problem (15.4) was circumvented by Kaup (1976) for cases where the three wave-envelopes are initially well separated. Then, the third-order problem may be replaced by three second-order ones, each applicable in regions where one wave amplitude is dominant. These three problems yield three scattering matrices,

$$a_{ij}^{(1)} = \begin{bmatrix} 1 & 0 & 0 \\ 0 & \bar{a}^{(1)} & -\bar{b}^{(1)} \\ 0 & b^{(1)} & a^{(1)} \end{bmatrix}, \quad a_{ij}^{(2)} = \begin{bmatrix} \bar{a}^{(2)} & 0 & -\bar{b}^{(2)} \\ 0 & 1 & 0 \\ b^{(2)} & 0 & a^{(2)} \end{bmatrix},$$

$$a_{ij}^{(3)} = \begin{bmatrix} \bar{a}^{(3)} & -\bar{b}^{(3)} & 0 \\ b^{(3)} & a^{(3)} & 0 \\ 0 & 0 & 1 \end{bmatrix},$$

with each $a^{(r)}, \bar{a}^{(r)}, b^{(r)}, \bar{b}^{(r)}$ known in terms of A_r. In fact, each $a_{ij}^{(r)}$ derives from a 'pump wave approximation' in which a single wave-mode A_r is dominant. For instance, (15.4) reduces to

$$\begin{rcases} (i\,d/dx - c_1\zeta)u_1 = V_{12}u_2, \\ (i\,d/dx - c_2\zeta)u_2 = V_{21}u_1, \\ (i\,d/dx - c_3\zeta)u_3 = 0, \end{rcases} \tag{15.7a}$$

when only A_3 is non-zero. Rescaling yields a corresponding second-order problem of the form

$$(d/dx + i\lambda)u_1 = qu_2, \quad (d/dx - i\lambda)u_2 = ru_1. \tag{15.7b}$$

Independent solutions $u_k^{(m)}$ $(k, m = 1, 2)$, incident from $x = -\infty$ as

$$u_k^{(1)} \sim \begin{bmatrix} e^{-i\lambda x} \\ 0 \end{bmatrix}, \quad u_k^{(2)} \sim \begin{bmatrix} 0 \\ e^{i\lambda x} \end{bmatrix} \quad (x \to -\infty),$$

behave as $x \to +\infty$ like

$$u_k^{(1)} \sim \begin{bmatrix} a(\lambda)\,e^{-i\lambda x} \\ b(\lambda)\,e^{i\lambda x} \end{bmatrix}, \quad u_k^{(2)} \sim \begin{bmatrix} -\bar{b}(\lambda)\,e^{-i\lambda x} \\ \bar{a}(\lambda)\,e^{i\lambda x} \end{bmatrix} \quad (x \to +\infty),$$

where $a, b, \bar{a}, \bar{b}$ and the eigenvalue λ depend on the potentials q and r. These correspond to the scattering matrix $a_{ij}^{(3)}$ above, (with superscript 3 added), the amplitude of u_3 being unaltered.

With the separate envelopes initially ordered 3, 2, 1 from left to right, the overall scattering matrix at $t = 0$ is just

$$a_{ij}(\zeta, 0) = a_{ik}^{(3)} a_{kl}^{(2)} a_{lj}^{(1)}.$$

If the envelopes again separate at sufficiently large times, say t_f, (which does not always happen!) the final scattering matrices $a_{ij}^{(r)}(\zeta, t_f)$ have similar form and

$$a_{ij}(\zeta, t_f) = a_{ik}^{(1)} a_{kl}^{(2)} a_{lj}^{(3)} \quad (t = t_f), \tag{15.8}$$

the relative positions of the envelopes being reversed in accordance with $c_1 < c_2 < c_3$. Results analogous to (15.6) relate each initial and final scattering matrix $a_{ij}^{(r)}$. Inversion of the scattering data yields the respective wave amplitudes after interaction has been completed: but this simpler treatment does not yield the solution *during* interaction.

The complete formal solution of the inverse-scattering problem, in terms of a linear integral equation, is not usually amenable to further analytic reduction. Nevertheless, a class of closed-form n-soliton solutions is available. These are associated with a discrete 'bound-state' spectrum of eigenvalues ζ (or λ) which corresponds to poles of the ratios $b^{(r)}/a^{(r)}$ and which exists only when the signs s_j differ. Each $a^{(r)}$ may be thought of (loosely) as a wave incident from $x = +\infty$ which yields a transmitted wave of unit amplitude as $x \to -\infty$ and a 'reflected' wave of amplitude $b^{(r)}$ (actually, $b^{(r)}$ is 'converted', not reflected, into another member of the resonant triad). The bound-state spectrum of ζ arises, analogously with resonant over-reflection, when 'transmitted' and 'reflected' waves occur in the absence of an incident mode.

In addition to these bound-state soliton solutions, there is a continuous spectrum of ζ with non-singular $b^{(r)}/a^{(r)}$. Since (14.10) are non-dispersive in the linear limit, the contribution from this continuous spectrum does not decay with time (as it does for the Korteweg–de Vries equation, for example) but remains on an equal footing with the solitons. Nevertheless, investigation of the properties of the soliton solutions provides valuable insight.

For a given eigenvalue ζ, a ratio $b^{(r)}/a^{(r)}$ may be singular at $t = 0$ but not at the final time t_f, or vice versa. Precisely, zeros of $a^{(r)}$ and $\bar{a}^{(r)}$ present at $t = 0$ persist at $t = t_f$ for the fast and slow envelopes $r = 1$ and 3, but all zeros of $a^{(2)}$ present at $t = 0$ are absent at $t = t_f$. This means that the middle envelope always loses any solitons it may originally possess, but that solitons are not usually lost from the fast and slow envelopes. A middle-envelope soliton with eigenvalue $\zeta^{(2)}$ is converted to solitons of the slow and fast envelopes with respective eigenvalues $\zeta^{(1)}$, $\zeta^{(3)}$ with

Figure 5.4. Soliton exchange interactions, computed by Kaup, Reiman & Bers (1979). Mode a_i (with intermediate group velocity) dominates at $t = 0$ but mode a_j (smallest group velocity) is also present. Subsequently, one, two or more pulses of a_j and the third mode a_k are emitted. Reference frame chosen with $\mathbf{v}_i = 0$, $\mathbf{v}_j = -\mathbf{v}_k$. (a) $|a_j/a_i| = 0.02$ and pulse length $L/L_c = 3.25$ initially; (b) $|a_j/a_i| = 0.002$ and $L/L_c = 6.4$ initially. (Courtesy of Amer. Phys. Soc.).

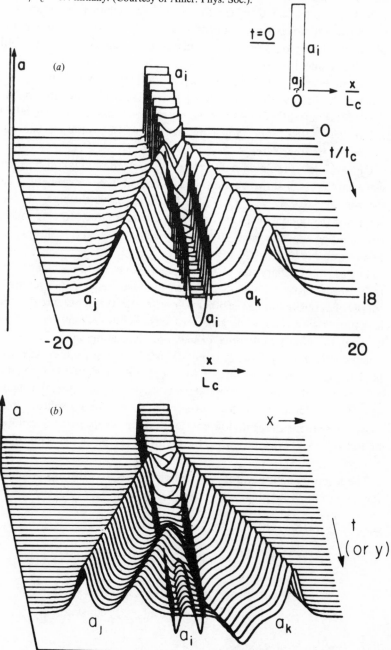

$\zeta^{(1)} + \zeta^{(3)} = \zeta^{(2)}$. These three solitons therefore form a resonant triad. If, exceptionally, the initial data provides fast and slow envelopes with solitons at such eigenvalues $\zeta^{(1)}$, $\zeta^{(3)}$ and with appropriate phases, these annihilate one another to produce a $\zeta^{(2)}$-soliton in the middle envelope, but this would subsequently disintegrate back into $\zeta^{(1)}$ and $\zeta^{(3)}$ solitons if it met any other disturbance. Such 'soliton exchange interaction' by nonlinear resonance is not confined to (14.10): it arises also for solutions of the Korteweg–de Vries and nonlinear Schrödinger equations (Miles 1977a, b; Kaup 1981b).

Soliton exchange interactions occur for (14.10) when the wave of greatest frequency has the middle group velocity and the signs s_j ($j = 1, 2, 3$) differ. Examples of such interaction, obtained numerically by Kaup *et al.* (1979), are shown in Figure 5.4. But soliton exchange is prohibited in other cases.

One such is 'stimulated back-scattering', where the highest-frequency wave has fastest or slowest group velocity and all three wave energies have the same sign (e.g. $s_1 = s_2 = 1, s_3 = -1$). Then, neither the high-frequency wave nor the middle envelope can ever possess solitons. Interaction of a laser and ion acoustic wave, to produce a backscattered (i.e. slower) laser pulse, is of this kind.

The assumed separation of the wave-envelopes at $t > t_f$ breaks down in 'explosive' cases. These occur when the high-frequency envelope has energy of opposite sign to the other two waves (i.e. the signs s_j are the same) and travels with the middle group velocity. Solutions may then develop a singularity after a finite time, or may not, depending on the initial data. Numerical examples, from Kaup *et al.*, are shown in Figures 5.5(*a*), (*b*): the latter explodes but the former does not. In Kaup's (1976) formulation, an earlier 'explosion' is indicated by inconsistencies in the scattering data at t_f: neither the fast nor slow envelope can ever contain solitons in this case, but neither can the middle envelope retain its solitons at t_f. Solutions bounded for all t exist only when the middle envelope contains no solitons at $t = 0$.

The n-soliton solution, for s_j of differing signs, has the following form for each (non-overlapping) envelope (Kaup *et al.*, Appendix B):

$$\left.\begin{array}{l} q^{(r)} = \sum_{j, k=1}^{n} \mathbb{D}_j S_{jk} \exp[-(\eta_j + \eta_k) x'] \quad (r = 1, 2, 3), \\[2ex] \lambda_j = i\eta_j, \quad S_{ij} = [(I + N^2)^{-1}]_{ij}, \quad \mathbb{D}_j = -ib_j[(\partial a_j / \partial \lambda)_{\pm \lambda = \lambda_j}]^{-1}, \\[2ex] N = N_{ij} = \mathbb{D}_j (\eta_i + \eta_j)^{-1} \exp[-(\eta_i + \eta_j) x'], \quad x' = x - c_r t - x_r \\[1ex] \hfill (x_r \text{ constant}), \end{array}\right\} \quad (15.9)$$

Figure 5.5. Collision of initially rectangular pulses in potentially explosive cases: (*a*) non-explosive, with $|a_j/a_i| = 0.25$ at $t = 0$; (*b*) explosive, with $|a_j/a_i| = 0.35$ at $t = 0$ (from Kaup, Reiman & Bers 1979, courtesy of Amer. Phys. Soc.).

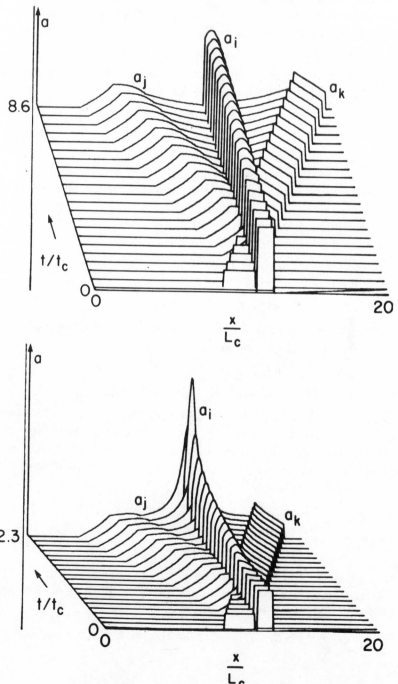

where $a_j = a_j^{(r)}$, $b_j = b_j^{(r)}$ ($r = 1, 2, 3$) are elements of the three scattering matrices $a_{k,l}^{(r)}$ corresponding to the jth eigenvalue-pair $\lambda = \pm\lambda_j$. These bound-state positive and negative eigenvalue pairs are pure imaginary and the summation yields two terms for each j, k. The scaled eigenvalues λ are related to the original eigenvalue ζ by

$$\lambda^{(1)} = \tfrac{1}{2}\zeta(c_3 - c_2), \quad \lambda^{(2)} = \tfrac{1}{2}\zeta(c_3 - c_1), \quad \lambda^{(3)} = \tfrac{1}{2}\zeta(c_2 - c_1)$$

for the respective envelopes (cf. equation 15.7a, b). Also, from (15.5) and (15.7), the amplitudes A_r are related to $q^{(r)}$ by

$$q^{(1)} = \frac{-s_2 s_3 A_1^*}{[(c_2 - c_1)(c_3 - c_1)]^{\frac{1}{2}}}, \quad q^{(2)} = \frac{-A_2}{[(c_2 - c_1)(c_3 - c_2)]^{\frac{1}{2}}},$$

$$q^{(3)} = \frac{-s_1 s_2 A_3^*}{[(c_3 - c_1)(c_3 - c_2)]^{\frac{1}{2}}}.$$

Each soliton travels with its linear group velocity c_r in the non-interacting region but a_j, b_j and so $\mathbb{D}_j$ take different values before and after interaction. Fuller details are given by Kaup *et al.*

The trivial one-soliton solution ($j = k = n = 1$) is

$$q^{(r)} = -2\eta_1 \operatorname{sgn}(-\mathbb{D}_1) \operatorname{sech}[2\eta_1(x' - x_0)]$$

with x_0 defined by

$$|\mathbb{D}_1| = 2\eta_1 \exp(2\eta_1 x_0).$$

The collision of two single envelope solitons is the next simplest solution (see Kaup *et al.*). At $t \to -\infty$ this is

$$q^{(1)} \sim 2\eta_1 \operatorname{sech}(2\eta_1 z_1) \, e^{i\phi_1}, \quad q^{(2)} \sim 0,$$

$$q^{(3)} \sim 2\eta_3 \operatorname{sech}(2\eta_3 z_3) \, e^{i\phi_3}, \quad z_j = x - c_j t - x_{0j}$$

where x_{0j}, ϕ_j are constants. Provided the eigenvalues $\lambda = i\eta_1$, $i\eta_3$ are such that the corresponding values ζ_1, ζ_3 of ζ are not equal, these solitons are non-resonant. They collide when $t \approx (x_{10} - x_{30})(c_3 - c_1)^{-1}$ and for a time produce a $q^{(2)}$ soliton. But, as $t \to \infty$, the $q^{(2)}$ soliton decays, replenishing the $q^{(1)}$ and $q^{(3)}$ solitons to

$$q^{(1)} \sim 2\eta_1 \operatorname{sech}(2\eta_1 z_1 - \delta) \, e^{i(\phi_1 + \theta)}, \quad q^{(2)} \sim 0,$$

$$q^{(3)} \sim 2\eta_3 \operatorname{sech}(2\eta_3 z_3 + \delta) \, e^{i(\phi_3 - \theta)},$$

$$\exp(-\delta + i\theta) \equiv (\eta_3 - \eta_1)(\eta_3 + \eta_1)^{-1}$$

(the eigenvalues λ being assumed imaginary).

When $\zeta_1 = \zeta_3$, the two solitons are resonant with a third, for then $\lambda^{(1)} + \lambda^{(3)} = \tfrac{1}{2}\zeta_1(c_3 - c_1) = \lambda^{(2)}$. In this case,

$$q^{(1)} \sim q^{(3)} \sim 0, \quad q^{(2)} \sim -2\eta_2 \operatorname{sech}(2\eta_2 z_2) \, e^{i\phi_2}$$

as $t \to \infty$, where $z_2 = x - c_2 t - x_{20}$ and

$$x_{20} = (x_{10} \eta_1 + x_{30} \eta_3) \eta_2^{-1}, \quad \phi_2 = \phi_1 + \phi_3.$$

Clearly, the $q^{(1)}$ and $q^{(3)}$ solitons have been converted to a $q^{(2)}$ soliton which completes the resonant triad.

The latter resonance occurs only in exceptional cases for which the initial data yield equal eigenvalues ζ_1 and ζ_3. Usually, the conversion process operates in reverse: a given middle envelope $q^{(2)}$ soliton interacting with an arbitrary small disturbance decays into fast and slow solitons which are resonant with it. This is normally referred to as a 'decay instability'.

Similar inverse scattering solutions have been found for three-wave interactions in inhomogeneous media. Then, a small wavenumber mismatch, say Δk denoting imperfect resonance, varies linearly with distance x; but the governing equations may still be transformed into (14.10) (Reiman 1979). Inverse-scattering and Bäcklund-transformation solutions for three-wave interaction with constant frequency or wavenumber mismatch were earlier given by Chu & Scott (1975) and Chu (1975a): these relate to cases where two of the three group velocities c_j are equal.

15.3   *Solutions in three and four dimensions*

For propagation in two or three space dimensions and time, the inverse scattering theory has been developed, along somewhat similar lines, by Zakharov (1976), Cornille (1979) and Kaup (1980). Zakharov first noted a broad class of explicit solutions, independently discovered using elementary analysis by Craik (1978). These and further classes of solutions have also been constructed, using a Bäcklund transformation, by Kaup (1981a, b).

With three, or four, independent variables $\mathbf{x}$, t available, it is possible to choose characteristic co-ordinates χ_i ($i = 1, 2, 3$) such that

$$\partial/\partial \chi_i = \partial/\partial t + \mathbf{v}_i \cdot \nabla \, (\mathbf{v}_1 \neq \mathbf{v}_2 \neq \mathbf{v}_3):$$

these are clearly linear combinations of $\mathbf{x}$, t. If there is dependence on three, rather than two, space co-ordinates and t, the fourth characteristic co-ordinate χ_4 is chosen so that

$$(\partial/\partial t + \mathbf{v}_i \cdot \nabla) \chi_4 = 0 \quad (i = 1, 2, 3).$$

In the amplitude equations, χ_4 is just a 'dummy variable' which acts like a fixed parameter and so is ignorable. The three amplitude equations (14.10) are then just

$$\partial A_i / \partial \chi_i = s_i A_j^* A_k^* \tag{15.10}$$

with i, j, k taken in cyclic order, equal to 1, 2 or 3.

Surprisingly, the inverse-scattering analysis reveals that the associated linear eigenvalue problem yields no discrete bound states. Accordingly, there are no three-dimensional n-solitons and the complete solution derives from the continuous spectrum.

On introducing moduli b_i and phases θ_i, where $A_i = b_i \exp(i\theta_i)$, (15.10) becomes

$$\left. \begin{aligned} \partial(s_i b_i^2)/\partial\chi_i &= 2b_1 b_2 b_3 \cos(\theta_1+\theta_2+\theta_3), \\ s_i b_i^2(\partial\theta_i/\partial\chi_i) &= -b_1 b_2 b_3 \sin(\theta_1+\theta_2+\theta_3). \end{aligned} \right\} \tag{15.11}$$

Solutions with constant phases $\theta_i = \theta_i^0$ have

$$b_i^2 = s_i \frac{\partial^2 F}{\partial\chi_j \partial\chi_k}, \qquad \theta_1^0+\theta_2^0+\theta_3^0 = N\pi,$$

$$\frac{\partial^3 F}{\partial\chi_1 \partial\chi_2 \partial\chi_3} = 2(-1)^N \left[s_1 s_2 s_3 \frac{\partial^2 F}{\partial\chi_1 \partial\chi_2} \frac{\partial^2 F}{\partial\chi_2 \partial\chi_3} \frac{\partial^2 F}{\partial\chi_3 \partial\chi_1} \right]^{\frac{1}{2}} \tag{15.12}$$

where $F(\chi_1,\chi_2,\chi_3)$ is any real solution of the last equation which yields positive values for each b_i^2. The problem lies in finding such functions F. That given by Craik (1978) is

$$F = -s_1 s_2 s_3 \ln[f_1(\chi_1)+f_2(\chi_2)+f_3(\chi_3)] \tag{15.13}$$

with $(-1)^N = -s_1 s_2 s_3$, the $f_i(\chi_i)$ being three not quite arbitrary differentiable functions. This class of explicit solutions also emerged as a special case of Zakharov's (1976) inverse-scattering representation and a somewhat similar solution, for propagation in x and t only, was given by Wilhelmsson, Watanabe & Nishikawa (1977).

For initially-bounded envelopes, the functions $f_i(\chi_i)$ at $t = 0$ must have non-zero sum $f_1+f_2+f_3$ at all points of space. Some further restrictions are necessary to ensure that the amplitudes b_j are real, but the class of solutions remains large. Initially-bounded envelopes 'explode' at a later time if $f_1+f_2+f_3$ becomes zero at any point in space. Craik (1978) showed this to be possible only when the three signs s_j are the same, as expected. He and also Kaup (1981a, b) give some particular examples of interacting localized wave-envelopes. They also give a 'bursting criterion', in terms of the available 'energies' of initial wave-envelopes. A typical configuration of interacting envelopes is shown in Figure 5.6. When one wave-mode is locally dominant, a 'pump-wave' approximation suffices (Craik & Adam 1978), but this cannot represent the explosive instability.

More general solutions were derived by Kaup (1981a), using a Bäcklund transformation. First, the linear scattering problem

$$\left. \begin{aligned} \partial\psi_i/\partial\chi_k &= s_k A_j^* \psi_k \\ \partial\psi_k/\partial\chi_i &= s_i A_j \psi_i \end{aligned} \right\} \quad (i, j, k = 1, 2, 3) \tag{15.14}$$

is defined, where i, j, k are cyclic. This comprises six equations for the three functions $\psi_i(\chi_1, \chi_2, \chi_3)$. These have three conservation laws,

$$\partial(s_i|\psi_i|^2)/\partial\chi_k = \partial(s_k|\psi_k|^2)/\partial\chi_i \quad (i \neq k)$$

and so there exists a function $D(\chi_1, \chi_2, \chi_3)$ with

$$\partial D/\partial\chi_i = -s_i|\psi_i|^2 \quad (i = 1, 2, 3). \tag{15.15}$$

If A_j $(j = 1, 2, 3)$ is any known solution of (15.10), the corresponding functions ψ_j and D may be found from these equations.

The Bäcklund transformation is

$$\tilde{A}_j = A_j + (\psi_i^* \psi_k / D), \tag{15.16}$$

with D real and cyclic i, j, k. It is readily verified that $\tilde{A}_j$ is then also a solution of (15.10)! The transformation therefore generates new solutions from old.

The trivial solution $A_j = 0$ yields

$$\tilde{A}_j = \psi_i^* \psi_k / D,$$

$$D = \sum_{i=1}^{3} s_i \left(\int_{\chi_i}^{\infty} |\psi_i(u)|^2 \, du \right) + \text{constant}$$

where each ψ_i is an arbitrary, differentiable, complex function of the corresponding single variable χ_i. This class of solutions, called 'one-lump' solutions, is precisely that of (15.12)–(15.13). 'Two-lump' solutions $\tilde{A}_j$ are next constructed by using the one-lump solution for A_j in (15.14) and (15.16) and calculating new functions ψ_j and D. Continuation of this process generates an arbitrary number of 'N-lump' solutions. Kaup

Figure 5.6. Configuration of localized disturbances (a) before and (b) after interaction. Each mode is contained within the domain D_j ($j = 1, 2, 3$) which moves in the direction indicated.

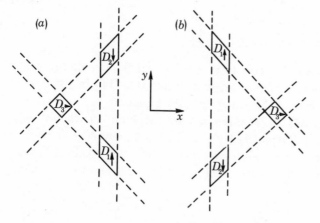

(1981a) gives only the one-lump and two-lump solutions explicitly, the latter containing six arbitrary functions compared with the one-lump solution's three. Kaup conjectures that the two-lump solution may be sufficiently general to describe the behaviour of *all* initially-localized wave-envelopes, but this has yet to be shown.

In one sense, the three-dimensional N-lump solutions may be regarded as counterparts of the two dimensional N-solitons: in inverse scattering theory, both derive from separable kernels of integral equations. In another, these solutions are very different, for the three-dimensional problem yields no discrete bound-state eigenvalues, such as give rise to N-solitons. A feature reminiscent of resonant soliton interactions is the 'decay instability' exhibited by some one-lump solutions: particular A_1 envelopes have been shown to convert completely into A_2 and A_3 envelopes on meeting a small amount of A_2 and A_3 (Case & Chiu 1977; Kaup 1981b).

15.4 Long wave–short wave interactions

Interactions among short waves and long waves are resonant when the group velocity c_g of the short-wave envelope equals the phase velocity c_p of the long waves. This is just the condition

$$\Delta k = k_l, \quad \Delta \omega = c_g \Delta k = \omega_l$$

where (k_l, ω_l) are the wavenumber and frequency of the long wave and $(\Delta k, \Delta \omega)$ are the limits $(k - k', \omega - \omega')$ as (k', ω') approaches the short-wave wavenumber and frequency (k, ω). Obviously, this is a special case of three-wave resonance.

With short-wave envelope $A_1(X, \tau)$ and long-wave envelope $B(X, \tau)$, the governing equations in many cases take the form

$$\left. \begin{aligned} iA_{1\tau} + \beta_1 A_{1XX} + \gamma_1 A_1 B &= 0, \\ B_\tau &= \lambda_1 (|A_1|^2)_X. \end{aligned} \right\} \tag{15.17a, b}$$

Here, $X = \epsilon^{\frac{2}{3}}(x - c_g t)$ and $\tau = \epsilon^{\frac{4}{3}}t$ (cf. 19.2a, b). Equality of c_g and c_p allows the dispersive term $\beta_1 A_{1XX}$ to be included in (15.17a), though this would otherwise be of higher order (see §19). Equations of this form were derived by Grimshaw (1977) for resonant interaction of a packet of short surface waves with a long internal wave. These were solved by Ma (1978) and Ma & Redekopp (1979), using inverse scattering, for real coefficients.

With *two* short-wave envelopes $A_1(X, \tau)$, $A_2(X, \tau)$ with differing wavenumbers and frequencies but with equal group velocities $c_{g1} = c_{g2} = c_p$, the corresponding equations are

$$
\left.
\begin{aligned}
iA_{1\tau} + \beta_1 A_{1XX} + \gamma_1 A_1 B &= 0, \\
iA_{2\tau} + \beta_2 A_{2XX} + \gamma_2 A_2 B &= 0, \\
B_\tau &= (\lambda_1 |A_1|^2 + \lambda_2 |A_2|^2)_X.
\end{aligned}
\right\}
\tag{15.18}
$$

These are discussed by Ma (1981), who gives their solution by inverse scattering for real coefficients with $\beta_1 = \beta_2$, $\gamma_1 = \gamma_2$ and sgn $\lambda_1 \lambda_2 = 1$. Such cases are conservative, with waves of like energy sign. There are, in effect, two degenerate resonant triads with the long wave in common. The restriction $c_{g1} = c_{g2}$ constrains the choice of modes.

With rescaled co-ordinates, (15.18) transform to

$$
\left.
\begin{aligned}
(i\,\partial/\partial\tau - \partial^2/\partial X^2)\,S_k &= -LS_k \quad (k = 1, 2), \\
\partial L/\partial\tau &= -2(|S_1|^2 + |S_2|^2)_X,
\end{aligned}
\right\}
\tag{15.19}
$$

and boundary conditions S_k, $L \to 0$ as $|X| \to \infty$ are imposed. The method of inverse scattering applies, in principle, also to interactions with N short waves $(k = 1, 2, ..., N)$. The linear eigenvalue problem associated with (15.19) is a fourth-order one, with complex eigenvalues $\zeta = -\xi + i\eta$ say, where ξ, $\eta > 0$. The one-soliton solution is

$$
\left.
\begin{aligned}
S_k &= i\eta(2\xi)^{\frac{1}{2}} c_k \frac{\exp[i(\xi^2 - \eta^2)\tau + i\xi X]}{\cosh[\eta(X - X_0) + 2\xi\eta\tau]}, \\
L &= -2\eta^2 \operatorname{sech}^2[\eta(X - X_0) + 2\xi\eta\tau], \quad c_1^2 + c_2^2 = 1,
\end{aligned}
\right\}
\tag{15.20}
$$

where X_0 and the 'polarization vector' $\mathbf{c} = (c_1, c_2)$ are known constants. All such solitons travel to the left, with velocity -2ξ, without change in form.

Interaction of two such solitons, with different eigenvalues ζ_1, ζ_2 yields solutions like (15.20) for both, as $t \to -\infty$ and $t \to +\infty$. But the respective values of $X_0^{(j)}$ and $\mathbf{c}^{(j)}$ $(j = 1, 2)$ differ before and after interaction. The centres of the two envelope solitons are shifted and the polarization vectors $\mathbf{c}^{(j)}$ are changed because of redistribution of energy among the S_1 and S_2 wave-modes in either soliton. Nevertheless, the total energy of each mode,

$$
\int_{-\infty}^{\infty} |S_1|^2 \, dX, \quad \int_{-\infty}^{\infty} |S_2|^2 \, dX, \quad \int_{-\infty}^{\infty} L \, dX,
$$

remains unchanged – a property not shared by other models of long wave–short wave interaction (Ma 1978; Newell 1978).

16   Linearly damped waves

16.1   *One wave heavily damped*

When the nonlinear interaction remains of conservative type, but the waves are linearly damped or amplified, the normalized amplitude equations resemble (14.10) but have an additional term $\sigma_j A_j$ $(j = 1, 2, 3)$ added to each left-hand side. A uniform wave-train, for given j, would then decay with exponential rate σ_j in the absence of the other two modes. (As before, the resonant triad is taken to have the form $\mathbf{k}_1 + \mathbf{k}_2 + \mathbf{k}_3 = 0$, $\omega_1 + \omega_2 + \omega_3 = 0$.)

The inverse-scattering solution exists only for $\sigma_j = 0$ and very few solutions incorporating damping are known, for amplitudes varying in both space and time. One relates to the case $\sigma_1 = \sigma_2 = 0$ and σ_3 very large. Then, the equation for A_3 yields the approximate solution

$$A_3 = \sigma_3^{-1} s_3 A_1^* A_2^*$$

and those for A_1 and A_2 yield

$$\left. \begin{array}{c} (\partial/\partial t + \mathbf{v}_1 \cdot \nabla) I_1 = I_1 I_2, \quad (\partial/\partial t + \mathbf{v}_2 \cdot \nabla) I_2 = I_1 I_2, \\ I_1 \equiv 2 s_2 s_3 \sigma_3^{-1} |A_1|^2, \quad I_2 \equiv 2 s_1 s_3 \sigma_3^{-1} |A_2|^2. \end{array} \right\} \tag{16.1}$$

Solutions of these are given by Hasimoto (1974), Chu (1975b) and Wilhelmsson, Watanabe & Nishikawa (1977) for amplitude variations in x and t. Chu & Karney (1977) consider variations in x, y and t: this extension derives from a straightforward change of reference frame since derivatives of A_3 are negligible compared with the large damping term. When I_1, I_2 depend on x and t, introduction of characteristic co-ordinates reduces (16.1) to

$$\left. \begin{array}{c} I_j = \partial \mathscr{F}/\partial \xi_j \quad (j = 1, 2), \quad \partial^2 \mathscr{F}/\partial \xi_1 \partial \xi_2 = -(\partial \mathscr{F}/\partial \xi_1)(\partial \mathscr{F}/\partial \xi_2), \\ \xi_1 = \dfrac{x - c_1 t}{c_1 - c_2}, \quad \xi_2 = \dfrac{x - c_2 t}{c_2 - c_1}, \end{array} \right\} \tag{16.2}$$

where c_j is the x-component of $\mathbf{v}_j$. The known class of solutions has

$$\mathscr{F}(\xi_1, \xi_2) = \ln\left[f_1(\xi_1) + f_2(\xi_2) \right].$$

This is reminiscent of the one-lump solutions of (15.10) in three dimensions.

16.2   *Waves dependent on t only*

(i) *Equal damping rates*

When the amplitudes depend on t only and each damping rate σ_j is the same, say σ, the governing equations may be reduced to the corresponding undamped equations. This is accomplished by the transformation

$$B_j = A_j \, e^{\sigma t}, \quad \tau = \sigma^{-1}(1 - e^{-\sigma t}), \tag{16.3a}$$

which yields

$$dB_i/d\tau = s_i B_j^* B_k^* \quad (i, j, k = 1, 2, 3 \text{ cyclically}). \tag{16.3b}$$

Solutions are then found as in §15.1. In explosive cases, the time of explosion is

$$t_\infty = -\sigma^{-1} \ln(1 - \sigma \tau_\infty)$$

where τ_∞ is the time of explosion in the undamped case. Obviously, the explosion is suppressed when $\sigma > \tau_\infty^{-1}$.

In non-explosive cases, the B_i vary periodically in τ, with period T, say; but corresponding recurrences in real time t do *not* have uniform periodicity. The intervals Δt_n between successive values $\tau = nT$ $(n = 0, 1, 2, \ldots)$ are

$$\Delta t_n = -\sigma^{-1} \ln\left(\frac{1 - (n+1)\sigma T}{1 - n\sigma T}\right), \quad (n+1)\sigma T < 1,$$

which increase with n. Only a finite number of recurrences of B_i can take place, since $n+1 < (\sigma T)^{-1}$; meanwhile, the actual amplitudes A_i incorporate the additional decay factor $\exp(-\sigma t)$.

When all three modes are *amplified* at the same rate, σ is negative and an identical transformation applies. Because of the change in sign, explosive instabilities appear sooner than in the unamplified case; while, in non-explosive cases, the recurrence times Δt_n of B_i are ever-decreasing and now infinite in number.

(ii) One wave damped

If only one wave, say A_3 is damped, and the phases satisfy $\eta_1 + \eta_2 + \eta_3 = N\pi$ (N integer), the three-wave equations are (cf. 15.2 with $\Delta\omega = 0$)

$$db_1/dt = (-1)^N s_1 b_2 b_3, \quad db_2/dt = (-1)^N s_2 b_1 b_3,$$
$$db_3/dt + \sigma b_3 = s_3 b_1 b_2.$$

A constant of motion is $\gamma \equiv s_1 b_1^2 - s_2 b_2^2$. When s_1, s_2 differ in sign, the substitution (Fuchs & Beaudry 1975)

$$b_1 = |\gamma|^{\frac{1}{2}} \sin \psi, \quad b_2 = |\gamma|^{\frac{1}{2}} \cos \psi \quad (0 \leqslant \psi \leqslant \tfrac{1}{2}\pi)$$
$$b_3 = (-1)^N s_1 \, \partial\psi/\partial t$$

yields

$$d^2y/dt^2 + \sigma \, dy/dt + r \sin y = 0,$$
$$y = 2\psi, \quad r = (-1)^N s_1 s_3 |\gamma|.$$

When $r > 0$, this is identical to the equation of a damped, nonlinear, simple pendulum.

(iii) Three different damping rates

The general case is more difficult to handle. Weiland & Wilhelmsson (1977) describe analytical approximations and numerical solutions both for equal and differing signs s_j. Miles (1976b) also does so for the degenerate case of two-mode interaction ($A_1 \equiv A_2$ say).

(iv) Three-wave strange attractor

When the wave of greatest frequency is linearly amplified and the other two are linearly damped, the temporal evolution of near-resonant triads with $\mathbf{k}_1 = \mathbf{k}_2 + \mathbf{k}_3$, $\omega_2 + \omega_3 - \omega_1 = \delta$ is described by normalized equations

$$\left.\begin{aligned}
\mathrm{d}b_1/\mathrm{d}t &= b_1 + b_2 b_3 \exp(\mathrm{i}\delta t), \\
\mathrm{d}b_2/\mathrm{d}t &= -\sigma_2 b_2 - b_1 b_3^* \exp(-\mathrm{i}\delta t), \\
\mathrm{d}b_3/\mathrm{d}t &= -\sigma_3 b_3 - b_1 b_2^* \exp(-\mathrm{i}\delta t),
\end{aligned}\right\} \tag{16.4}$$

(corresponding to $s_2 = s_3 = -s_1$). The normalized damping rates σ_2, σ_3 are positive.

Wersinger, Finn & Ott (1980a, b) discovered that these equations have a surprisingly rich structure, with solutions which exhibit bifurcations to successively more complex periodic orbits and also the 'chaotic' behaviour of a 'strange attractor'. On setting

$$b_1 = a_1 \exp(\mathrm{i}\phi_1), \quad b_k = a_k \exp(\mathrm{i}\phi_k - \tfrac{1}{2}\mathrm{i}\delta t) \quad (k = 2, 3)$$

and restricting attention to the special case $a_2 = a_3$, $\sigma_2 = \sigma_3 \equiv \sigma$, (16.4) become

$$\mathrm{d}a_1/\mathrm{d}t = a_1 + a_2^2 \cos\phi, \quad \mathrm{d}a_2/\mathrm{d}t = -a_2(\sigma + a_1 \cos\phi),$$

$$\mathrm{d}\phi/\mathrm{d}t = -\delta + a_1^{-1}(2a_1^2 - a_2^2)\sin\phi, \quad \phi = \phi_1 - \phi_2 - \phi_3.$$

These depend on just two parameters, δ and σ. Wersinger *et al.* undertook a numerical study for $\delta = 2$ and $1 \leqslant \sigma \leqslant 25$. When the initial phase ϕ was chosen in the range $0 < \phi < \pi$, it remained so at all subsequent times. Wersinger *et al.* recorded their results by plotting a_1 against a_2 each time the solution orbit crossed $\phi(t) = \tfrac{1}{2}\pi$ with $\mathrm{d}\phi/\mathrm{d}t < 0$.

For small damping rates, $1 < \sigma < 3$, a_1 and a_3 grow without bound; while, for $3 \leqslant \sigma \leqslant 8.5$, all solutions approach a simple limit cycle about a single fixed-point solution. At $\sigma \approx 8.5$, bifurcation occurs from the single fixed point to one unstable and two stable fixed points: solutions then approach a 2-point periodic limit cycle at large t. Further bifurcations, to stable 4-point, 8-point, 16-point and 32-point limit cycles occur at $\sigma \approx 11.9$, 12.8, 13.15 and 13.16 respectively. For $13.16 < \sigma < 13.20$ and $13.20 < \sigma < 16.8$, solutions had no observable periodicity and showed the typical chaotic behaviour of a strange attractor, but a stable 48-point periodic cycle was observed at $\sigma = 13.20$. For $16.8 < \sigma < 17.4$, periodicity reappeared as a stable 3-point limit cycle, but at $\sigma \approx 17.4$ this bifurcated to a 6-point cycle. Apparently chaotic behaviour resumed at $\sigma \approx 18.5$ and persisted to $\sigma = 25$.

(v)   Equilibrium solutions with external forcing

With three damped waves and external forcing, equations for the moduli and phases of the amplitudes are (cf. 14.9)

$$\tfrac{1}{2}d(b_i^2)/dt = b_1 b_2 b_3(Jk_i/\omega_i)\sin\eta - \sigma_i b_i^2 + b_i F_i \cos(\gamma_i - \eta_i),$$

$$b_i^2 \, d\eta_i/dt = b_1 b_2 b_3(Jk_i/\omega_i)\cos\eta + b_i F_i \sin(\gamma_i - \eta_i)$$

$$\eta = \sum_{i=1}^{3} \eta_i \quad (i = 1, 2, 3).$$

Here, the damping rates are σ_i and F_i, γ_i represent the respective magnitudes and phases of the external forcing. If F_i and γ_i are constants, steady-state solutions exist with $\eta_i = \gamma_i$ when $\gamma_1 + \gamma_2 + \gamma_3 = (N - \tfrac{1}{2})\pi$ (N integer). The amplitudes b_i then satisfy

$$\sigma_i b_i^2 - F_i b_i = (-1)^N b_1 b_2 b_3 (Jk_i/\omega_i) \quad (i = 1, 2, 3),$$

and the number of permissible solutions varies. They have the form

$$\left.\begin{aligned}
b_j &= b_j^{\pm}(K) = \tfrac{1}{2}[\sigma_j^{-1}F_j \pm (\sigma_j^{-2}F_j^2 + KB_j)^{\frac{1}{2}}], \\
B_j &\equiv (-1)^N Jk_j/\omega_j \sigma_j,
\end{aligned}\right\} \tag{16.5}$$

where K is a root of

$$b_1^{\pm} b_2^{\pm} b_3^{\pm} = \tfrac{1}{4}K.$$

McEwan, Mander & Smith (1972) examined such triads of standing internal waves in stratified fluid, both experimentally and theoretically, when only the wave of greatest absolute frequency is forced. Then, $F_2 = F_3 = 0$ and the only steady-state solutions are

$$\eta_1 = \gamma_1, \quad b_1 = \sigma_1^{-1}F_1, \quad b_2 = b_3 = 0,$$

$$\eta_2, \eta_3 \text{ arbitrary,}$$

and

$$\left.\begin{aligned}
&\eta_1 = \gamma_1, \quad b_1 = b_{10} = (B_2 B_3)^{-\frac{1}{2}}, \quad \eta_2 = (N - \tfrac{1}{2})\pi - \gamma_1 - \eta_3, \\
&b_2(-B_1 B_3)^{\frac{1}{2}} = b_3(-B_1 B_2)^{\frac{1}{2}} = (-1 + F_1 \sigma_1^{-1} b_{10}^{-1})^{\frac{1}{2}}, \quad \eta_3 \text{ arbitrary.}
\end{aligned}\right\} \text{(16.6a, b)}$$

The latter solution exists only if $B_2 B_3 > 0$ – when $B_1 B_2$ and $B_1 B_3$ are both negative by (16.5) – and if $\sigma_1^{-1} F_1 > (B_2 B_3)^{-\frac{1}{2}}$. With weaker forcing than this, only the former solution occurs.

As F_1 is increased from zero, at first only the single wave b_1 appears, but this solution is unstable for $F_1 \geqslant \sigma_1 (B_2 B_3)^{-\frac{1}{2}}$ and resonating waves b_2 and b_3 appear. As F_1 increases beyond the critical value, the amplitudes of b_2 and b_3 increase while that of the driven wave remains constant at $(B_2 B_3)^{-\frac{1}{2}}$. This behaviour was confirmed experimentally by McEwan *et al.* Figure 5.7 shows an example of their experimental and theoretical results. A single standing-wave mode was first produced and then a paddle, which directly excited a second mode, was turned on. The results show establishment of a three-wave equilibrium solution, comprising these and a third, resonating, mode.

Chester (1968) and Miles (1976b) have analysed resonant oscillations of water waves in closed containers subjected to periodic forcing, and related experiments are described by Chester & Bones (1968). When the forcing frequency ω and the frequencies ω_1, ω_2 of standing-wave modes satisfy

Figure 5.7. Forced triad interaction among standing internal waves (from McEwan, Mander & Smith 1972). Mode $(1, 2)$ was present initially and mode $(3, 1)$ was driven by an oscillatory paddle. Three-wave equilibrium, with these and mode $(2, -1)$, was eventually established. Q denotes amplitudes, N the number of paddle cycles; ●, □ show measurements and solid lines theoretical results.

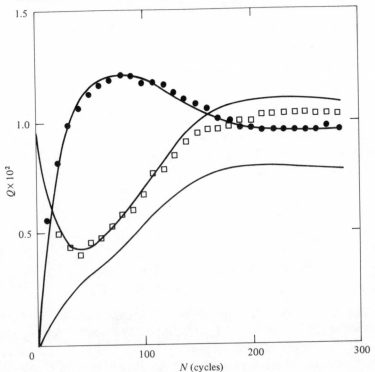

$\omega \approx \omega_1 \approx \frac{1}{2}\omega_2$, and when the wavenumbers satisfy $k_1 = \frac{1}{2}k_2$, Miles' coupled amplitude equations are

$$dA/dt = -\sigma_1 A + BA^* + 1,$$
$$dB/dt = -\sigma_2 B + A^2.$$

This is a degenerate case of three-wave resonance with damping and periodic forcing. Similar resonances of elastic structures containing a liquid, under periodic forcing, were investigated both theoretically and experimentally by Ibrahim & Barr (1975a, b). For harmonic two-mode interactions, equations like Miles' were recovered; while three-mode interactions satisfied equations like those of McEwan, Mander & Smith.

Waves in oscillating containers were also studied theoretically by Mahony & Smith (1972) and Huntley & Smith (1973), but with imposed vibrations of much higher frequency Ω than those of standing water waves. The latter are driven unstable by a limiting form of three-wave resonance, $\omega_3 = \omega_1 - \omega_2$ say, where $\omega_1 \approx \omega_2 \approx \Omega \gg \omega_3$. The modulations of forced components with frequencies ω_1, ω_2 may then occur on the same $O(\omega_3^{-1})$ timescale as the standing-wave oscillations. Huntley's (1972) experiments on waves in a rapidly-vibrated glass beaker and Franklin, Price & Williams' (1973) acoustically-excited water waves gave dramatic demonstrations of the phenomenon. With cubic nonlinearities retained in the amplitude equations, the theoretical model successfully reproduced observed features of hysteresis and nonlinear detuning.

It is appropriate also to mention parallel work on resonance and forced oscillations in mechanical systems with just a few degrees of freedom: Rott's (1970) multiple pendulum and Barr & Ashworth's (1977) weighted elastic beams provide particularly elegant demonstrations of parametric and internal resonances.

(vi) Coupled triads

When there exist *two* resonant triads with one member in common, say

$$\mathbf{k}_1 + \mathbf{k}_2 + \mathbf{k}_3 = 0, \quad \omega_1 + \omega_2 + \omega_3 = 0,$$
$$\mathbf{k}_1 + \mathbf{k}_4 + \mathbf{k}_5 = 0, \quad \omega_1 + \omega_4 + \omega_5 = 0,$$

there are five coupled amplitude equations, each with quadratic nonlinearities. Undamped cases are discussed by Wilhelmsson & Pavlenko (1973) (see also Weiland & Wilhelmsson 1977) for potentially explosive interactions. Analytic solutions are found, in particular cases, in terms of elliptic functions.

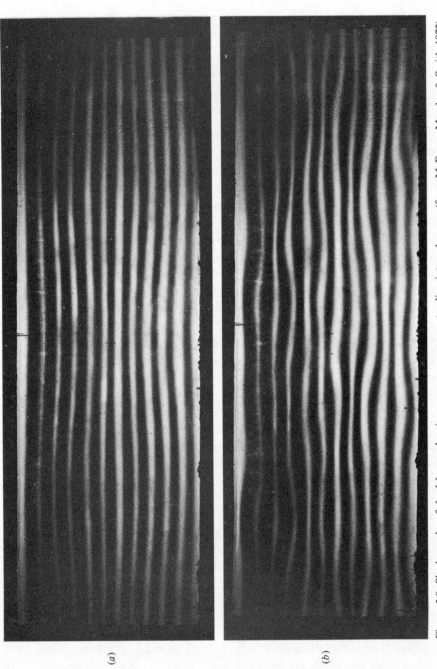

Figure 5.8. Shadographs of dyed layers showing resonance among standing internal waves (from McEwan, Mander & Smith 1972). (a) shows initial single-mode state with dimensionless wavenumber (4, 2). (b) shows state after 430 cycles of paddle driving the (5, 1) mode. This mode resonates both with (4, 2), (1, −1) and with (3, 3), (8, 4) the latter of which have grown to prominence.

McEwan, Mander & Smith (1972) considered five damped internal waves, with the common wave b_1 externally forced and having the greatest absolute frequency. They showed that the steady state $b_1 = \sigma_1^{-1} F_1$, $b_j = 0$ ($j = 2, 3, 4, 5$) is stable when $\sigma_1^{-1} F_1$ is below the threshold value for both triads, regarded separately. Otherwise, the only stable equilibrium state corresponds to (16.6b) for the triad with lowest threshold, the remaining two waves having zero amplitude.

Figure 5.8, from McEwan, Mander & Smith (1972), shows just such a case. The initial configuration (a) is a single standing-wave mode, say $\mathbf{k}_2$. Direct forcing of mode $\mathbf{k}_1$ at first led to growth of this and the resonant $\mathbf{k}_3$ component; but, after sufficient time, modes $\mathbf{k}_4$ and $\mathbf{k}_5$ grew to prominence as shown in (b).

Explosive instability of triads and coupled pairs of triads of baroclinic waves was recently discussed by Merkine & Shtilman (1984): their model equations incorporate linear growth or damping and also cubic nonlinearities. The mutual interaction of *many* coupled resonant triads was investigated numerically by Loesch & Domaracki (1977); see also §19.3.

16.3   *Higher-order effects*

The influence of higher-order terms on three-wave resonance is described, in part, by Weiland & Wilhelmsson (1977); see also Goncharov (1981). Their equations have the general form

$$\left. \begin{aligned} \mathrm{d}A_1/\mathrm{d}t + \sigma_1 A_1 &= c_{23} A_3 A_2^* - \mathrm{i}A_1 \sum_{k=1}^{3} \alpha_{1k} |A_k|^2, \\ \mathrm{d}A_2/\mathrm{d}t + \sigma_2 A_2 &= c_{13} A_3 A_1^* - \mathrm{i}A_2 \sum_{k=1}^{3} \alpha_{2k} |A_k|^2, \\ \mathrm{d}A_3/\mathrm{d}t + \sigma_3 A_3 &= c_{12}^* A_1 A_2 - \mathrm{i}A_3 \sum_{k=1}^{3} \alpha_{3k} |A_k|^2, \end{aligned} \right\} \quad (16.7\mathrm{a,b,c})$$

for resonance with $\mathbf{k}_3 = \mathbf{k}_1 + \mathbf{k}_2$, $\omega_3 = \omega_1 + \omega_2$. In addition to the three second-order coupling coefficients, there are nine third-order coefficients α_{ij} ($i, j = 1, 2, 3$). The real parts of α_{ij} contribute amplitude-dependent frequency shifts and the imaginary parts yield extra growth or damping terms.

For conservative systems, $\sigma_j = \mathrm{Im}\{\alpha_{ij}\} = 0$ and (16.7) transform to scaled equations like (15.2), namely

$$\left. \begin{aligned} \mathrm{d}b_i/\mathrm{d}t &= s_i b_j b_k \cos \eta \quad (i, j, k \text{ cyclic}), \\ \mathrm{d}\eta/\mathrm{d}t &= \Delta\omega - b_1 b_2 b_3 \sin \eta \sum_{j=1}^{3} s_j b_j^{-2}. \end{aligned} \right\} \quad (16.8)$$

But the frequency shift $\Delta\omega$ is now an additional variable,

$$\Delta\omega = \sum_{j=1}^{3} \beta_j b_j^2, \quad \beta_j = \mathrm{Re}\{\alpha_{j3} - \alpha_{j1} - \alpha_{j2}\}.$$

A solution with all three waves equal and each $s_i = 1$ is

$$b_j = b(t) = [\gamma^2 + (t_1 - t)^2]^{-\frac{1}{2}} \quad (j = 1, 2, 3),$$

$$\sin\eta = -\gamma b(t), \quad \gamma \equiv \tfrac{1}{4}\sum_{j=1}^{3}\beta_j, \quad t_1 \equiv b^{-1}(0)[1 - \gamma^2 b^2(0)]^{\frac{1}{2}}.$$

When $\gamma = 0$, this is the explosive solution iv(c) of §15.1; but, for finite γ, $b_j(t)$ reach a maximum of γ^{-1} before decaying to zero. The explosion is suppressed by nonlinear detuning of the resonance.

More generally, the conservative equations (16.8) may be cast in the form (15.3)$'$ since they, like (15.2), have constants of motion. But the function $\pi(x)$ is now a fourth, not third, degree polynomial. This again yields solutions in terms of elliptic functions: the sn function when all four roots of $\pi(x)$ are real and the cn function when two are real and two complex. Since $\pi(x) \sim 2\gamma^2 x^4 > 0$ as $|x| \to \infty$, unbounded solutions cannot occur. Except for a few limiting cases, like that above, the solutions are periodic. Cases which are explosive in the absence of third-order terms normally exhibit 'repeated stabilized explosions' as shown schematically in Figure 5.9.

In non-conservative cases, only approximate analytic solutions are known when c_{ij} are non-zero; but it is a simple matter to solve any given

Figure 5.9. Repeated stabilized explosions (after Weiland & Wilhelmsson 1977).

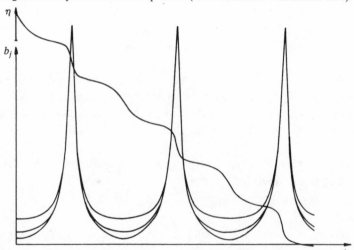

initial-value problem numerically. These, too, are discussed by Weiland & Wilhelmsson. Bifurcating equilibrium solutions of (16.7) were considered by Craik (1975): when only one wave, A_3, is present, (16.7c) reduces to a Landau-type equation (cf. equation 8.8),

$$\mathrm{d}A_3/\mathrm{d}t + \sigma_3 A_3 = -\mathrm{i}\alpha_{33} |A_3|^2 A_3.$$

This may have an equilibrium-amplitude solution with

$$|A_3| = [\mathrm{Re}\,\sigma_3/\mathrm{Im}\,\alpha_{33}]^{\frac{1}{2}} \equiv a(R) \quad (R \geqslant R_\mathrm{c})$$

where R denotes a variable flow parameter which yields $a(R_\mathrm{c}) = 0$ at the critical value R_c. As R and $a(R)$ increase away from R_c, initially-infinitesimal, linearly-damped, wave modes A_1 and A_2 may eventually be driven unstable by terms containing A_3 in (16.7a, b).

A fuller discussion of three-mode interactions with cubic nonlinearities is postponed until §§25–26, where examples of thermal convection and shear-flow instability are treated.

17   Non-conservative wave interactions

17.1   *Resonant triads in shear flows*

Until now, we have mostly considered energy-conserving interactions: even for the linearly-damped modes of §§16.1 and 16.2, the interaction coefficients were of conservative form. For waves in shear flows,

Figure 5.10. Curves of constant c_r and c_i for three-dimensional disturbances with wavenumber components (α, β) in Blasius flow at $R = 882$. The curves of constant c_i are shown only for $\beta \leqslant 0$ and those of constant c_r for $\beta \geqslant 0$. Both curves are symmetrical about $\beta = 0$. The arrows designate resonant triads (from Craik 1971).

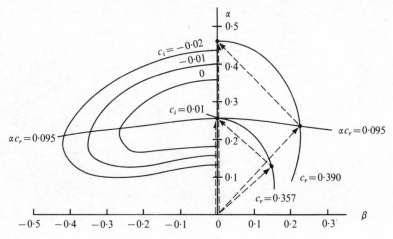

Figure 5.11. Smoke-streak flow visualization photographs, by Saric and co-workers, of growing disturbances in the Blasius boundary layer in air. Photographs (a), (b) and (c) show the same location, for $U_\infty = 6.8$ m s^{-1} and disturbances of differing amplitudes but same frequency (39 Hz) introduced upstream by a vibrating ribbon. Case (a) with $|u'|/U_\infty = 2.5 \times 10^{-3}$ at onset of linear instability, shows two-dimensional waves. Case (b), with $|u'|/U_\infty = 5.0 \times 10^{-3}$ shows three-dimensional disturbances with 'staggered peak-valley' arrangement characteristic of oblique-wave resonance.

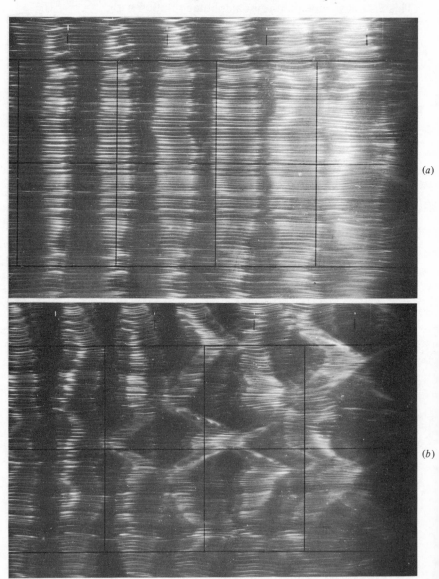

(a)

(b)

Case (c), with $|u'|/U_\infty = 9.0 \times 10^{-3}$ shows coexisting 'staggered' and 'aligned' peak-valley structures. Case (d) shows a typical 'aligned' configuration with 'Λ-vortices' such as was observed by Klebanoff *et al.* (1962). Flow is from left to right. (Photographs kindly supplied by Professor Saric.)

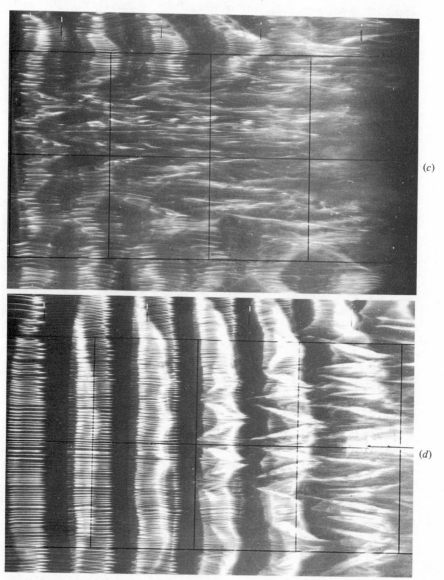

(c)

(d)

this is not usually so and nonlinearity may be responsible for additional energy transfer to or from the primary flow.

A *weak* shear flow may participate in three- or four-wave resonance just as if it were a 'wave' with zero frequency (cf. Phillips 1966). But this approach is unsuitable for strong shear flows. Raetz (1959, 1964) and Stuart (1962b) first showed that resonant triads of Tollmien–Schlichting waves can occur in boundary-layer flows and Kelly (1967, 1968) examined subharmonic and three-wave resonance of inviscid modes for jet and shear-layer profiles. In all this work, attention was confined to wave-modes which are neutrally stable according to linear theory, although linearly unstable modes also exist in these flows.

Craik (1968) showed that resonant triads of surface gravity waves and interfacial gravity waves exist in sufficiently strong uniformly sheared flows. When such waves share a common critical layer, there is a surprisingly large $O(R)$ *viscous* contribution to the quadratic interaction coefficients, where R is the Reynolds number. The same was found to be true for waves in general parallel shear flows, provided the critical layer is far from boundaries (Craik 1971, Usher & Craik 1974). These triads consist of one two-dimensional and two oblique wave-modes with periodicities

$$\exp[i\alpha(x - c_r t)], \quad \exp[i(\tfrac{1}{2}\alpha x \pm \beta y - \tfrac{1}{2}\alpha c_r' t)]$$

which are in resonance provided $c_r = c_r'$. These are best located from curves of constant c_r plotted in the α–β wavenumber plane at given Reynolds numbers R. Such curves may be deduced from the Orr–Sommerfeld eigenvalues $c_r + i c_i = c(\alpha, R)$ by using Squire's transformation. Examples, from Craik (1971), are shown in Figure 5.10 for Blasius flow at $R = 882$. Since c_r and c_i are even functions of β they are shown only for $\beta \geqslant 0$ and $\beta \leqslant 0$ respectively. The arrows denote two separate resonant triads. That with $\alpha = 0.254$ has the linearly-most-unstable mode as its two-dimensional component; that with $\alpha = 0.46$ has linearly-unstable oblique waves with frequency equal to that of the linearly-most-unstable mode. Such resonant triads can cause preferential amplification of particular oblique modes.

The classic boundary-layer experiments of Klebanoff, Tidstrom & Sargent (1962) and the plane Poiseuille flow experiments of Nishioka, Iida & Ichikawa (1975) showed the spontaneous growth of oblique modes with frequency close to that of the fundamental, but provided no evidence of the generation of subharmonic oblique waves with half that frequency. The first such evidence in boundary layers was found by Kachanov, Kozlov & Levchenko (1977) and this was later substantiated by Thomas & Saric (1981), Kachanov & Levchenko (1984) and Saric & Thomas (1984). Saric

& Thomas' work employed an ingenious smoke-wire technique for flow visualization; typical results are shown in Figure 5.11 where the development of spanwise-periodic oblique disturbances is clear. Kozlov & Ramazanov's (1984) similar study of plane Poiseuille flow shows structures as in Figure 5.11(d).

The work of Levchenko and co-workers records the evolution downstream of a vibrating ribbon, of the complete frequency spectrum of disturbances. At low excitation amplitudes, the development of a broad band of rather low frequencies was noticed. When the ribbon was excited with two frequencies ω_0, ω_1 simultaneously, sum and difference frequencies $\omega_0 \pm \omega_1$ could be observed along with other higher-order combinations. Judicious variation of the ribbon amplitudes and frequencies provided clear evidence of three-wave resonance at downstream positions: see Figure 5.12. When $\omega_1 = \frac{1}{2}\omega_0$, the oblique wave resonance is symmetric;

Figure 5.12. Measured frequency spectra of Kachanov & Levchenko (1984), at a fixed location with $R = 633$ and fixed primary (ω_0) oscillation frequency 120 Hz, for various secondary excitation frequencies (ω_1) detuned from symmetric resonance at $\frac{1}{2}\omega_0$. Curves labelled 1 to 5 correspond to detuning of -30, -10, -5, 0, $+10$ Hz respectively. The twin low-frequency peaks of 2, 3, 5 are associated with non-symmetric resonance, the single peak of 4 with symmetric resonance. Note contributions of higher harmonics and of broadband low-frequency components deriving from resonantly amplified non-controlled disturbances.

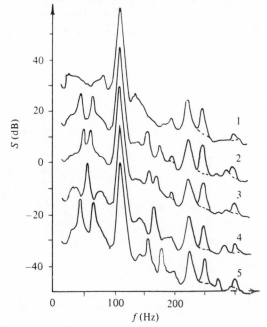

when not, there are two non-symmetric triads of the form $(k_0, 0)$, $(k_1, \pm l)$, $(k_0 - k_1, \mp l)$ as well as a range of near-resonant triads. This rapid growth of oblique modes, without diminution of the two-dimensional mode, is the clearest – indeed virtually the only – indication of 'explosive' three-wave interaction in fluids yet obtained. This is discussed further in §26.1.

Subharmonic resonance has also been detected in unstable shear layers (Miksad 1972, 1973). Miksad's anti-symmetric shear layer was excited by a loudspeaker at one or two chosen frequencies. Though evolution from linear instability to turbulent breakdown covered only about five wavelengths, he could discern various sub-, super-, and sum-and-difference harmonics of the imposed frequencies.

For symmetric wakes, Sato (1970) found no subharmonic resonance with imposed disturbances of a single frequency. Miksad *et al.* (1982) investigated similar wakes with two-frequency excitation and recorded the amplitude and phase modulations of the various sum-and-difference frequency combinations driven by nonlinear interactions. They, and also Miksad, Jones & Powers (1983), suggest that phase modulations may play a major rôle in bringing about the spectral broadening characteristic of transition to turbulence.

17.2   *The interaction equations*

The $O(a^2)$ interaction equations may be derived as outlined in §8 with spatially-uniform wave amplitudes (cf. Craik 1971). With a primary shear flow $\mathbf{u} = [U(z), 0, 0]$, the two-dimensional wave has linearized velocity components (u_3, v_3, w_3) given by

$$u_3 = \partial \Psi_3 / \partial z, \quad v_3 = 0, \quad w_3 = -\partial \Psi_3 / \partial x$$

with streamfunction

$$\Psi_3 = \mathrm{Re}\{A_3(t)\,\phi_3(z)\exp[\mathrm{i}\alpha(x-ct)]\}.$$

The eigenfunction ϕ_3 and eigenvalue $c = c_r + \mathrm{i}c_i$ are those of the Orr–Sommerfeld equation

$$L[\phi_3] \equiv \mathrm{i}\alpha[(U-c)(\phi_3'' - \alpha^2\phi_3) - U''\phi_3] - R^{-1}(\phi_3^{\mathrm{iv}} - 2\alpha^2\phi_3'' + \alpha^4\phi_3) = 0 \tag{17.1}$$

and appropriate homogeneous boundary conditions.

The oblique-wave velocity components (u_j, v_j, w_j) $(j = 1, 2)$ are best expressed as

$$\gamma u_j = \tfrac{1}{2}\alpha\hat{u}_j \mp \beta\hat{v}_j, \quad \gamma v_j = \pm\beta\hat{u}_j + \tfrac{1}{2}\alpha\hat{v}_j,$$

$$\gamma \equiv (\tfrac{1}{4}\alpha^2 + \beta^2)^{\frac{1}{2}},$$

with upper signs chosen for $j = 1$ and lower for $j = 2$. The new horizontal velocity components $\hat{u}_j$, $\hat{v}_j$ are those perpendicular and parallel to the respective wave crests. Accordingly,

$$\hat{u}_j = \partial\Psi_j/\partial z, \quad w_j = -(2\gamma/\alpha)\,\partial\Psi_j/\partial x \quad (j = 1, 2),$$

$$\Psi_j = \text{Re}\{A_j(t)\,\phi_j(z)\,\exp[\mathrm{i}(\tfrac{1}{2}\alpha x \pm \beta y - \tfrac{1}{2}\alpha\tilde{c}t)]\}$$

where $\phi_j(z)$, $\tilde{c} = c_r + \mathrm{i}\tilde{c}_i$ are eigenfunction and eigenvalue of the modified Orr–Sommerfeld equation

$$L'[\phi_j] \equiv \tfrac{1}{2}\mathrm{i}\alpha[(U-c)(\phi_j'' - \gamma^2\phi_j) - U''\phi_j] - R^{-1}(\phi_j^{\mathrm{iv}} - 2\gamma^2\phi_j'' + \gamma^4\phi_j) = 0$$

$$(17.2)$$

and given homogeneous boundary conditions. At resonance, the eigenvalues c and $\tilde{c}$ have equal real parts and this requirement selects a particular value of β. The remaining velocity components $\hat{v}_j$ satisfy

$$\tfrac{1}{2}\mathrm{i}\alpha(U - \tilde{c})\hat{v}_j - R^{-1}(\hat{v}_j'' - \gamma^2\hat{v}_j) = \mp\mathrm{i}U'\beta\phi_j \qquad (17.3)$$

(cf. equation 7.3).

On retaining terms of order $O(a^2)$ in the vorticity equations for each Fourier component, one finds that

$$A_1(t)\,L'[\phi_1] = -(\mathrm{d}A_1/\mathrm{d}t)(\phi_1'' - \gamma^2\phi_1) + A_3 A_2^* \,\mathrm{e}^{\alpha c_i t}\,G_1(z),$$

$$A_2(t)\,L'[\phi_2] = -(\mathrm{d}A_2/\mathrm{d}t)(\phi_2'' - \gamma^2\phi_2) + A_3 A_1^* \,\mathrm{e}^{\alpha c_i t}\,G_2(z),$$

$$A_3(t)\,L[\phi_3] = -(\mathrm{d}A_3/\mathrm{d}t)(\phi_3'' - \alpha^2\phi_3) + A_1 A_2 \,\mathrm{e}^{\alpha(\tilde{c}_i - c_i)t}\,G_3(z).$$

The functions $G_i(z)$ $(i = 1, 2, 3)$, which are omitted for brevity, involve products of the ϕ_k, $\hat{v}_j$ and their derivatives and so may be evaluated from linear theory. When $(\phi_1, \hat{v}_1)$ and $(\phi_2, \hat{v}_2)$ satisfy the same boundary conditions – as is usual – G_1 and G_2 are identical, by symmetry.

The orthogonality conditions necessary for the existence of solutions of these equations are obtained on multiplication by the respective adjoint functions $\psi_i(z)$ $(i = 1, 2, 3)$ and integration across the flow from boundaries at, say, z_1 and z_2. That is,

$$\mathrm{d}A_1/\mathrm{d}t \int_{z_1}^{z_2} \psi_1(\phi_1'' - \gamma^2\phi_1)\,\mathrm{d}z = A_3 A_2^* \,\mathrm{e}^{\alpha c_i t} \int_{z_1}^{z_2} G_1\psi_1\,\mathrm{d}z,$$

$$\mathrm{d}A_2/\mathrm{d}t \int_{z_1}^{z_2} \psi_2(\phi_2'' - \gamma^2\phi_2)\,\mathrm{d}z = A_3 A_1^* \,\mathrm{e}^{\alpha c_i t} \int_{z_1}^{z_2} G_2\psi_2\,\mathrm{d}z,$$

$$\mathrm{d}A_3/\mathrm{d}t \int_{z_1}^{z_2} \psi_3(\phi_3'' - \alpha^2\phi_3)\,\mathrm{d}z = A_1 A_2 \,\mathrm{e}^{\alpha(\tilde{c}_i - c_i)t} \int_{z_1}^{z_2} G_3\psi_3\,\mathrm{d}z,$$

where ψ_i are eigenfunctions of linear equations and boundary conditions adjoint to those of ϕ_i. The functions ϕ_i, ψ_i may be normalized in any convenient manner. (Note that the notation differs from that of §8.2.)

In terms of the amplitudes $a_{1,2} = A_{1,2} \exp(\tfrac{1}{2}\alpha\tilde{c}_i t)$, $a_3 = A_3 \exp(\alpha c_i t)$, these are

$$
\left.
\begin{aligned}
\mathrm{d}a_1/\mathrm{d}t &= \tfrac{1}{2}\alpha\tilde{c}_i\, a_1 + \lambda_1 a_3 a_2^*, \\
\mathrm{d}a_2/\mathrm{d}t &= \tfrac{1}{2}\alpha\tilde{c}_i\, a_2 + \lambda_2 a_3 a_1^*, \\
\mathrm{d}a_3/\mathrm{d}t &= \alpha c_i\, a_3 + \lambda_3 a_1 a_2,
\end{aligned}
\right\} \tag{17.4}
$$

where each λ_i is a ratio of the above integrals and terms of $O(a^3)$ are omitted. Normally, $\lambda_1 = \lambda_2$ by symmetry. The omission of the higher-order terms is formally justified only in the limit $a \to 0$ where a is a measure of the wave amplitudes. The $O(a^2)$ nonlinear terms are then comparable with the linear growth terms if $|\tfrac{1}{2}\alpha\tilde{c}_i/\lambda_{1,2}|$ and $|\alpha c_i/\lambda_3|$ are $O(a)$. But it is only in exceptional cases that all three waves of a resonant triad are neutrally stable, as the formal limit $a \to 0$ requires. However, typical linear growth rates are numerically small, and the λ_i may be quite large. Accordingly, it is reasonable to hope that the truncated equations (17.4) may yield satisfactory approximations at sufficiently small but finite wave amplitudes; but formal justification for omitting higher-order terms is lacking. The next order $O(a^3)$ terms were considered by Usher & Craik (1975); these have the form of those in (16.7) (see also §26.1).

Consideration of the asymptotic structure of the linear eigenfunctions allows approximate evaluation of the interaction coefficients λ_i for arbitrary primary flows $U(z)$ at large Reynolds numbers R. When the critical layer is located far from the boundaries z_1 and z_2 and the wavenumber α is $O(1)$ – which is not usually so as $R \to \infty$ but may nevertheless be true at fairly large R – Craik (1968, 1971) found that $|\lambda_{1,2}|$ is $O(R)$ and $|\lambda_3|$ is $O(1)$ relative to R for near-neutral waves and standard normalizations of the ϕ_j. The rather unexpected $O(R)$ contributions derive from the immediate vicinity of the common critical layer, and depend crucially on three-dimensionality through the velocity components $\hat{v}_j$. Large interaction coefficients for the next order $O(a^3)$ terms were also predicted by Usher & Craik (1975). Though lacking validity in the formal limit $R \to \infty$, these results provided an early indication of the remarkable strength of nonlinear interactions in shear flows at fairly large, finite, Reynolds numbers.

Large interaction coefficients for oblique waves may lead to enhanced growth of three-dimensional modes, even when these are linearly damped. The coefficients λ_i were evaluated explicitly by Craik (1971) for boundary layers approximated by straight-line profiles, but computer evaluation is normally necessary. Computations by Hendricks, reported in Usher & Craik (1975), give values of λ_i for Blasius flow at $R = 882$ and various α. The ratio of $|\lambda_{1,2}/\lambda_3|$ increases rapidly with α, from near 1 at $\alpha = 0.1$ (when the asymptotic theory is certainly invalid) to about 30 at $\alpha = 0.5$.

Later, Volodin & Zel'man (1979) calculated the λ_i for R ranging from 650 to 1300 and confirmed that the ratio $|\lambda_{1,2}/\lambda_3|$ remains fairly large. Their analysis of three-wave resonance in Blasius flow considers spatial, rather than temporal, wave evolution and attempts to incorporate the effects of downstream variation of the primary flow. Then, spatial derivatives are associated with the respective group velocities v_j of each wave, as in (14.10), but each v_j may be complex.

An alternative derivation of (17.4) is given by Usher & Craik (1974). In this, they extend to viscous flows the variational approach used by Simmons (1969, see §14.2 above) for inviscid water waves. To do this, they employed a little-used variational formulation of the Navier–Stokes equations, given by Dryden, Murnahan & Bateman (1956). Since the Navier–Stokes equations are not self-adjoint, this variational formulation involves additional functions, of no direct physical significance, which may be regarded as a 'pseudo-velocity vector' and 'pseudo pressure'. In the linear approximation, these extra functions are identified as the solutions of the adjoint Orr–Sommerfeld problems. Apart from this additional complexity, the derivation of the interaction equations proceeds much as in Simmons' case and it entails rather less algebraic manipulation than does the direct method described above.

Exceptionally, for plane Poiseuille flow between rigid boundaries, the interaction coefficients λ_i are identically zero, by symmetry, when the three participating waves have eigenfunctions ϕ_i which are all even with respect to the centre of the channel. Accordingly, the least damped modes do not interact resonantly at this order: see §26.1. Resonant interactions in a two-layer channel flow, with fluids of differing viscosity but the same density, were considered by Hame & Muller (1975). Lekoudis (1980) has examined the boundary layer on a swept wing for similar resonance.

Wave interactions in Blasius boundary-layer flows were further analysed by Nayfeh & Bozatli (1979a, b, 1980). The first and third of these studies confirm that exact resonance does not occur between two-dimensional waves with frequencies ω_1 and $\omega_2 = \frac{1}{2}\omega_1$ and that the spatial mismatch $\Delta = k_1 - 2k_2$ is normally sufficiently large to suppress parametric instability of this subharmonic. Moreover, because the primary flow develops with downstream distance, modes which are resonant at one downstream location become progressively detuned with distance from it. Their 1979b paper concerns oblique interactions rather like those above, but with the additional ingredient that the two-dimensional second harmonic with perodicity $\exp[2i\alpha(x - c_r t)]$ lies close to a linear mode. In effect, they consider two coupled near-resonant triads, one comprising the two-dimensional wave and its harmonic with wavenumbers $(\alpha, 0)$, $(2\alpha, 0)$, the other consisting of the wave $(2\alpha, 0)$ and the oblique waves (α, β), $(\alpha, -\beta)$. These oblique waves have downstream wavenumbers and frequencies close to that of the two-dimensional wave $(\alpha, 0)$ – cf. Figure 5.10.

17.3   *Some particular solutions*

Although numerical solutions of (17.4) are easily obtained, no general analytic solutions, like those for conservative interactions, are known. When $\lambda_1 = \lambda_2 \equiv \lambda$, the change of variables

$$b_3 = \lambda a_3, \quad b_{1,2} = (\lambda_3 \lambda)^{\frac{1}{2}} a_{1,2}$$

yields

$$\left. \begin{array}{l} db_1/dt = \tilde{\sigma} b_1 + b_3 b_2^*, \quad db_2/dt = \tilde{\sigma} b_2 + b_3 b_1^*, \\[1mm] db_3/dt = \sigma b_3 + e^{i\phi} b_1 b_2, \quad e^{i\phi} \equiv \lambda_3 \lambda / |\lambda_3 \lambda|, \\[1mm] \sigma \equiv \alpha c_i, \quad \tilde{\sigma} \equiv \tfrac{1}{2} \alpha \tilde{c}_i. \end{array} \right\} \qquad (17.5)$$

Conservative interactions with $d(|b_1|^2 + |b_2|^2 \pm |b_3|^2)/dt = 0$ exist when $\exp i\phi = \pm 1$ but not otherwise. With arbitrary ϕ, Craik (1971) gives two particular solutions of the system (17.5). The first is periodic, with

$$b_{1,2} = [\sigma\tilde{\sigma}/\cos\theta \, \cos(\phi-\theta)]^{\frac{1}{2}} \exp[i(\theta_{1,2} - \tilde{\sigma} t \tan\theta)],$$

$$b_3 = (-\tilde{\sigma}/\cos\theta) \exp[i(\theta_3 - 2\tilde{\sigma} t \tan\theta)]$$

where the real phases θ_j $(j = 1, 2, 3)$ satisfy $\theta_3 - \theta_2 - \theta_1 = \theta$ and

$$\tan(\phi - \theta) \cot\theta = 2\tilde{\sigma}/\sigma.$$

This solution exists provided $(\sigma + 2\tilde{\sigma})^2 \cot^2\phi + 8\sigma\tilde{\sigma} > 0$; in particular, when all three waves are damped or amplified according to linear theory.

When $\tilde{\sigma} = \sigma$, a transformation like (16.3a) may be used. Equations like (16.3b) are obtained, namely

$$dB_1/d\tau = B_3 B_2^*, \quad dB_2/d\tau = B_3 B_1^*, \quad dB_3/d\tau = e^{i\phi} B_1 B_2 \quad (17.6)$$

with

$$B_j = b_j e^{-\sigma t}, \quad \tau = \sigma^{-1}(e^{\sigma t} - 1).$$

A further class of solutions of (17.6) is then

$$B_1 = \mathbb{B} e^{i\theta_1}, \quad B_2 = \mathbb{B} e^{i\theta_2}, \quad B_3 = \mathbb{B}_3 e^{i\theta_3},$$

$$\mathbb{B} \left[\frac{\cos(\phi-\theta)}{\cos\theta} \right]^{\frac{1}{2}} = \mathbb{B}_3 = \frac{K}{1 - \tau K \cos\theta}, \quad \theta = \theta_3 - \theta_2 - \theta_1,$$

where K is an arbitrary positive constant, θ is a root of

$$\tan(\phi - \theta) \cot\theta = 2$$

and any two of the phases θ_j $(j = 1, 2, 3)$ may be assigned arbitrarily.

Roots with $\cos \theta < 0$ yield solutions B_j which decay algebraically with τ. When $\sigma > 0$, the corresponding b_j approach equilibrium values

$$b_k = -\sigma[\cos \theta \cos (\phi - \theta)]^{-\frac{1}{2}} e^{i\theta_k} \quad (k = 1, 2), \quad b_3 = (-\sigma/\cos \theta) e^{i\theta_3}$$

as $t \to \infty$; when $\sigma < 0$ they decay to zero. Roots with $\cos \theta > 0$ yield explosive growth of B_j with a singularity appearing at $\tau = (K \cos \theta)^{-1}$. Provided $\exp i\phi \neq -1$, initial phases θ_j may always be found which lead to this singularity. Corresponding singularities of b_j are attained at the finite real time

$$t = \sigma^{-1} \log (1 + \sigma/K \cos \theta)$$

provided $\sigma > -K \cos \theta$, but sufficient linear damping causes the b_j to decay. Such explosive growth may often be expected in shear flows, for the interaction coefficients are then complex, with no simple phase-relation between their arguments. Of course, the assumptions underlying the truncated equation (17.5) must fail before the singularity is reached.

Wang (1972) has derived upper and lower bounds for the solutions of three-wave interacting systems with linear damping and frequency mismatch and arbitrary coupling coefficients. These yield a sufficient condition for the non-existence of explosive instability. Weiland & Wilhelmsson (1977) also discuss the qualitative nature of solutions, with particular regard to the presence, or not, of explosive growth and phase locking.

It seems that no exact solutions of the corresponding equations with both spatial and temporal amplitude variations have yet been published, for non-conservative interactions. The 'pump-wave' approximation was developed by Craik & Adam (1978) in such cases with amplitudes that vary in x, y and t. The system may then be transformed into the telegraph equation, valid in regions where a single mode is dominant.

However, very recently, Craik (1985, to be published) has re-examined equations of the form (cf. 15.10)

$$\partial A_i/\partial \chi_i = s_i A_j^* A_k^* \quad (i, j, k = 1, 2, 3 \text{ cyclically})$$

where the s_i are *complex* coupling coefficients $s_i = \exp(i\phi_i)$. A new class of solutions has been found which closely resembles the class of 'one-lump' solutions described in §15.3. These, like the conservative 'one-lump' solutions, are phase-locked, with constant values of $\theta = \theta_1 + \theta_2 + \theta_3$ where $\theta_i \equiv \text{ph} A_i$. But now θ must be a root of

$$\tan (\phi_1 - \theta) + \tan (\phi_2 - \theta) + \tan (\phi_3 - \theta) = 0.$$

Such roots always exist and solutions for A_i may exhibit explosive instability after a finite time much as already described.

Chapter six

EVOLUTION OF A NONLINEAR
WAVE-TRAIN

18 Heuristic derivation of the evolution equations

Evolution equations for linear wave-packets were discussed in §7.3 above. The form of such equations is determined by the linear dispersion relation. For weakly nonlinear wave-packets centred on a single wavenumber and frequency, the appropriate governing equations may be inferred heuristically by similar means.

Let a wave-packet have the form $\mathrm{Re}\{A(\mathbf{x}, t)\exp \mathrm{i}(\mathbf{k}_0 \cdot \mathbf{x} - \omega_0 t)\}$ where $\omega = \omega_0$ is a real root of the linear dispersion relation

$$F(\omega, \mathbf{k}, V) = 0 \tag{18.1}$$

when $\mathbf{k} = \mathbf{k}_0$ and $V = V_0$. V is some flow parameter, typically the Reynolds number R or, for inviscid flows, a dimensionless measure of the velocity scale. When the spatial and temporal modulations in wave amplitude $A(\mathbf{x}, t)$ are slow compared with the characteristic wavelength $2\pi/|\mathbf{k}_0|$ and wave period $2\pi/\omega_0$, the linearized evolution equation for A is found by replacing ω, $\mathbf{k}$, V in (18.1) by

$$\omega_0 + \mathrm{i}\,\partial/\partial t, \quad \mathbf{k}_0 - \mathrm{i}\,\partial/\partial\mathbf{x}, \quad V_0 + V' \tag{18.2}$$

and applying the resultant partial differential operator to $A(\mathbf{x}, t)$. Here, $\mathbf{x}$ denotes only the propagation space of the waves; non-periodic dependence of the linear eigenmode on any other co-ordinate direction, such as the depth for gravity waves, is removable from a suitably-defined amplitude equation. The replacement (18.2), suggested by the work of Whitham (1965), is not a strictly rigorous procedure, but it is both instructive and fruitful. An early application was given by Benney & Newell (1967) and the method was subsequently developed by Whitham (1974), Davey (1972), Benney & Maslowe (1975), Weissman (1979) and others.

172

When $F(\omega, \mathbf{k}, V) = 0$ takes the simple form $\omega - \Omega(\mathbf{k}, V) = 0$ characteristic of a single wave-mode, the amplitude equation is readily found to be

$$\left(\frac{\partial}{\partial t} + \frac{\partial \Omega}{\partial k_i} \frac{\partial}{\partial x_i}\right) A - \frac{i}{2} \frac{\partial^2 \Omega}{\partial k_i \partial k_j} \frac{\partial^2 A}{\partial x_i \partial x_j} + i \frac{\partial \Omega}{\partial V} V' A = 0. \tag{18.3}$$

Here, k_i denote the Cartesian components of the wavenumber vector $\mathbf{k}$, x_i the propagation space co-ordinates and summation over repeated indices is implied. All partial derivatives of $\Omega(\mathbf{k}, V)$ are evaluated at $\mathbf{k} = \mathbf{k}_0$ and $V = V_0$.

For weakly nonlinear wave-packets, the same linear operator must arise, but the right-hand side is no longer zero. If $|A(\mathbf{x}, t)|$ is $O(\epsilon)$, the $O(\epsilon^2)$ self-interaction of the wave normally yields a second harmonic in $A^2 \exp 2i(\mathbf{k}_0 \cdot \mathbf{x} - \omega_0 t)$ and a mean flow contribution such as was found in Chapter 4. Sometimes, as for inviscid surface gravity waves in deep water, the $O(\epsilon^2)$ mean flow is identically zero (cf. §11.2). In other cases, typically those with two-dimensional modulations where A depends on t and only that space co-ordinate parallel to $\mathbf{k}_0$, the mean flow is directly proportional to $|A|^2$. In both these situations, the interaction of the fundamental wave with its own second harmonic and mean flow (if present) yields $O(\epsilon^3)$ terms with the same periodicity as the fundamental, which are directly proportional to $|A|^2 A$.

Such terms are of magnitude comparable with those of (18.3) when weak modulations of $A = \epsilon \tilde{A}$ are appropriately described by the scalings

$$\tau = \epsilon^2 t, \quad \xi = \epsilon[\mathbf{x} - (\partial \Omega / \partial \mathbf{k}) t], \quad V' = \epsilon^2 \mathcal{V}. \tag{18.4}$$

At leading order, $O(\epsilon^2)$, (18.3) then shows that the wave-packet propagates as a whole with the group velocity $\mathbf{c}_g = \partial \Omega / \partial \mathbf{k}$; while, at $O(\epsilon^3)$, it evolves according to

$$\frac{\partial \tilde{A}}{\partial \tau} - \frac{i}{2} \frac{\partial^2 \Omega}{\partial k_i \partial k_j} \frac{\partial^2 \tilde{A}}{\partial \xi_i \partial \xi_j} + i \frac{\partial \Omega}{\partial V} \mathcal{V} \tilde{A} = \Lambda |\tilde{A}|^2 \tilde{A}. \tag{18.5}$$

The usually complex constant Λ must be determined by detailed study of each particular case. Clearly, it is the Landau constant and (18.5) reduces to Landau's equation (8.8) in cases of purely temporal modulation, with $-i(\partial \Omega / \partial V) \mathcal{V}$ identified as the linear growth rate. Equation (18.5) is a *nonlinear Schrödinger equation*, the properties of which are discussed in §21 below.

For three-dimensional wave-packets, the $O(\epsilon^2)$ mean flow is not usually directly proportional to $|A|^2$. For instance, with $\tilde{A} = \tilde{A}(\xi_1, \xi_2, \tau)$, the $O(\epsilon^2)$

mean flow $u^{(2)}$ associated with inviscid water waves satisfies an equation of the form (11.17); that is,

$$u^{(2)} = \epsilon^2 \partial \phi_{01}/\partial \xi_1 \equiv \epsilon^2 \tilde{B}(\xi_1, \xi_2, \tau),$$

$$\left(\alpha \frac{\partial^2}{\partial \xi_1^2} + \frac{\partial^2}{\partial \xi_2^2} \right) \tilde{B} = \kappa \frac{\partial^2}{\partial \xi_1^2} |\tilde{A}|^2 \qquad (18.6)$$

where α, κ are constants. The nonlinear equation for $\tilde{A}$ then resembles (18.5) but has a right-hand side $(\Lambda_1 |\tilde{A}|^2 + \Lambda_2 \tilde{B}) \tilde{A}$ with suitable constants Λ_1, Λ_2. This and (18.6) form a coupled system, the *Davey–Stewartson equations*.

The equation for $\tilde{A}$ may also be deduced from the *nonlinear* dispersion relation, $\omega = \Omega(\mathbf{k}, V, |A|^2, B)$ say, where $B = \epsilon^2 \tilde{B}$, if this is known. It is readily seen that $\Lambda_1 = -i \partial \Omega/\partial |A|^2$ and $\Lambda_2 = -i \partial \Omega/\partial B$ with partial derivatives evaluated at $|A|^2 = B = 0$.

When the linear dispersion relation has the more general form (18.1), similar arguments normally apply. The linear group velocity $\mathbf{c}_g$ is then

$$\mathbf{c}_g = -(\partial F/\partial \mathbf{k})/(\partial F/\partial \omega)$$

and no difficulties arise unless $\partial F/\partial \omega = 0$ at $(\mathbf{k}_0, \omega_0, V_0)$. But the latter case is of genuine interest, for $\partial F/\partial \omega$ vanishes whenever modes coalesce: for instance, this is so at every point of the stability boundary of inviscid Kelvin–Helmholtz instability. At the critical point for instability, $\partial F/\partial \mathbf{k}$ also vanishes but at other points of the stability boundary it does not. The evolution equation then differs from those given above.

Following Weissman (1979), we set $\mathbf{k} = (k, l)$ and let $\mathbf{k}_0 = (k_c, 0)$, $V_0 = V_c$ denote the critical wavenumber and flow velocity for onset of a Kelvin–Helmholtz type instability. The linear dispersion relation and the substitution (18.2) then yields an evolution equation

$$F\left(\omega_c + i \frac{\partial}{\partial t}, \; k_c - i \frac{\partial}{\partial x}, \; -i \frac{\partial}{\partial y}, \; V_c + V' \right) A(x, y, t) = \text{n.l.t.}$$

where the right-hand side comprises appropriate nonlinear terms. When $\partial F/\partial \omega = \partial F/\partial k = \partial F/\partial l = 0$ at $(\omega_c, k_c, 0, V_c)$ and $\partial^2 F/\partial \omega \, \partial l = \partial^2 F/\partial k \, \partial l = 0$ also, by symmetry, this becomes

$$-\tfrac{1}{2} F_{\omega\omega} \frac{\partial^2 A}{\partial t^2} + F_{\omega k} \frac{\partial^2 A}{\partial t \, \partial x} - \tfrac{1}{2} F_{kk} \frac{\partial^2 A}{\partial x^2} - \tfrac{1}{2} F_{ll} \frac{\partial^2 A}{\partial y^2} + F_V \, V' A = \text{n.l.t.} \quad (18.7)$$

the subscripts denoting partial derivatives evaluated at the critical point. The nonlinear terms are typically similar to those described above. For unbounded Kelvin–Helmholtz flows, there is no mean-flow distortion and these terms have the form $N \,|\, A \,|^2 A$, where N is some constant. On writing

$$A = \epsilon \tilde{A}, \quad X = \epsilon(x - ct), \quad Y = \epsilon y, \quad T = \epsilon t, \quad V' = \epsilon^2 \mathscr{V},$$

$$d = 2 F_V \mathscr{V} / F_{\omega\omega}, \quad c = -F_{\omega k} / F_{\omega\omega}, \quad b = -F_{ll} / F_{\omega\omega},$$

$$a = (F_{\omega k}^2 - F_{\omega\omega} F_{kk}) / F_{\omega\omega}^2, \quad n = 2N / F_{\omega\omega},$$

equation (18.7) yields

$$\frac{\partial^2 \tilde{A}}{\partial T^2} - a \frac{\partial^2 \tilde{A}}{\partial X^2} - b \frac{\partial^2 \tilde{A}}{\partial Y^2} - d\tilde{A} = n \,|\, \tilde{A} \,|^2 \tilde{A} + O(\epsilon^2). \tag{18.8}$$

Akylas & Benney (1980) give a rather similar derivation, but without Y-dependence.

The group velocity $\mathbf{c}_g$ is now multiple-valued, with x and y components $(c \pm a^{\frac{1}{2}}, \pm b^{\frac{1}{2}})$: the chosen reference frame moves with the 'mean' group velocity $(c, 0)$ relative to the original co-ordinates. Equation (18.8) is a *nonlinear Klein–Gordon equation*, which arises in the buckling of elastic shells (Lange & Newell 1971) as well as in hydrodynamic instability: see Weissman (1979) for other occurrences. Its linear counterpart was discussed in §7.3. Another evolution equation with double-valued group velocity is discussed in §19.3.

At other points on the neutral curve of a Kelvin–Helmholtz type of instability, $\partial F / \partial \omega = 0$ but $\partial F / \partial k, \; \partial F / \partial l \neq 0$: the latter derivatives vanish only at the critical point. Wave-packets centred on such points turn out to have evolution equations of the same form as the nonlinear Schrödinger equation (18.5), but with the rôles of x and t interchanged (Watanabe 1969; Nayfeh & Saric 1972, Ma 1984b). However, such wave-packets are of less interest since they are usually dominated by more rapidly-growing modes with wavenumbers in the linearly-unstable range.

Exceptionally, more than two modes may coalesce at a given wave-number. Then, $\partial^2 F / \partial \omega^2 = 0$ and third or higher time derivatives must arise in the corresponding evolution equation (see equation 19.5 below).

For long inviscid gravity waves in shallow water, yet another evolution equation arises. For two-dimensional waves with frequency ω and wave-number k, the linear dispersion relation is (cf. §11.2)

$$\omega = \pm (gh)^{\frac{1}{2}} k[1 - \tfrac{1}{6}(kh)^2 + O\{(kh)^4\}]$$

and $\omega \to 0$ in the long-wave approximation $kh \to 0$. In effect, the modulations are about $k_0 = 0$ and $\omega_0 = 0$ and the corresponding weakly-nonlinear amplitude equation is obviously

$$\left[\pm \frac{\partial}{\partial t} + (gh)^{\frac{1}{2}} \frac{\partial}{\partial x} \left(1 + \tfrac{1}{6} h^2 \frac{\partial^2}{\partial x^2} \right) \right] A = \text{n.l.t.}$$

Here, A is the actual surface-wave elevation, not the modulated amplitude. Moreover, the nonlinear terms are found to be *quadratic* in A, namely $-\frac{3}{2}A\,\partial A/\partial x$ with suitably defined A (see e.g. Peregrine 1972). This is because the nonlinearities 'felt by the wave' need no longer match its periodicity: the second-harmonic and mean flow are themselves 'long waves' like the fundamental. The resultant equation

$$\pm\frac{\partial A}{\partial t}+(gh)^{\frac{1}{2}}\frac{\partial A}{\partial x}+\frac{3}{2}A\,\frac{\partial A}{\partial x}+\frac{1}{6}h^2(gh)^{\frac{1}{2}}\frac{\partial^3 A}{\partial x^3}=0 \qquad (18.9a)$$

is the celebrated Korteweg–de Vries equation, more simply expressed in rescaled form as

$$\frac{\partial A}{\partial \tau}+\sigma A\,\frac{\partial A}{\partial \xi}+\frac{\partial^3 A}{\partial \xi^3}=0. \qquad (18.9b)$$

A generalization of (18.9b) is

$$\frac{\partial A}{\partial \tau}+\sigma A\,\frac{\partial A}{\partial \xi}+\alpha\,\frac{\partial^2 A}{\partial \xi^2}+\beta\,\frac{\partial^3 A}{\partial \xi^3}+\gamma\,\frac{\partial^4 A}{\partial \xi^4}=0. \qquad (18.10)$$

The corresponding linear dispersion relation for small-amplitude waves with periodicity $\exp[\mathrm{i}(k\xi-\omega\tau)]$ is

$$\omega \equiv \omega_r+\mathrm{i}\omega_i = -\beta k^3+\mathrm{i}(\alpha k^2-\gamma k^4);$$

this gives instability for $0 < k < (\alpha/\gamma)^{\frac{1}{2}}$ when α, β, γ are real and $\alpha, \gamma > 0$. Equations of the form (18.10) have arisen in a variety of physical contexts. For fluids, cases with $\beta = 0$ approximate long waves in thin viscous layers flowing down an incline (Benney 1966) and on the interface between two viscous layers (Hooper & Grimshaw 1985). In the former case, the destabilizing α-term derives from gravity or downstream pressure gradient, but in the latter it may also be due to differing viscosities (cf. Yih 1967). The stabilizing γ-term represents the effect of surface tension. When $\beta = \gamma = 0$, (18.10) reduces to a form of Burgers' equation, first proposed as a model of turbulence (Burgers 1946). This may be transformed into the heat equation and solved for arbitrary initial conditions (see e.g. Whitham 1974). A three-dimensional counterpart of (18.10) with $A = A(\xi, \eta, \tau)$ is considered by Lin & Krishna (1977).

19   Weakly nonlinear waves in inviscid fluids

19.1   *Surface and interfacial waves*

The evolution of weakly nonlinear surface waves in inviscid fluid has been much studied. Hasimoto & Ono (1972) considered gravity waves with slow amplitude variations in x and t; Benney & Roskes (1969) and Davey & Stewartson (1974) incorporated modulations in x, y and t;

Djordjevic & Redekopp (1977) and Ablowitz & Segur (1979) added surface tension effects. All employed perturbation theory but a parallel development using Whitham's method is practicable (Yuen & Lake 1975, 1982). For linearly-stable interfacial waves in inviscid Kelvin–Helmholtz flow, the analysis follows similar lines; but differences occur for modes located near the stability boundary, for reasons outlined above.

For gravity waves on a free surface, the velocity potential satisfies Laplace's equation within the fluid, together with the surface conditions (11.13), (11.14) and the bottom boundary condition $\partial\phi/\partial z = 0$. Only a brief outline of the perturbation analysis need be given here. The wave-train has the same form as in §11.2, with fundamental periodicity $E \equiv \exp i(kx - \omega t)$ and amplitude modulations described by the scaled co-ordinates (11.2). The velocity potential ϕ and free surface displacement ζ are expanded in ascending powers of the wave-slope parameter ϵ. Any $O(\epsilon^2)$ mean flow driven by amplitude modulations is precisely as given in (11.17). In the series expansions for ϕ and ζ, there are also $O(\epsilon^2)$ terms with periodicity E, forced by the modulation of the fundamental.

A second harmonic is driven by the self-interaction of the fundamental: this is $O(\epsilon^2)$ and proportional to $A^2E^2 + $c.c. except when 2ω is close to the natural frequency of a linear mode with wavenumber $2k$. This exceptional case is one of second-harmonic resonance, first identified by Wilton (1915) and later elucidated by Nayfeh (1970) and McGoldrick (1970a, b, 1972) for capillary–gravity waves and by Nayfeh & Saric (1972) and Weissman (1979) for unbounded Kelvin–Helmholtz flows. This degenerate example of three-wave resonance was discussed in §14.2 above; here we consider non-resonant cases.

The perturbation analysis leads to a set of equations for the various terms proportional to $\epsilon^p E^q$ ($p = 1, 2, 3, ..., q = 0, \pm 1, \pm 2, \pm ...$), and these may be solved *seriatim*. The interaction of the fundamental $O(\epsilon)$ wave with the $O(\epsilon^2)$ mean flow and second harmonic yields $O(\epsilon^3)$ terms in the nonlinear boundary conditions (11.13) and (11.14), some of which have the same periodicity $E^{\pm 1}$ as the fundamental wave-train. Such terms cause slow amplitude modulations of the fundamental waves, on the time scale $\tau = \epsilon^2 t$. Once the solutions for the mean flow and second harmonic have been found, evaluation of these terms is a straightforward, but rather tedious task. Third-harmonic terms in $\epsilon^3 E^{\pm 3}$ are also present but play no part in modifying the fundamental at this order. These may be disregarded except in special cases of resonance between the third harmonic and fundamental: for the latter, see Nayfeh (1970) and §22.2.

The expansions of ϕ and ζ contain respective terms in $\epsilon^3 E\phi_{13}(z)$ and

$\epsilon^3 E \zeta_{13}$. These enter linearly into the boundary conditions at $O(\epsilon^3)$. Terms in $\epsilon^3 E$ may then be extracted from (11.13) and (11.14), evaluated at the mean surface $z = 0$ by a Taylor expansion. These yield two algebraic equations for $\phi_{13}(0)$ and ζ_{13}, in the form

$$\mathbb{A} \begin{pmatrix} \phi_{13} \\ \zeta_{13} \end{pmatrix} = \begin{pmatrix} b_1 \\ b_2 \end{pmatrix}$$

where $\mathbb{A}$ is a known 2×2 matrix. The right-hand side contains all nonlinear interaction terms with periodicity E, also linear terms involving partial derivatives of $A(\xi, \eta, \tau)$ which arise from the modulation of the wave-train. Since the *linear* approximation is $\mathbb{A}(\phi, \zeta)^{tr} = 0$, the linear dispersion relation is just $\det \mathbb{A} = 0$. The quantities b_1 and b_2 must therefore satisfy a compatibility condition in order that a solution (ϕ_{13}, ζ_{13}) exists. It is this compatibility condition which yields the governing equation for the evolution of the amplitude modulations.

For surface capillary–gravity waves and stable waves in Kelvin–Helmholtz flow, this amplitude evolution equation and the $O(\epsilon^2)$ mean-flow equation take the form derived heuristically above:

$$\left. \begin{aligned} iA_\tau + \lambda A_{\xi\xi} + \mu A_{\eta\eta} &= \chi |A|^2 A + \chi_1 A \Phi_\xi, \\ \alpha \Phi_{\xi\xi} + \Phi_{\eta\eta} &= -\beta(|A|^2)_\xi, \end{aligned} \right\} \tag{19.1a, b}$$

where $\Phi \equiv \phi_{01}(\xi, \eta, \tau)$, subscripts ξ, η, τ denote partial differentiation and

Figure 6.1. Dependence of α, λ, χ and ν on kh and $\tilde{T}$ (from Ablowitz & Segur 1979). Curves indicate where the various coefficients change sign.

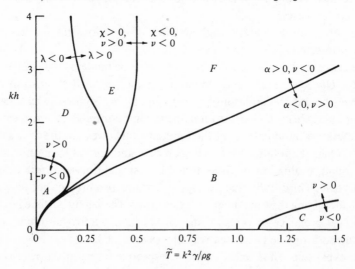

λ, μ, χ, χ_1, α and β are known real constants. These equations were first derived by Davey & Stewartson (1974). For capillary–gravity waves, expressions for the various constants are given by Djordjevic & Redekopp (1977) and Ablowitz & Segur (1979). In particular, we note that

$$\alpha = 1 - (c_g^2/gh),$$

that μ, χ_1 and β are non-negative and bounded, and that α, λ, χ and $\nu \equiv \chi - \chi_1(\beta/\alpha)$ change sign as shown in Figure 6.1. The two axes represent dimensionless wavenumber kh and the surface tension parameter $\tilde{T} = k^2\gamma/\rho g$. Each line denotes a simple zero of the designated coefficient except those bounding region F, which denote singularities of ν and χ. These singularities arise where

$$\tilde{T} = \sigma^2(3-\sigma^2)^{-1}, \quad \sigma \equiv \tanh kh:$$

this is the condition for second-harmonic resonance, for which the present perturbation expansion breaks down.

Cases where $\alpha = 0$ are also singular. For these, $c_g = \pm(gh)^{\frac{1}{2}}$ and the group velocity of the fundamental waves coincides with the phase speed of much longer shallow water waves. This is also a degenerate three-wave interaction: a long wave with small wavenumber $\kappa \ll k$ and frequency $\Omega = \kappa(gh)^{\frac{1}{2}}$ resonates with the fundamental wave of wavenumber and frequency (k, ω), and either of its 'sidebands' $(k \pm \kappa, \omega \pm \Omega)$. This long-wave resonance has been treated by Djordjevic & Redekopp (1977) for amplitude modulations in x and t only. A new co-ordinate scaling

$$\xi = \epsilon^{\frac{2}{3}}(x - c_g t), \quad \tau = \epsilon^{\frac{4}{3}}t$$

is required and the surface displacement and x-velocity associated with the long wave are taken to be $O(\epsilon^{\frac{4}{3}})$. The governing equations are then

$$\left.\begin{aligned} iA_\tau + \lambda A_{\xi\xi} &= \delta\Phi_\xi A, \\ \Phi_{\tau\xi} &= -\alpha_1(|A|^2)_\xi, \end{aligned}\right\} \tag{19.2a, b}$$

with λ as above and

$$\alpha_1 = -\tfrac{1}{2}k^2\operatorname{cosech}^2 kh, \quad \delta = k[1 + \tfrac{1}{2}(c_g k/\omega)(1 + \tilde{T})\operatorname{cosech}^2 kh]$$

for capillary–gravity waves. Here, $\epsilon^{\frac{2}{3}}\Phi(\xi, \tau)$ is the leading-order velocity potential of the long wave.

A similar pair of coupled equations was derived by Grimshaw (1977) and Koop & Redekopp (1981) for the resonant interaction of a packet of internal gravity waves and a long internal wave-mode in stratified fluid. Grimshaw also covers the case of near resonance, when an extra term in $\Phi_{\xi\xi}$ enters equation (19.2b). Such equations are likely to arise in many other configurations where the group velocity of a modulated wave-train matches the phase velocity of a much longer wave. However, in water of infinite depth, the equations governing such interaction of short and long capillary–gravity waves do *not* take this form (Benney 1976): this is because the second-order mean flow generated by self-interaction of the short waves vanishes in the deep-water limit.

Solution of (19.2a, b) by the method of inverse-scattering is given by Ma (1978) and Ma & Redekopp (1979) as mentioned in § 15.4. Complementary numerical work is described by Bryant (1982). Newell (1978) proposed a different model of long-wave/short-wave interactions,

$$A_t = 2\sigma(|B|^2)_x,$$

$$B_t - iB_{xx} = -A_x B + iA^2 B - 2i\sigma |B|^2 B$$

in scaled form, for which exact solutions can also be found.

The interesting experiments of Koop & Redekopp (1981) concern interacting waves on the two interfaces of a three-layer configuration. In these, packets of relatively short gravity waves on the upper interface resonate with long waves on the lower interface. Examples of their results are shown in Figure 6.2, the long wave being generated by the short-wave modulations. Their results are generally in fair agreement with theoretical estimates which (somewhat empirically) incorporate viscous effects. However, long time-scale recurrent modulations predicted by inviscid theory could not be confirmed in their apparatus.

For stable waves in Kelvin–Helmholtz flow of unbounded extent in z, and for waves in water of infinite depth, the constants χ_1 and β of (19.1) are zero. The vanishing of β is shown by (11.17), there being no $O(\epsilon^2)$ mean flow in infinitely deep inviscid fluid. The wave-packet is then governed by the nonlinear Schrödinger equation

$$iA_\tau + \lambda A_{\xi\xi} + \mu A_{\eta\eta} = \chi |A|^2 A, \tag{19.3}$$

the coefficient χ resulting solely from the interaction of the fundamental and the second harmonic. For unbounded Kelvin–Helmholtz flow, Weissman (1979) gives χ explicitly, in the form

$$\chi = -N_1(\partial F/\partial \omega)^{-1}$$

where $F(\omega, k, l, U)$ is the dispersion function. The value of χ may be positive or negative and it is singular, due to the vanishing of $\partial F/\partial \omega$, on the neutral curve of Kelvin–Helmholtz instability.

Weakly nonlinear marginally-unstable waves near the critical point for Kelvin–Helmholtz instability were first examined by Drazin (1970), for purely temporal modulation. Spatial and temporal evolution of such waves is discussed by Nayfeh & Saric (1972) and Weissman (1979), the latter of whom gives the governing equation (18.8).

Figure 6.2. Examples of resonant long-wave/short-wave interaction, from experiments of Koop & Redekopp (1981). Modulated short waves generated at upper interface cause long waves at lower interface of a three layer fluid. (*a*) periodic modulations; (*b*) single wave packet; (*c*) 'dark-pulse' modulation.

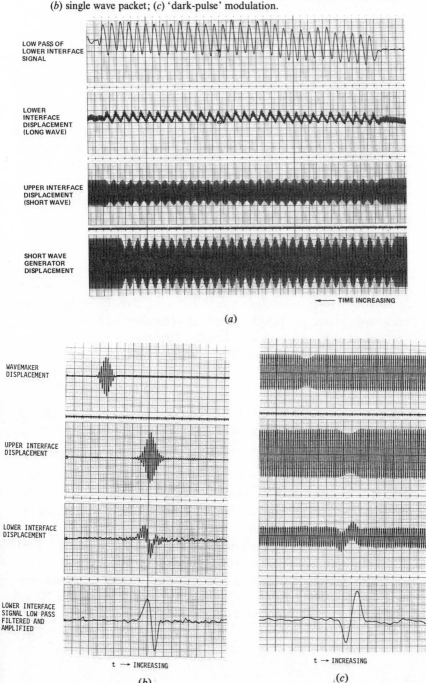

LOW PASS OF
LOWER INTERFACE
SIGNAL

LOWER
INTERFACE
DISPLACEMENT
(LONG WAVE)

UPPER INTERFACE
DISPLACEMENT
(SHORT WAVE)

SHORT WAVE
GENERATOR
DISPLACEMENT

◄—— TIME INCREASING

(*a*)

WAVEMAKER
DISPLACEMENT

UPPER INTERFACE
DISPLACEMENT

LOWER INTERFACE
DISPLACEMENT

LOWER INTERFACE
SIGNAL LOW PASS
FILTERED AND
AMPLIFIED

t —► INCREASING

t —► INCREASING

(*b*)

(*c*)

For wave-packets with amplitude-dependence on ξ and τ only, the appropriate mean-flow solution of (19.1b) is just

$$\Phi_\xi = (-\beta/\alpha)|A|^2.$$

The system (19.1a, b) then also reduces to a nonlinear Schrödinger equation for A of the form (19.3), but with χ replaced by

$$\nu = \chi - \chi_1(\beta/\alpha).$$

For deep-water gravity-wave packets, Dysthe (1979), Janssen (1983) and Lo & Mei (1985) consider a higher-order approximation to the evolution equation. In this, the nonlinear Schrödinger equation is replaced by another containing additional terms which, under the scaling (11.2), would be vanishingly small. For water of finite depth, Prasad & Krishnan (1978) propose a generalization of the Davey–Stewartson equations (19.1a, b) to account for more strongly curved wavefronts.

Further discussion of the properties of these various governing equations and of related experiments is deferred till §21.

19.2 Internal waves

Grimshaw's (1977) investigation of internal gravity waves in stratified fluid at rest also led to a modification of the Davey–Stewartson equations. A further nonlinear term SA must then be added to the right-hand side of (19.1a). S denotes the contribution from a discrete spectrum of extremely long-wavelength internal-wave modes which are excited by the amplitude modulations of the shorter waves. It has the form

$$S = \sum_{j=1}^{\infty} \delta_j W_j(\xi, \eta, \tau)$$

with known constants δ_j, and each W_j relates to a single long-wavelength mode. The functions W_j satisfy a set of equations

$$[1 - (c_g/c_j)^2]\, \partial^2 W_j/\partial \xi^2 + \partial^2 W_j/\partial \eta^2 = \nu_j \partial^2 |A|^2/\partial \eta^2 \quad (j = 1, 2, \ldots)$$

where c_g is the group velocity of the fundamental packet, c_j is the phase velocity of the jth long-wave mode and each ν_j a known constant. The resonant case $c_j = c_g$ requires separate treatment and was mentioned above.

The linear theory of over-reflecting internal gravity waves was discussed in §4.2, for continuous density stratification with constant N and the discontinuous Helmholtz velocity profile (4.3). For resonant over-reflection, the linear dispersion relation (4.6) gives three modes, with Kelvin–Helmholtz instability at wavenumbers $k^2 > 2N^2/U^2$. Grimshaw (1976, 1979b) has developed a weakly nonlinear analysis of such cases, for temporal, but not spatial, amplitude modulation.

At wavenumbers below the cutoff value for linear instability, his amplitude equation is

$$\partial A/\partial \tau = \beta |A|^2 A + I \qquad (19.4)$$

where I is a known forcing term representing waves incident from $z = \pm \infty$. For the symmetric mode with $c = \frac{1}{2}U$, β is found to be zero. Though this mode and its harmonic are in resonance when $k^2 < N^2/U^2$, the resonant coupling is so weak that the harmonic remains $O(\epsilon^2)$. The vanishing of β suggests that unforced steady waves may indeed exist at the interface, with internal waves radiating off to $z = \pm \infty$ in accordance with linear theory.

For the other pair of modes, $\beta \neq 0$ and $\mathrm{Re}\,\beta > 0$, indicating a destabilizing effect. The interaction of the fundamental wave and the $O(\epsilon^2)$ mean flow is found to contribute only to $\mathrm{Im}\,\beta$: accordingly, the destabilization of these linearly stable modes derives solely from the interaction of the fundamental with its second harmonic. If the velocity discontinuity is replaced by a very thin shear layer, equation (19.4) is modified only by addition of a linear term γA to the right-hand side. The value of γ is negative for the symmetric mode and positive for the asymmetric modes (Grimshaw 1981a).

At the cutoff value $k_c = 2^{\frac{1}{2}}N/U$ for onset of linear instability, all three modes of the vortex-sheet profile coalesce. Close to this value, with $k = k_c(1 + \epsilon^2 K)$, the dispersion relation has the renormalized form

$$\omega'(\omega'^2 + \epsilon^2 K) = 0$$

and the heuristic replacement (18.2) is just $\omega' = i\,\partial/\partial t$. The corresponding nonlinear equation for $A(T)$, where $T \equiv \epsilon t$, is

$$-\frac{\partial^3 A}{\partial T^3} + K\frac{\partial A}{\partial T} = \tfrac{9}{2}|A|^2\frac{\partial A}{\partial T} + 2A^2\frac{\partial A^*}{\partial T} \qquad (19.5)$$

with no forcing. Grimshaw (1976) derives this equation and discusses some of its properties.

The value of investigating such stable or marginally unstable modes is open to question when, as here, other strongly-unstable linear modes are present. On the other hand, it is instructive to know what influence weak nonlinearities may have on over-reflection, even though the model incorporates unrealistic features.

19.3   Baroclinic waves

Baroclinic waves occur in rotating, stratified systems: they are important in large-scale atmospheric and other geophysical flows and have been studied in many laboratory experiments. They derive from the fact that equilibrium states in rotating systems have non-coincident surfaces of constant gravitational potential and constant density. Mathematical models of baroclinic phenomena have been developed both for superposed layers of immiscible fluids with differing densities and for fluid with continuous density stratification. Baroclinic waves are linearly unstable in the presence of sufficiently strong mean-flow variations. Weakly-nonlinear analyses have been carried out for conditions close to those for onset of linear instability. Reviews of baroclinic flow phenomena are given by Hide & Mason (1975), Drazin (1978) and Hart (1979).

Spatial and temporal evolution of marginally unstable finite-amplitude waves was studied by Pedlosky (1970, 1971, 1972) for a two-layer model and corresponding analyses of continuously-stratified shear flows were undertaken by Drazin (1972), Pedlosky (1979) and Moroz & Brindley (1981, 1984).

At the critical point of linear inviscid instability, the group velocity c_g becomes double-valued, as in (18.8); but the form of the nonlinear terms now involves wave-driven variations of the mean flow. As in the Davey–Stewartson equations (19.1), a pair of coupled equations arises for the wave amplitude A and mean-flow modification B. For modulations in x and t only, these turn out to have the form

$$\left.\begin{array}{l}\left(\dfrac{\partial}{\partial T}+c_1\dfrac{\partial}{\partial X}\right)\left(\dfrac{\partial}{\partial T}+c_2\dfrac{\partial}{\partial X}\right)A = \pm\alpha A-\beta AB,\\[4mm]\left(\dfrac{\partial}{\partial T}+c_2\dfrac{\partial}{\partial X}\right)B = \left(\dfrac{\partial}{\partial T}+c_1\dfrac{\partial}{\partial X}\right)|A|^2\end{array}\right\} \qquad (19.6)$$

with suitably scaled variables. Here, c_1, c_2 are the two group velocities, $\pm\alpha$ is the (small) linear growth or decay rate and β a real coupling coefficient. Pedlosky (1972) gave solutions depending on a single variable, $X-VT$, with constant V.

A derivation of these equations from a more general standpoint is given by Gibbon & McGuinness (1981): the coefficients α, β may then be complex. These equations are solvable by inverse scattering (Gibbon, James & Moroz 1979) and have soliton-type solutions. If A (and so α, β) is restricted to be real, the substitutions

$$A = (2\beta)^{-\frac{1}{2}}\left(\frac{\partial}{\partial T}+c_2\frac{\partial}{\partial X}\right)\Phi, \quad B = \pm(\alpha/\beta)(1-\cos\Phi)$$

transform (19.6) into the sine–Gordon equation

$$\left(\frac{\partial}{\partial T}+c_1\frac{\partial}{\partial X}\right)\left(\frac{\partial}{\partial T}+c_2\frac{\partial}{\partial X}\right)\Phi = \pm\alpha\sin\Phi,$$

which has well-known properties (see, e.g., Dodd *et al.* 1982).

In the purely time-dependent case, Gibbon & McGuinness (1982) have argued that the counterparts of (19.6) in the presence of weak viscosity should be

$$\left.\begin{array}{l} d^2A/dT^2+\Delta_1\,dA/dT = \pm\alpha A-\beta AB,\\ dB/dT+\Delta_2\,B = d\,|A|^2/dT+\Delta_3\,|A|^2, \end{array}\right\} \tag{19.7}$$

the new Δ_j-terms representing viscous damping. (But care is necessary: Δ_1^2/α must remain small as $\alpha\to 0$ if the neutral curve is to be only slightly affected – cf. Newell 1972.) With α, Δ_1 allowed to be complex, but with the other parameters real, the transformation

$$\tau = \Omega T,\quad \Omega = \text{Re}\,(\Delta_1)-\tfrac{1}{2}\Delta_3,$$
$$\mathscr{X} = (2\beta)^{\frac{1}{2}}\Omega^{-1}A,\quad \mathscr{Z} = 2\beta\Omega^{-1}\Delta_3^{-1}B$$

enables (19.7) to be rewritten as

$$\left.\begin{array}{l} d\mathscr{X}/d\tau = -\sigma\mathscr{X}+\sigma\mathscr{Y},\\ d\mathscr{Y}/d\tau = r\mathscr{X}-a\mathscr{Y}-\mathscr{X}\mathscr{Z},\\ d\mathscr{Z}/d\tau = -b\mathscr{Z}+\tfrac{1}{2}(\mathscr{X}^*\mathscr{Y}+\mathscr{X}\mathscr{Y}^*) \end{array}\right\} \tag{19.8}$$

where

$$\sigma = \tfrac{1}{2}\Omega^{-1}\Delta_3,\quad b = \Omega^{-1}\Delta_2,\quad a = 1+i\,\text{Im}\,(\Delta_1)\Omega^{-1},$$
$$r = a+2\alpha(\Delta_3\,\Omega)^{-1}.$$

When α and Δ_1 (and so a and r) are real, so also may be A and B (and so $\mathscr{X}$ and $\mathscr{Y}$). Equations (19.8) then reduce to the Lorenz equations (Lorenz 1963; Yorke & Yorke 1981), which constitute the first studied strange-attractor system. Properties of the complex system (19.8) are discussed by Fowler, Gibbon & McGuinness (1983).

With stronger dissipation, equations (19.7) are inappropriate. Linear instability is then no longer due to a coalescence of two neutral modes, but to the crossing of a single previously-damped mode into the amplified régime. Accordingly, second-order time derivatives should not arise in the nonlinear evolution equations (cf. §18). The weakly-nonlinear evolution of such viscous baroclinic waves was treated by Pedlosky (1976), allowing for cross-stream variations in y as well as time t. His coupled equations for (scaled) wave and mean-flow components ϕ, u have the form

$$\left.\begin{array}{l} \partial\phi/\partial t-\partial^2\phi/\partial y^2 = s^2\phi(1+u)\quad (s\text{ constant})\\ \partial u/\partial t-\partial^2u/\partial y^2 = \partial^2\,|\phi|^2/\partial y^2, \end{array}\right\} \tag{19.9}$$

where ϕ is associated with the periodicity $\exp[ik(x-ct)]$. Pedlosky imposed 'sidewall' boundary conditions $u = \phi = 0$ at $y = 0$ and 1, which restrict the cross-stream structure: there is then a discrete set of linear modes, as $\phi, u \to 0$, given by

$$\phi_m = A \exp(\sigma_m t) \sin(m\pi y), \quad \sigma_m = s^2 - (m\pi)^2$$

where m is integer. The mth mode is unstable if $s > m\pi$.

When the first mode is slightly supercritical, $s^2 = \pi^2 + \delta$ say, its weakly-nonlinear evolution is found from (19.9) to be given by

$$\phi = \delta^{\frac{1}{2}} A(\tau) \sin \pi y + \text{smaller harmonics},$$
$$u = -|\phi|^2, \quad \tau \equiv \delta t \quad (0 < \delta \ll 1),$$
$$dA/d\tau = A - (3\pi^2/4)|A|^2 A.$$

This evolves to a finite-amplitude equilibrium, $|A| = 2(3^{\frac{1}{2}}\pi)^{-1}$.

For arbitrarily large s, (19.9) admits time-independent solutions with

$$u(y) = -\phi^2(y), \quad d^2\phi/dy^2 - s^2\phi(1 - \phi^2) = 0, \quad \phi \text{ real}.$$

Those satisfying $\phi(0) = 0$ are Jacobian elliptic functions

$$\phi(y) = a \operatorname{sn}[\eta, \nu], \quad \eta \equiv asy/2^{\frac{1}{2}}\nu, \quad a \equiv 2^{\frac{1}{2}}\nu(\nu^2 + 1)^{-\frac{1}{2}}$$

with modulus ν. The remaining boundary condition $\phi(1) = 0$ is satisfied for values ν such that $as/2^{\frac{1}{2}}\nu$ coincides with a zero of the elliptic function. Pedlosky showed that, when $j\pi < s < (j+1)\pi$ with integer j, there are exactly j such steady states. He also undertook a numerical investigation of the temporal evolution of finite-amplitude disturbances, based upon equations (19.9), and employing truncated Fourier-series representation in y. The results indicated eventual equilibration at the steady state with largest available amplitude, usually after an interval of decaying oscillations. On the other hand, with weaker dissipation, Hart (1973) and Pedlosky (1977) found finite-amplitude permanent oscillations and so the issue remains unresolved. It is also worth remarking that, for s substantially above π, it seems unduly restrictive to admit just a single Fourier mode in x, but several in y.

Resonant triads of *inviscid* baroclinic waves were studied by Loesch (1974) and Pedlosky (1975), for cases where one of the three waves is marginally unstable. The governing equations for temporal evolution are not then of the standard form (15.1) since the linear inviscid instability is due to coalescence of modes. Rather, they have the (scaled) form

$$\left.\begin{array}{l} d^2A_1/dT^2 - \sigma^2 A_1 = N_1 A_1(N_0|A_1|^2 + N_2|A_2|^2 + N_3|A_3|^2), \\ dA_2/dT = iN_2 A_1^* A_3^*, \\ dA_3/dT = iN_3 A_1^* A_2^*, \end{array}\right\} \quad (19.10\mathrm{a,b,c})$$

when $\omega_1 + \omega_2 + \omega_3 = 0$, $\mathbf{k}_1 + \mathbf{k}_2 + \mathbf{k}_3 = 0$ and subscripts 1 designate the marginally unstable mode. Here, σ is related to the linear growth rate and the constants N_j are real. Loesch gave several numerical solutions while Pedlosky made analytical progress in cases with $N_0 = 0$. The latter correspond to waves A_1 having no meridional (north–south) wavenumber component: these are of particular interest because, when $A_2 = A_3 = 0$, there are no cubic terms to suppress the linear wave growth. In such cases, $N_1 < 0$ while N_2, $N_3 > 0$.

A further scaling

$$T = \tau/\sigma, \quad A_1 = \sigma(N_2 N_3)^{-\frac{1}{2}} r_1 \, e^{i\theta_1}, \quad A_2 = \sigma(-N_1 N_3)^{-\frac{1}{2}} r_2 \, e^{i\theta_2},$$
$$A_3 = \sigma(-N_1 N_2)^{-\frac{1}{2}} r_3 \, e^{i\theta_3}$$

then yields three real equations

$$\mathrm{d}^2 r_1/\mathrm{d}\tau^2 = r_1(1 - r_2^2 - r_3^2), \quad \mathrm{d}r_2/\mathrm{d}\tau = r_1 r_3, \quad \mathrm{d}r_3/\mathrm{d}\tau = r_1 r_2,$$

provided $\theta_1 + \theta_2 + \theta_3$ is chosen to equal $\frac{1}{2}\pi$. Pedlosky showed that the solutions are oscillatory, with $r_2^2 - r_3^2$ constant. His rather cumbersome account is simplified by the transformation

$$r_2 = A \cosh u, \quad r_3 = A \sinh u, \quad r_2^2 - r_3^2 = A^2 \geqslant 0.$$

This readily leads to

$$r_1 = \mathrm{d}u/\mathrm{d}\tau = \pm(u^2 + au + b - \tfrac{1}{2}A^2 \cosh 2u)^{\frac{1}{2}},$$

where a and b are constants determined by the initial data. The solution is obviously periodic, with period

$$2 \int_{u_1}^{u_2} (u^2 + au + b - \tfrac{1}{2}A^2 \cosh 2u)^{-\frac{1}{2}} \, \mathrm{d}u,$$

where u_1 and u_2 are the values of u for which $r_1 = 0$.

Loesch & Domaracki (1977) extended this work to investigate the effect of *several* co-existing resonant triads on a marginally unstable mode A_1. Their numerical results for 5, 9 then 13 participating modes show increasingly complex modulations; but these apparently remain quasi-periodic, with two or three superposed periods. In contrast, Merkine & Shtilman (1984) found instances of explosive growth among (three-wave) triads and coupled (five-wave) triads, in the presence of weak viscous effects: since their marginally unstable mode satisfies a first-order equation, their systems resemble those discussed in Chapter 5.

20 Weakly nonlinear waves in shear flows

20.1 Waves in inviscid shear flows

The simplest such flows to handle are constant-density flows with broken-line velocity profiles, which yield explicit expressions for the linear dispersion relation and eigenfunctions. The absence of curvature of the velocity profile, except at 'joins' between layers of differing vorticity, means that the linear eigenfunctions of two-dimensional waves are non-singular at any critical layer.

Benney & Maslowe (1975) investigate weakly nonlinear waves in the presence of the primary velocity profiles

$$\bar{u} = \begin{cases} z & (0 \leqslant z \leqslant 1) \\ 1 & (1 < z < \infty), \end{cases} \tag{20.1a}$$

$$\bar{u} = \begin{cases} z & (|z| \leqslant 1) \\ z/|z| & (1 < |z| < \infty), \end{cases} \tag{20.1b}$$

the former of which has a plane rigid boundary at $z = 0$. For profile (20.1a), inviscid linear modes with periodicity $\exp[ik(x-ct)]$ have (cf. Tietjens 1925)

$$c = 1 - k^{-1} e^{-k} \sinh k.$$

A perturbation analysis, similar to that of §19.1, again leads to result (19.3), with the Landau constant χ given by

$$\chi = \frac{k(1-3\sigma)(1-2\sigma-\sigma^2)}{2\sigma(1+\sigma)}, \quad \sigma \equiv \tanh k.$$

In deriving this result, Benney & Maslowe ignored any contribution due to modifications of the mean flow at $O(\epsilon^2)$. This is apparently justified for their unbounded constant-density flows, but it would not normally be so for bounded fluid layers of differing densities, under gravity. Then, mean flow modifications are driven by spatial modulation of the wave-train and coupled equations of the form (19.1a, b) must result.

Benney & Maslowe (1975) restrict their study of profile (20.1b) to wave-modes with $c = 0$. By symmetry, this case corresponds to coalescence of right and left propagating modes centred on the respective interfaces (cf. §2). This is an instance of marginal stability, with neighbouring unstable modes, and the governing amplitude equation is then

$$\frac{\partial^2 A}{\partial \tau^2} = 0.201 \, i \left(\frac{\partial A}{\partial X} + 2.11 \, iA^2 A^* \right) \tag{20.2}$$

with suitably scaled variables X, τ. This has the form typical of wave-modes near the neutral curve of Kelvin–Helmholtz instability, but not at the critical wavenumber (see §18). Again, mean flow modifications may be ignored in deriving this result since gravitational effects are absent. The second time derivative also arises, for similar reasons, in the evolution equation derived by Engevik (1982) for waves in some stratified shear flows.

The linear stability of the inviscid shear-layer profile

$$\bar{u} = \tanh z \quad (-\infty < z < \infty) \tag{20.3}$$

was described in §3.1. The behaviour of weakly-nonlinear marginally unstable modes was treated by Schade (1964) and Stuart (1967). Schade considered purely temporal wave growth and Stuart confined attention from the outset to nonlinear constant-amplitude disturbances. Any mean flow modification was ignored. This is obviously permissible for constant-amplitude waves in any inviscid parallel flow: the mean flow may be chosen as (20.3) in the presence of such waves, without regard to the original velocity profile in their absence. But the neglect of mean flow variations requires justification when temporal or spatial amplitude changes are admitted. The perturbation solutions are valid for sufficiently small linear growth rates αc_i or, equivalently, for wavenumbers sufficiently close to that of the linear neutral mode, here $\alpha = 1$. Stuart continued his analysis of equilibrium disturbances to higher order, finding that disturbances with the streamfunction (cf. §3.1)

$$\psi = \epsilon \operatorname{sech} z \cos\left[\alpha(x - ct)\right] + O(\epsilon^2), \quad c = 0,$$

have

$$\alpha^2 = 1 - \tfrac{4}{3}\epsilon^2 + \tfrac{16}{3}(1 + \tfrac{1}{3}\ln 2)\,\epsilon^4 + O(\epsilon^6). \tag{20.4}$$

Correspondingly, the equilibrium amplitude ϵ is

$$\epsilon = \pm\left[\tfrac{3}{4}(1 - \alpha^2)\right]^{\frac{1}{2}}\left[1 + \tfrac{1}{2}(3 + \ln 2)(1 - \alpha^2) + O(|1 - \alpha^2|^2)\right],$$

the leading-order approximation of which agrees with Schade's result.

Benney & Maslowe (1975) show that the latter generalizes, with spatial and temporal variation, to the amplitude equation

$$\frac{\partial A}{\partial \tau} = \frac{2i}{\pi}\left(\frac{\partial A}{\partial X} + \frac{8i}{3}|A|^2 A\right), \tag{20.5}$$

the linear group velocity $\partial \omega / \partial \alpha = -2i/\pi$ being purely imaginary and deriving from the influence of the linear (viscous) critical layer. In contrast, for the piecewise-linear profile (20.1b), there is no critical layer contribution and the linear group velocity is infinite for waves with $c = 0$, so leading to (20.2). Benney & Maslowe further show that, for a *nonlinear* critical layer (see §22.4), which sustains no jump in Reynolds stress, the tanh z profile leads to an equation similar to (20.2). Attempts to incorporate viscous effects for the profile (20.3) are discussed in §20.3 below.

20.2   Near-critical plane Poiseuille flow

We here outline the derivation of the amplitude equation of a weakly nonlinear wave-packet in plane Poiseuille flow (11.5), taking into account amplitude modulations in x, y and t. The treatment broadly follows that of Davey, Hocking and Stewartson (1974): this incorporates as special cases the work of Stuart (1960), Watson (1960, 1962) and Chen & Joseph (1973) on purely temporal and purely spatial growth and also that of Stewartson & Stuart (1971) for modulations in x and t. Weinstein (1981) has demonstrated that some differences of method in these papers do not affect the results.

The flow is taken to be marginally unstable, at a Reynolds number R just above the critical value $R_c = 5772$ for onset of linear instability. A localized small disturbance initially centred on $x = y = t = 0$ at first evolves according to linear theory, as discussed in §7.3, into a slowly-varying wave-train with streamfunction

$$\psi = \text{Re} \{\epsilon A(\xi, \eta, \tau) \psi_1(z) \exp[i\alpha_0(x - c_0 t)] + O(\epsilon^2)\}. \tag{20.6}$$

Here, α_0 and c_0 are the real wavenumber and phase speed of the neutral mode at $R = R_c$ and $\psi_1(z)$ its corresponding (normalized) eigenfunction. The scaled co-ordinates ξ, η, τ are as defined in (11.2) where ϵ is a small parameter and A is regarded as $O(1)$.

Since R is close to R_c, we set

$$R = R_c + \epsilon^2/d_{1r}$$

and denote the precise linear eigenvalue c at this value of R with $\alpha = \alpha_0$ by

$$c = c_r + ic_i = c_0 + \epsilon^2 i d_1 (d_{1r} \alpha_0)^{-1}, \quad d_1 = d_{1r} + i d_{1i}$$

for suitable $O(1)$ real constants d_{1r}, d_{1i}. This choice fixes ϵ^2 as the small linear growth rate $\alpha_0 c_i$. (This differs from Davey, Hocking & Stewartson's choice of $\epsilon = \alpha_0 c_i$: throughout, their $\epsilon^{\frac{1}{2}}$ is our ϵ.)

Successive approximations are constructed by expressing the velocity components (u, v, w) and pressure p as

$$u = u_0(\xi, \eta, \tau, z, \epsilon) + \text{Re}[Eu_1(\xi, \eta, \tau, z, \epsilon) + E^2 u_2 + \ldots],$$

$$E \equiv \exp[i\alpha_0(x - c_0 t)],$$

and similar expressions. Each function u_m may be expanded in powers of ϵ as

$$u_0 = 1 - z^2 + \qquad \epsilon^2 u_{02}(\xi, \eta, \tau, z) + \epsilon^3 u_{03} + \ldots$$

$$u_1 = \epsilon u_{11}(\xi, \eta, \tau, z) + \epsilon^2 u_{12}(\xi, \eta, \tau, z) + \epsilon^3 u_{13} + \ldots$$

$$u_2 = \qquad\qquad \epsilon^2 u_{22}(\xi, \eta, \tau, z) + \epsilon^3 u_{23} + \ldots$$

$$u_3 = \qquad\qquad\qquad\qquad \epsilon^3 u_{33} + \ldots$$

together with similar expansions for the corresponding components of v, w and p. Those for v_0 and w_0 contain no $O(1)$ term in $1 - z^2$; also,

$$p_0 = -2R^{-1}x + \text{constant} + \epsilon P_{01} + \epsilon^2 P_{02} + \dots,$$

the leading-order pressure gradient $-2R^{-1}$ being that which maintains the primary flow.

The various coefficients of $\epsilon^n E^m$ $(n, m = 0, 1, 2, \dots)$ which arise in the governing Navier–Stokes equations (1.1a, b)' must then be equated to zero. That in ϵE yields the linear solution

$$u_{11} = A(\xi, \eta, \tau)\, D\psi_1(z), \quad v_{11} = 0, \quad w_{11} = -\mathrm{i}\alpha_0 A\psi_1(z),$$

where $D \equiv \mathrm{d}/\mathrm{d}z$ and $\psi_1(z)$ is the Orr–Sommerfeld eigenfunction at wavenumber α_0 and $R = R_c$. For definiteness, the normalization $\psi_1(0) = 2$ may be adopted (this accords with Davey et al.'s normalization).

The $O(\epsilon^2)$ mean flow components (u_{02}, v_{02}, w_{02}) are just as found in §11.1 on applying the condition of constant mass flux: these are given by (11.6) and (11.7). The other $O(\epsilon^2)$ components in E and E^2 are also readily found (see Davey et al., §2).

The evolution equation for $A(\xi, \eta, \tau)$ is deduced from the $O(\epsilon^3 E)$ equations. These are three momentum equations and the equation of continuity, connecting u_{13}, v_{13}, w_{13}, P_{13} and lower-order functions. By elimination of u_{13}, v_{13} and P_{13}, these reduce to a single equation for w_{13} of the form

$$\mathscr{L}[w_{13}] = \mathscr{F}(\xi, \eta, \tau, z)$$

where $\mathscr{L}$ is just the Orr–Sommerfeld operator of (3.1b) and $\mathscr{F}$ is known in terms of A, its partial derivatives, and the lower-order functions already found. As described in §8.2, the function $\mathscr{F}$ must be orthogonal to the adjoint function $\Phi(z)$ of $\mathscr{L}$. Accordingly, multiplication by $\Phi(z)$ and integration across the flow domain yields

$$\int_{-1}^{1} \mathscr{F} \Phi \, \mathrm{d}z = 0.$$

This gives the equation for A in the form

$$\frac{\partial A}{\partial \tau} - a_2 \frac{\partial^2 A}{\partial \xi^2} - b_2 \frac{\partial^2 A}{\partial \eta^2} = \frac{d_1}{d_{1r}} A + k\,|A|^2 A + qAB, \tag{20.7}$$

as already indicated in §18. Here,

$$B(\xi, \eta, \tau) = |A|^2 - \tfrac{1}{3}R_c \left(\int_0^1 S(z)\,\mathrm{d}z \right)^{-1} \partial P_{01}/\partial\xi$$

and the $O(\epsilon^2)$ mean flow modification u_{02} is related to B by

$$u_{02} = |A|^2 S(z) + \tfrac{3}{2}(1 - z^2) \int_0^1 S(z)\,\mathrm{d}z\,(B - |A|^2)$$

in agreement with (11.6). Also, from (11.7),

$$\frac{\partial^2 B}{\partial \xi^2} + \frac{\partial^2 B}{\partial \eta^2} = \frac{\partial^2 |A|^2}{\partial \eta^2}. \tag{20.8}$$

(The substitution $\tilde{B} = \kappa(B + |\tilde{A}|^2)$ in (18.5) and (18.6) yields equations of the same form.)

In agreement with (18.5), the complex constants a_2, b_2 of (20.7) are $\frac{1}{2}i\,\partial^2\omega/\partial\alpha^2$, $\frac{1}{2}i\,\partial^2\omega/\partial\beta^2$ respectively, evaluated at $\alpha = \alpha_0$, $\beta = 0$, $R = R_c$, where $\omega(\alpha, \beta, R) = \alpha c$ is the complex frequency of oblique linear modes with periodicity $\exp i(\alpha x \pm \beta y - \omega t)$. The constants d_{1r}, d_{1i} may also be deduced from the linear dispersion relation, while

$$q = \tfrac{3}{2}(R_c\,d_1 - i\alpha_0\,c_0) \int_0^1 S(z)\,\mathrm{d}z.$$

The Landau constant k is an integral of a rather lengthy expression which need not be reproduced here.

Numerical integration at the neutral point for plane Poiseuille flow gives

$R_c = 5772.22, \quad \alpha_0 = 1.020\,55, \quad c_0 = 0.264, \quad c_g = 0.383$

$a_2 = 0.187 + 0.0275\,i, \quad b_2 = 0.004\,66 + 0.0808\,i,$

$d_1 = (0.168 + 0.811\,i)\,10^{-5}, \quad q = -1.27 + 29.1\,i, \quad k = 30.8 - 173\,i.$

These values for k and q may be deduced from results of Reynolds & Potter (1967); the others were computed by Davey *et al.*

For amplitude modulations independent of η, u_{02} is directly proportional to $|A|^2$ and B is identically zero. The coupled system (20.7), (20.8) then reduces to a nonlinear Schrödinger equation, with Landau constant k. Since the real part of k is positive, the nonlinear terms exert a destabilizing influence upon spatially-uniform waves. Periodic solutions with $\alpha = \alpha_0$ therefore bifurcate subcritically from $R = R_c$.

Ikeda (1977, 1978) has developed a higher order approximation to the evolution equation for waves in plane Poiseuille flow. For wave amplitudes independent of η, the nonlinear Schrödinger equation is replaced by

$$\frac{\partial A}{\partial \tau} - a_2 \frac{\partial^2 A}{\partial \xi^2} - \frac{d_1}{d_{1r}}\,A - k|A|^2 A = \epsilon\left[\theta_1 A \frac{\partial |A|^2}{\partial \xi} + \theta_2 |A|^2 \frac{\partial A}{\partial \xi}\right] + O(\epsilon^2)$$

where θ_1, θ_2 are complex constants. This equation is simpler than Dysthe's (1979) higher-order equation for deep-water waves, the latter being influenced by the free-surface boundary conditions. Ikeda's higher-order equations for wave amplitudes dependent on ξ, η and τ form a coupled system substantially more complex than (20.7) and (20.8).

20.3   Non-critical (nearly) parallel flows

For wave-modes at other values of (α, R) *sufficiently close to the neutral curve* for linear stability, the governing equations are similar to (20.7) and (20.8), but with different numerical coefficients. For these modes, the linear growth or decay rate may again be characterized by an arbitrarily small parameter ϵ. But such cases are of less interest than the critical one because of the presence of neighbouring more-strongly amplified modes.

One might hope that similar equations would give acceptable approximations at values of (α, R) some distance from the neutral curve, provided the linear growth or decay rate remains numerically small; but such hopes appear to be unduly optimistic. There is an inherent difficulty: for, if the linear eigenvalue is $c = c_r + ic_i$, with finite c_i, one cannot consistently choose αc_i to be $O(\epsilon)$ since $c = c_r$ is not an eigenvalue of the linear problem. But retention of c_i at $O(1)$ negates the assumption that the linear growth or decay is of no greater importance than the nonlinear contribution.

Certainly, a self-consistent series expansion may be constructed with wave amplitude A, but not αc_i, characterized by a small parameter, as was indicated in §8.3(ii). But truncation after a few terms is unlikely to succeed at amplitudes Q large enough that $|Q|^2$ is $O(\alpha c_i)$.

Reynolds & Potter (1967) and Pekeris & Shkoller (1967) sought to overcome this dilemma by including c_i in the linear Orr–Sommerfeld equation but omitting it elsewhere. Reynolds & Potter omitted it from both linear operators arising in the $O(\epsilon^2)$ harmonic and mean-flow equations; Pekeris & Shkoller omitted it only from the latter. Reynolds & Potter argued that their procedure should yield correct results for finite amplitude equilibrium states, for then the $O(\epsilon^2)$ forcing terms are of constant amplitude. With truncation at $O(|Q|^3)$, the resulting Landau equation for purely temporal growth is (cf. 8.8) $dQ/dt - \alpha c_i Q = k|Q|^2 Q$. Many values of the Landau constant k were computed, for a range of α and R, both for plane Poiseuille flow and for plane Couette–Poiseuille flows with relative motion of the channel walls. Since both authors considered only temporal modulations, they were free to impose either constant mass flux or constant pressure gradient; Reynolds & Potter chose the former and Pekeris & Shkoller the latter. Despite these differences, their results for k are in quite close agreement. Figure 6.3, from Pekeris & Shkoller, shows the curve on which $\operatorname{Re} k = k_r = 0$, together with the neutral curve of linear theory, for plane Poiseuille flow. Sub-critical equilibrium may be expected only when $c_i < 0$ and $k_r > 0$ and supercritical equilibrium when $c_i > 0$ and $k_r < 0$. When both k_r and c_i are negative, both terms cause the wave to decay; while, when both are positive, they promote wave growth. Corresponding results for two superposed horizontal layers of fluid with differing densities and viscosities, flowing between parallel walls, have been obtained by Blennerhassett (1980).

Figure 6.4 shows, for the first time, a direct comparison of predictions of the subcritical equilibrium amplitude in plane Poiseuille flow. These results, obtained by the methods of Reynolds & Potter and of Pekeris & Shkoller, were kindly provided by A. Davey. Clearly, substantial discrepancies result from the differing treatments of growth terms. Series truncation is a further source of error at larger $|Q|$.

The omission (or inclusion) of the growth terms from the harmonic and mean-flow equations is of no consequence provided $|\alpha c_i|$ is sufficiently small. But the necessary restriction on $|\alpha c_i|$ is severe and only those results for (α, R) close to the linear neutral curve should be trusted. For $\alpha \geqslant 1.0$ the magnitude of k increases rather rapidly with both α and R, reaching several thousands at $\alpha = 1.2$ and $R = 2.4 \times 10^4$ for example. This is in rough accord with an asymptotic result of Usher & Craik (1975), that $|k|$ should vary as $R^{\frac{5}{4}}$ for large R, provided α and c_r remain $O(1)$ and $|\alpha c_i|R \leqslant O(1)$. Unfortunately, the latter restrictions are not normally met for channel flows as $R \to \infty$; but suitably compliant walls would admit such behaviour. The increasing magnitude of $|k|$ with R suggests that the radius of convergence of the amplitude expansion decreases as R increases, as

Figure 6.3. Curves $k_r = 0$ and $c_i = 0$ in α–R plane for plane Poiseuille flow (from Pekeris & Shkoller 1967).

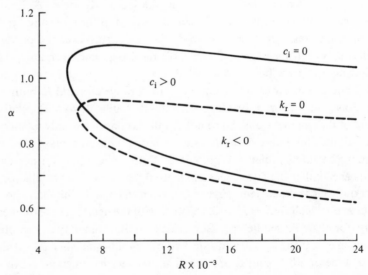

some inverse power of R. This partly explains the great sensitivity to nonlinear effects of those flows which remain linearly stable up to fairly large values of R.

A formally-valid asymptotic theory for $R \to \infty$ has been constructed by Hall & Smith (1982) for wave-modes situated indefinitely close to the lower branch of the neutral curve of plane Poiseuille and boundary-layer flows.

Figure 6.4. Subcritical equilibrium amplitude $|Q| = (-\alpha c_i/k_r)^{\frac{1}{2}}$ versus frequency αc_r for plane Poiseuille flow at $R = 5000$. Solid line, Pekeris & Shkoller's method (constant pressure gradient). Dashed line, Reynolds & Potter's method (constant mass flux). ●, Reynolds & Potter's method but with constant pressure gradient. (Results of A. Davey.) △ and ▽ are 3rd and 4th order results of Zhou (1982).

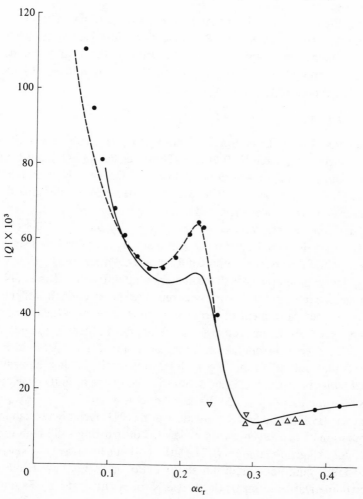

Since $\alpha \to 0$ and $c \to 0$ as $R \to \infty$ on *both* branches, attention is thereby confined to very long waves. Near the lower branch (but not the upper), the critical layer and viscous wall layer merge into one. The Landau constant k is then found to vary as $R^{\frac{1}{3}}$. Surprisingly, for plane Poiseuille flow, the lower branch of the curve $k_r = 0$ does not remain below that of $c_i = 0$, as suggested by Figure 6.3; consequently, linearly neutral modes on the lower branch are destabilized by nonlinearity at sufficiently large R. In contrast, k_r is stabilizing for boundary layers. Hall & Smith also consider the influence of forced wavelike distortions of a boundary wall. Unfortunately, enormous Reynolds numbers seem to be required for the validity of their asymptotic analysis, since $R^{-\frac{1}{3}}$ must be small compared with unity.

From Figure 6.3, it is seen that both αc_i and k_r vanish at a point on the neutral curve with $\alpha \approx 0.91$, $R \approx 6600$. Struminskii & Skobelev (1980) develop a higher-order approximation for waves in the neighbourhood of this point, finding two equilibrium solutions (in addition to $A = 0$) for an amplitude evolution equation of the form

$$dA/dt = A(\alpha c_i + a_1 |A|^2 + a_2 |A|^4).$$

Attempts to treat the nonlinear stability of flows which have no linear neutral curve were made by Ellingsen, Gjevik & Palm (1970) and Coffee (1977), who examined plane Couette flow, and Davey & Nguyen (1971), Itoh (1977a, b) and Davey (1978), who considered Poiseuille pipe flow. Ellingsen *et al.* and Coffee predicted finite-amplitude equilibrium solutions for plane Couette flow, but recent computations by Orszag & Kells (1980) suggest otherwise. Contradictory results were also obtained by Davey & Nguyen, who followed the method of Reynolds & Potter, and by Itoh, who treated the troublesome growth terms somewhat differently. Davey (1978) examined the reasons for the latter discrepancy, interpreting the difference as due to a rearrangement of terms of an infinite series. When the radii of convergence of the respective series are unknown, there is no apparent reason for favouring one rather than the other. Davey pragmatically suggests that only when the two methods give results in close agreement should *either* be trusted. Cowley & Smith (1985) have recently developed an asymptotic theory for such flows, valid as $R \to \infty$.

Herbert (1983a) and Sen & Venkateswarlu (1983) recently re-examined the question of convergence with a view to constructing valid high-order amplitude expansions at finite R. The latter authors, like Davey & Nguyen, found finite-amplitude equilibrium states for Poiseuille pipe flow. However, direct numerical solution of the Navier–Stokes equations by Patera &

Orszag (1981) yielded only damped disturbances at values of α and R for which equilibrium solutions are predicted by amplitude-expansion methods. The issue is accordingly unresolved.

Weakly nonlinear, spatially-varying waves are discussed by Watson (1962) and Itoh (1974b, c) for plane Poiseuille flow and for the Blasius boundary layer. Linear modes are then spatially amplified or damped as $\exp(-\alpha_1^0 x)$ where x denotes distance downstream of a continuous source of disturbance, such as a vibrating ribbon. Itoh finds coupled equations for the complex wave amplitude $A(x)$ and a measure $B(x)$ of the mean-flow modulation, in the form

$$\left. \begin{aligned} \mathrm{d}\,|\,A\,|^2/\mathrm{d}x &= |\,A\,|^2 (-2\alpha_1^0 + \lambda B + \mu\,|\,A\,|^2), \\ \mathrm{d}B/\mathrm{d}x &= -\kappa B - \gamma\,|\,A\,|^2, \end{aligned} \right\} \tag{20.9a, b}$$

higher-order terms in $|\,A\,|$ and B having been neglected. Here, λ, μ, κ and γ are real constants. These yield constant-amplitude solutions

$$|\,A\,|^2 = \frac{2\alpha_1^0 \kappa}{\mu\kappa - \lambda\gamma}, \quad B = \frac{-2\alpha_1^0 \gamma}{\mu\kappa - \lambda\gamma} \tag{20.10}$$

which agree with those of Watson (1962) as $\alpha_1^0 \to 0$. But the range of validity of the truncated equations (20.9) remains unclear for finite α_1^0.

A further recent attempt to resolve the difficulties associated with finite growth rates was made by Zhou (1982). In this, the primary flow is regarded as perturbed from its actual form to another notional one which yields a linearly-neutral mode. The (hopefully) small $O(\epsilon)$ modification of the primary flow then reappears at higher order in the perturbation expansion. Zhou claims better agreement with the experiments of Nishioka *et al.* (1975) than that hitherto obtained for the amplitudes of equilibrium solutions. Some of Zhou's results are shown on Figure 6.4: agreement with previous computations of k is not good. Furthermore, the data of Nishioka *et al.* at larger αc_r – with which better agreement was claimed – are now thought to be influenced by three-dimensional effects (see Figure 6.13).

Despite the fact that the 'tanh' shear-layer velocity profile (20.3) is not a solution of the Navier–Stokes equations, the nonlinear stability of a viscous fluid with this profile has been studied by various authors. Maslowe (1977a, b) investigated both constant-density and stratified flows, neglecting any mean flow distortion as the wave amplitude changes; but this omission is hardly justifiable. Huerre (1980) avoided some of the difficulties of dealing with the mean flow modification by incorporating an artificial, unphysical, body force which keeps the mean flow parallel. In any case, a homogeneous fluid with profile (20.3) is linearly unstable at all moderately large values of R, having a range of wavenumbers with $O(1)$ growth rates. The study of weakly nonlinear marginally unstable modes is therefore of reduced physical interest.

The experiments of Miksad (1972) on an unstable free shear layer reinforce this view. The complete process, from spatial amplification of small wavelike disturbances to ultimate transition to turbulence takes place within a short downstream distance of around five fundamental wavelengths. Within this distance, Miksad discerns no fewer than six distinct regions of behaviour, including linear growth, generation of harmonics and subharmonics, development of three-dimensional longitudinal vortices, secondary instability and breakdown into turbulence. Though Miksad compares his results with those of various weakly-nonlinear theories, any agreement can only be qualitative for such rapidly growing disturbances (cf. Ho & Huerre 1984).

Brown, Rosen & Maslowe (1981) give a more satisfactory treatment of stratified shear flows. Inclusion of stratification permits the study of a marginally unstable mode that is also the least stable of all available wavenumbers. Weakly nonlinear theory, with purely temporal modulations, leads to a Landau–Stuart equation of the type (8.8).

Brown *et al.* find that, when $R \gg 1$, the Landau constant k is real and $O(R)$ (except when the Prandtl number Pr is unity). They also present numerical results at several values of R showing that k changes sign as Pr varies: the disturbance is supercritically stable ($k < 0$) at smaller values of Pr, and unstable ($k > 0$) at larger. At larger amplitudes, nonlinear terms within the critical layer greatly modify the evolution equation. This is discussed in § 22.4 below. In all the above mentioned work, the critical layer region is dominated by viscosity, rather than nonlinearity.

Further difficulties are met in treating nearly, but not quite, parallel flows such as jets, wakes and boundary layers. Liu's (1969) early treatment of a plane compressible jet ignored these by assuming parallel flow and neglecting the mean flow distortion; but Itoh (1974c), Herbert (1975), Corner, Barry & Ross (1974), Smith (1979b) and others have variously treated non-parallelism and mean flow distortion in discussing the Blasius boundary layer.

The asymptotic suction boundary layer is simpler to treat, being independent of x and t with primary flow

$$\mathbf{u} = \{U_0[1 - \exp(-W_0 z/\nu)], \; 0, \; -W_0\}$$

where W_0 is a constant suction velocity. For this, Hocking (1975) has shown that marginally stable weakly nonlinear waves of the form (20.6) obey a nonlinear Schrödinger equation, like (20.7) but with $q = 0$. He gives numerical values for the coefficients at $\alpha = 0.1555$ and $R = 54370$, these being the values at the critical point of linear instability. The real part of k is then positive, indicating that nonlinearity has a destabilizing effect.

Benney (1984) has recently proposed a yet-incomplete inviscid theory for the development of three-dimensional disturbances in shear flows. Consider an initially parallel shear flow $\bar{u}_0(z)$ which sustains $O(\epsilon)$ two-dimensional constant-amplitude waves with periodicity $\exp [i\alpha(x - ct)]$. Let this be given an infinitesimal spanwise-periodic disturbance comprising mean-flow components

$$[u'_0, v'_0, w'_0] = \delta\, e^{iK(X-CT)} [u_0 \cos\beta y,\ \epsilon v_0 \sin\beta y,\ \epsilon w_0 \cos\beta y],$$

$$X \equiv \epsilon x,\quad T \equiv \epsilon t,\quad \delta \ll 1$$

and a corresponding $O(\delta\epsilon)$ perturbation of the wave field. Here, u_0, v_0, w_0 depend only on the normal-to-wall co-ordinate z. The governing equations for the $O(\delta\epsilon)$ wave field involve a term due to coupling of the original $O(\epsilon)$ waves and the $O(\delta)$ mean-flow perturbation u'_0; and the equations governing the mean flow contain terms which represent the interaction, at $O(\epsilon^2\delta)$, of the original and perturbed wave fields. This set of equations is linear in δ and, together with appropriate boundary conditions, defines an eigenvalue problem for $C = C(K, \alpha, \beta)$. Spanwise-periodic disturbances are unstable, for a given real wavenumber α, if there exist roots with $\operatorname{Im}(C) > 0$ for any real values K and β. Benney speculates that such instability may explain the experimental observations of Klebanoff, Tidstrom & Sargent (1962): but see also §26.2. Though developed from an entirely different standpoint, Benney's theory is similar in essence to that earlier described by Craik (1982d): see §13.2.

21 Properties of the evolution equations

21.1 *Nonlinear Schrödinger equation with real coefficients*

Equation (19.3), which has real coefficients λ, μ, χ, describes amplitude modulations in time τ and two space co-ordinates ξ and η. It is not amenable to solution by inverse scattering, except in degenerate cases where only one space co-ordinate is relevant. For this reason, no exact three-dimensional solutions are known.

When modulations depend on a single space co-ordinate, say $\xi_1 = \xi \cos\phi + \eta \sin\phi$ where ϕ is a fixed angle of orientation, (19.3) reduces to

$$iA_\tau + \lambda_1 A_{\xi_1\xi_1} = \chi\,|A|^2 A \tag{21.1}$$

where $\lambda_1 = \lambda \cos^2\phi + \mu \sin^2\phi$. In similar circumstances, the Davey–

Stewartson equations (19.1a, b) also reduce to this form. This equation is solvable by inverse scattering (Zakharov & Shabat 1972). With $\lambda_1 \chi < 0$, initially localized wave-packets evolve into discrete 'envelope solitons' and a dispersive 'tail'. Each soliton propagates with unchanged form and survives intact any interactions with other solitons. Such solitons are also unaffected by non-resonant interaction with wave-packets centred on other wavenumbers and frequencies. Recent accounts of the inverse scattering theory for this equation are given by Ablowitz & Segur (1981) and Dodd et al. (1982).

The simplest, one-soliton, solution for $\lambda_1 \chi < 0$ is

$$A = a \,|\, 2\lambda_1/\chi \,|^{\frac{1}{2}} \operatorname{sech}\left[a(\xi_1 - 2b\tau)\right] \exp\left[ib\xi_1 + i\lambda_1(a^2 - b^2)\tau\right], \quad (21.2)$$

where b denotes a small arbitrary shift in fundamental wavenumber. The width of the envelope is $O(a^{-1})$ and so decreases with increasing amplitude a. The actual number of solitons which emerge from an initial disturbance increases with the 'strength' of that disturbance.

In contrast, there are no localized soliton solutions when $\lambda_1 \chi > 0$. Smooth, initially-localized disturbances are then completely described by the remaining dispersive part of the solution: the amplitude decays as $\tau^{-\frac{1}{2}}$ as the disturbance spreads in space.

Hui & Hamilton (1979) give exact solutions of (21.1), for gravity waves in deep water, which elucidate the directional properties of the envelope modulations. For such waves with wavenumber components (k_1, k_2), the frequency is $\Omega = g^{\frac{1}{2}}(k_1^2 + k_2^2)^{\frac{1}{4}}$. Accordingly, from (18.5), wave-trains with $(k_1, k_2) = (k, 0)$ have

$$\lambda = -\tfrac{1}{8}\Omega/k^2, \quad \mu = \tfrac{1}{4}\Omega/k^2.$$

Renormalization of A and τ allows the choice $\lambda = -\tfrac{1}{8}, \mu = \tfrac{1}{4}, \chi = \tfrac{1}{2}$ of Hui & Hamilton. Clearly, $\lambda_1 \chi$ passes through zero when $\tan\phi = \pm 2^{-\frac{1}{2}}$ and so solitary-wave solutions exist only for $|\phi|$ or $|180° - \phi|$ less than $35°$. Within these sectors, there are also periodic solutions in the form of cn and dn elliptic functions. Solutions of permanent form – but not localized ones – also exist outside these sectors: these are periodic sn elliptic functions and solitary tanh solutions. The latter wave-envelopes have a single depression, with $A = 0$ at its centre, and a finite constant wave amplitude as $\xi_1 \to \pm\infty$. Hui & Hamilton also discuss wave groups propagating along the characteristic directions with $|\tan\phi| = 2^{-\frac{1}{2}}$, with application to Kelvin ship waves. These and other solutions of (21.1) are usefully reviewed by Peregrine (1983). Modifications which take account of slowly-varying depth or current are discussed by Turpin, Benmoussa & Mei (1983).

Good experimental agreement with the 'sech' soliton (21.2) was found by Hammack, reported in Ablowitz & Segur (1979), for the envelope of deep-water gravity waves in a narrow channel. Results at two distances downstream of a suitably modulated wavemaker are reproduced in

Figure 6.5. Measured surface displacements, by Hammack, of envelope soliton at two downstream locations in a narrow channel. Dashed curves show theoretical envelope shapes (from Ablowitz & Segur 1979).

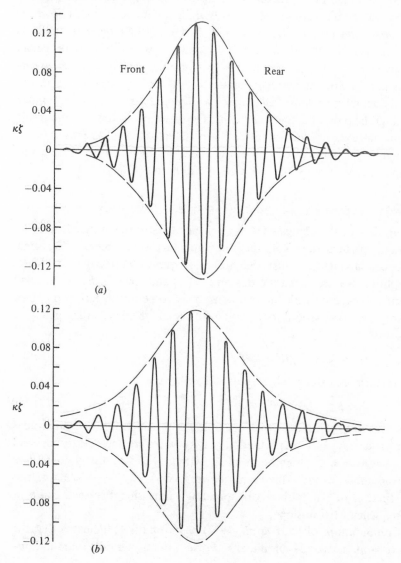

Figure 6.5. Owing to viscous dissipation, the wave heights slowly decrease with distance. The theoretical envelope breadth consequently increases in accord with (21.2), but the experimental data display less broadening. Further experiments on the evolution of localized disturbances in narrow channels and on the interaction of envelope solitons are described by Yuen & Lake (1975, 1982): again, agreement with the solutions of (21.1) is satisfactory.

Another interesting feature described by (21.1) is the instability and subsequent modulation of initially-uniform wave-trains, known as *Benjamin–Feir instability.* Benjamin & Feir (1967) demonstrated the progressive disintegration of initially-uniform waves as they propagate downstream from a wavemaker performing sinusoidal oscillations. Benjamin (1967) showed theoretically that energy passes from the fundamental wave into 'sidebands' with slightly different wavenumber. Though he did not employ (21.1), this equation leads to the same result, as the more general analysis of Zakharov (1968) showed. The result is also in agreement with somewhat earlier analyses by Lighthill (1965), who employed Whitham's method, and by Zakharov (1966).

Let

$$A = [a + b_+ \exp(iK\xi_1 + i\Omega\tau) + b_- \exp(-iK\xi_1 - i\Omega^*\tau)] \exp(-ia^2\chi\tau)$$

where Ω, with its conjugate Ω^*, is an eigenvalue to be determined and $\pm\epsilon K$ denote perturbations of the fundamental wavenumber. The final exponential factor is just the Stokes frequency correction for finite-amplitude waves. Substitute this into (21.1) and consider $b_\pm$ to be small enough compared with the real constant a to permit linearization. Two homogeneous equations for b_+ and b_- result and these yield the eigenvalue relation

$$\Omega^2 = \lambda_1 K^2(\lambda_1 K^2 + 2\chi a^2). \tag{21.3a}$$

Imaginary, and so unstable, roots arise whenever

$$0 < K^2 < (-2\chi/\lambda_1)a^2; \tag{21.3b}$$

that is to say, there is a range of unstable wavenumbers K for *all* amplitudes a, provided $\lambda_1 \chi < 0$. In particular, gravity waves are unstable to sideband modulations with orientation of ξ_1 within $\pm 35°$ of the fundamental wavenumber vector. Maximum growth occurs at $K = (-\chi/\lambda_1)^{\frac{1}{2}}a$; this is $\mathrm{Im}\,\Omega = \chi a^2$ and the corresponding dimensional growth rate is $\frac{1}{2}$ frequency × (wave-slope)^2.

Computations of Lake *et al.* (1977) based on (21.1) indicated that the wave and sidebands ultimately evolve into a periodic modulation–

demodulation cycle with repeated recurrence of a virtually 'pure' wave-train. Such behaviour, known as Fermi–Pasta–Ulam recurrence, was first discovered in lattice dynamics (Fermi *et al.* 1955) but was unexpected for water waves. Experiments by Lake *et al.* provided some evidence for this phenomenon, but results were restricted to just one cycle by the length of their wave tank. They also found a reduction in carrier frequency rather than true recurrence. A fuller discussion is given by Yuen & Lake (1982) and Thyagaraja (1983). When more than one pair of sidebands is admitted, the temporal evolution may become much more complicated (Caponi, Saffman & Yuen 1982; see also §23.5).

Also important are three-dimensional modulations of the form

$$A = [a + b_+ \exp(iK\xi + iL\eta + i\Omega\tau)$$
$$+ b_- \exp(-iK\xi - iL\eta - i\Omega^*\tau)] \exp(-ia^2\chi\tau) \quad (21.4)$$

where A satisfies the three-dimensional nonlinear Schrödinger equation (19.3). Here, $\epsilon(\pm K, \pm L)$ denote prescribed wavenumber perturbations in the ξ and η directions. Linearization with respect to $b_\pm$ yields the eigenvalue relation

$$\Omega^2 = (\lambda K^2 + \mu L^2)(\lambda K^2 + \mu L^2 + 2\chi a^2), \quad (21.5a)$$

which reveals instability whenever

$$0 < K^2 < (-\mu/\lambda)L^2 + (-2\chi/\lambda)a^2. \quad (21.5b)$$

For gravity waves, $-\mu/\lambda = 2$ and $\lambda\chi < 0$. Then, unlike result (21.3b), unstable modes are *not* confined within narrow sidebands of the fundamental, since both L and K may be large while satisfying (21.5b). Further, the maximum growth rate is attained everywhere along a curve with $L/K = \tan\phi = \pm 2^{-\frac{1}{2}}$ as asymptote. Of course, when ϵK and ϵL are of the same order as the fundamental wavenumber k_0, the Schrödinger equation (19.3) is no longer a valid approximation. Because such modes can grow, disturbances initially confined to a narrow wavenumber band near $(k_0, 0)$ – and so well represented by (19.3) – eventually pass out of its range of validity. Wave-trains in wide tanks are susceptible to these and other three-dimensional modulations, as shown by Melville (1982) and Su *et al.* (1982). Crawford *et al.* (1981) describe a more general theory of such three-dimensional instabilities, based on that of Zakharov (1968), which permits the interacting modes to have widely differing wavenumbers. In effect, this is a four-wave interaction problem and so is deferred to the following chapter.

The stability of spatially-varying solutions of (21.1) has also received attention. Vakhitov & Kolokolov (1973) first showed that the envelope soliton (21.2) is stable to small perturbations which depend on ξ_1 and τ only: that is, the Benjamin–Feir instability of uniform wave-trains is suppressed by the spatial variation. But spanwise-periodic perturbations akin to (21.4) yield instability (Zakharov & Rubenchik 1974, Martin *et al.* 1980, Yuen & Lake 1982), for (21.2) and also for periodic 'dn' solutions of (21.1). A more general discussion of the transverse instability of solitons, for governing equations other than (19.3), is given by Ablowitz & Segur (1981, §3.8). Unpublished observations of Hammack, briefly reported by Ablowitz & Segur (1981, §4.3), show evidence of the transverse instability of water wave-packets. Only when the channel is sufficiently narrow are such instabilities suppressed and permanent two-dimensional envelopes attained.

More dramatically, the three-dimensional nonlinear Schrödinger equation (19.3) may exhibit focusing at a finite time, when singularities develop: this occurs when λ, μ are both positive and χ is negative. These conditions are met for gravity waves (cf. Ablowitz & Segur 1979, equation 2.25 and note misprint in their 2.24d). In finite depth, the corresponding equations are (19.1a, b), which exhibit similar behaviour mentioned below.

The breakdown of localized solutions with radial symmetry was elegantly demonstrated by Zakharov (1972). Then, (19.3) rescales to

$$i\,\frac{\partial A}{\partial t}+\frac{\partial}{\partial\rho}\left(\frac{1}{\rho^2}\frac{\partial}{\partial\rho}(\rho^2 A)\right)+|A|^2 A = 0$$

where ρ denotes the radial co-ordinate. This has constants of motion

$$I_1 = \int_0^\infty \rho^2 |A|^2\,\mathrm{d}\rho,\quad I_2 = \int_0^\infty \left(|\partial(\rho A)/\partial\rho|^2+2|A|^2-\tfrac{1}{2}\rho^2|A|^4\right)\mathrm{d}\rho$$

and it is readily shown that

$$\frac{\mathrm{d}^2}{\mathrm{d}t^2}\int_0^\infty \rho^4|A|^2\,\mathrm{d}\rho < 6I_2.$$

Accordingly,

$$0 \leqslant \int_0^\infty \rho^4|A|^2\,\mathrm{d}\rho < 3I_2\,t^2+C_1\,t+C_2 \quad (C_1, C_2 \text{ constant})$$

and solutions exist only for a finite time if the initial envelope $A(\rho, 0)$ is sufficiently large that $I_2 < 0$. Zakharov extends this argument to more general disturbances.

21.2 *Davey–Stewartson equations with real coefficients*

The three-dimensional Davey–Stewartson equations (19.1a, b) have not been solved by inverse scattering for general real values of the coefficients. But for capillary–gravity waves in the shallow-water limit $kh \to 0$, $\epsilon \ll (kh)^2$ they rescale to

$$\left.\begin{array}{l} iA_T - sA_{XX} + A_{YY} = s\,|A|^2\,A + A\Phi_X, \\ s\Phi_{XX} + \Phi_{YY} = -2(|A|^2)_X, \end{array}\right\} \quad s \equiv \operatorname{sgn}\left[\tfrac{1}{3} - (\gamma k^2/\rho g)\right]. \tag{21.6}$$

These latter are of inverse-scattering type and various two- and three-dimensional soliton solutions are given by Ablowitz & Haberman (1975), Anker & Freeman (1978a) and Satsuma & Ablowitz (1979). Particular solutions for intermediate depth are noted by Kirby & Dalrymple (1983).

The stability of uniform wave-trains in finite depth is governed by (19.1a, b), in place of (19.3) for infinite depth. Disturbances A representing weak envelope modulations may be taken in similar form to (21.4), but with a modified frequency correction which (by a change of reference frame) incorporates any uniform part $\Phi_0 = B\xi$ of the mean flow. Modifications to the mean flow take the corresponding form

$$\Phi = f \exp\left(iK\xi + iL\eta + i\Omega\tau\right) + f^* \exp\left(-iK\xi - iL\eta - i\Omega^*\tau\right).$$

Substitution in (19.1a, b) and linearization with respect to $b_\pm$ and f yields

$$\Omega^2 = (\lambda K^2 + \mu L^2)\{\lambda K^2 + \mu L^2 + 2a^2[\chi - \beta\chi_1 K^2(\alpha K^2 + L^2)^{-1}]\}. \tag{21.7}$$

This reveals the possibility of unstable modes at some (K, L) whenever

$$(\lambda K^2 + \mu L^2)[\chi - \beta\chi_1 K^2(\alpha K^2 + L^2)^{-1}] < 0:$$

this result, derived by Davey & Stewartson (1974), was earlier deduced by Hayes (1973) using Whitham's formulation.

For gravity waves, such unstable wavenumbers (K, L) always exist. For capillary–gravity waves, the kh, $k^2\gamma/\rho g$ parameter space divides into regions shown in Figure 6.1 (cf. Djordjevic & Redekopp 1977; Ablowitz & Segur 1979). Instability is present in all but region E. For plane modulations with $L = 0$, the instability criterion reduces to (21.3) with λ_1, χ replaced by λ, $\nu \equiv \chi - \chi_1(\beta/\alpha)$. In regions A and D (where $\lambda\mu < 0$) and in regions B and C (where $\alpha < 0$) the unstable domain extends to large values of (K, L): in these, the developing disturbance presumably departs from the range of validity of (19.1a, b) at large enough times τ, just as happens with (19.3). In region F, unstable wavenumbers are restricted to sidebands with K, L of order a.

In region F, breakdown of equations (19.1a, b) can take a different form: some solutions are 'self-focusing', with initially-smooth modulations developing a singularity after a finite time (Ablowitz & Segur 1979). No experimental demonstration of this has yet been given for water waves; but a similar phenomenon has been observed in nonlinear optics, where the nonlinear Schrödinger equation (19.3) applies (Vlasov, Petrishchev & Talanov 1974).

21.3   Nonlinear Schrödinger equation with complex coefficients

For two-dimensional wave-packets with $A = A(\xi, \tau)$, the coupled system (20.7), (20.8) reduces to the nonlinear Schrödinger equation

$$\partial A/\partial \tau - a_2 \, \partial^2 A/\partial \xi^2 = (d_1/d_{1r}) A + k \, |A|^2 A \qquad (21.8)$$

with *complex* coefficients a_2, d_1/d_{1r} and k. An equation of similar form results when A depends on just one spatial direction, $\xi_1 = \xi \cos \phi + \eta \sin \phi$ say. In this form, the wave-mode is taken to be linearly unstable, since $\mathrm{Re} \, (d_1/d_{1r}) = 1$.

If a_{2r} is also positive, (21.8) rescales without loss to

$$\partial A/\partial \tau - (1 + ia_i) \, \partial^2 A/\partial \xi^2 = A + (\delta_r + i\delta_i) \, |A|^2 A, \qquad (21.9)$$

where $\delta_r = 1$, 0 or -1 according as the nonlinear term is destabilizing, neutral or stabilizing. A particular exact solution, independent of ξ, has

$$|A| = \frac{|A_0| \, e^\tau}{[1 + \delta_r |A_0|^2 (1 - e^{2\tau})]^{\frac{1}{2}}} \qquad (21.10)$$

where $A = A_0$ at $\tau = 0$ (Hocking & Stewartson 1972). Result (21.10) is independent of δ_i and 'bursts' at a finite time whenever $\delta_r > 0$. Exact solutions with periodicity $\exp iK\xi$ behave similarly provided $K^2 < 1$.

(i)   Soliton solutions

When $\delta_r < 0$, bursting of spatially-uniform solutions cannot occur because the nonlinear term is stabilizing. Equilibrium of $|A|$ is then possible, with $|A| = |A_0| = (-\delta_r)^{-\frac{1}{2}}$. Also, Pereira & Stenflo (1977) give an equilibrium solution localized in ξ, in the form

$$A = p(\mathrm{sech} \, q\xi)^{1+ir} \, e^{-is\tau} \qquad (21.11)$$

for particular real constants p, q, r, s determined by the coefficients of (21.8). This yields the soliton solution (21.2, with $b = 0$) as a special case; but, unlike the latter, (21.11) normally represents a single solution, not a family with an arbitrary amplitude parameter.

Interesting special cases of (21.8) are

$$\left. \begin{array}{l} iA_T + i\epsilon A + A_{XX} + 2 \, |A|^2 A = 0, \\[4pt] iA_T + (1 - i\epsilon) A_{XX} + 2 \, |A|^2 A = 0, \end{array} \right\} \qquad (21.12a, b)$$

obtainable from (21.9) on rescaling when $\delta_r = 0$. Both represent decaying waves when $\epsilon > 0$. When $0 < \epsilon \ll 1$, approximate solutions may be developed by perturbation of the conservative equation (21.1), in an alliance of inverse-scattering theory and 'two-timing' (Kaup 1976; Keener & McLaughlin 1977). Numerical solutions of (21.12a, b) by Pereira (1977) and Pereira & Chu (1979) display damped single and double solitons.

Segur (1981) investigated the viscous decay rate of a gravity-wave packet initially of envelope soliton form. He assumed that the solution remained close to (21.2) as the amplitude decreased, a supposition supported at leading order in ϵ by Pereira (1977) and also by experiment. Segur's result, that the exponential decay rate of $A(X, T)$ is twice as great as for uniform wave-trains, is readily understood. The breadth of the envelope soliton is of order $[\mathrm{Max}_X | A(X, T) |]^{-1} \equiv a^{-1}(T)$ and so the total energy E within the envelope is $O(a)$ rather than $O(a^2)$ as for uniform wave-trains. With energy dissipated at the same rate $\Lambda = E^{-1} \, \mathrm{d}E/\mathrm{d}T$, one immediately obtains $a^{-1} \, \mathrm{d}a/\mathrm{d}T = \Lambda$ for the soliton and $a^{-1} \, \mathrm{d}a/\mathrm{d}T = \frac{1}{2}\Lambda$ for the uniform train. In other words, exactly half the rate of reduction in amplitude of the soliton is attributable to spreading of the packet.

(ii) Sideband modulations

Equations (21.8) or (20.7) and (20.8) may be employed, as were (21.1) and (19.1a, b) above, to investigate the modulational stability of uniform wave-trains. Stuart & DiPrima (1978) did so for (21.8), taking

$$A = A_0 \, e^{i(\mu_0 \xi + \gamma_0 \tau)} + B(\xi, \tau)$$

where A_0 is constant and the small disturbance B has the form

$$B(\xi, \tau) = a(\tau) \, e^{i(\mu_1 \xi + \gamma_1 \tau)} + b(\tau) \, e^{i(\mu_2 \xi + \gamma_2 \tau)},$$

$$\mu_1 + \mu_2 = 2\mu_0, \quad \gamma_1 + \gamma_2 = 2\gamma_0.$$

For equilibrium,

$$| A_0 |^2 = (\mu_0^2 \, a_{2r} - d_1/d_{1r}) k_r^{-1}.$$

The linearized equations for $a(\tau)$, $b(\tau)$ admit exponential growth, and so instability, when an explicit but algebraically complex criterion involving $a_2, d_1/d_{1r}$ and k is met. Stuart & DiPrima investigated several special cases, including those of Taylor-vortex and Bénard flows: a relationship with earlier work of Eckhaus (1963) is made clear.

Equation (21.9) (under the alias of the 'Ginzburg–Landau equation') has been investigated numerically by Moon, Huerre & Redekopp (1983). They examined cases with $\delta_r = -1$, $a_i = \delta_i \equiv c_0^{-1}$ where c_0 is a variable parameter, and chose initial conditions

$$A(x, 0) = 1 + 0.2 \cos q\xi$$

representative of a uniform wave-train modulated by a sideband. They imposed periodic boundary conditions at $x = \pm \pi/q$ (which inhibit the development of subharmonics $\frac{1}{2}q$ etc.) and investigated a range of q with various fixed c_0 but mainly $c_0 = \frac{1}{4}$. Note that $|A| = 1$ is the equilibrium amplitude of temporal oscillations with no ξ-dependence and that infinitesimal sidebands $\propto \exp(\pm iq\xi)$ are unstable for a range of q. The following was observed for $c_0 = \frac{1}{4}$ as q was decreased.

Motion was simply periodic, with a single frequency ω_1 and harmonics, for q above 0.60 (though more complex motion apparently sets in near $q = 1$); doubly periodic, with independent frequencies ω_1, ω_2 and combinations, for $0.52 < q < 0.60$; triply periodic for $0.49 < q < 0.52$. Phase locking into a limit cycle solution was found at $q = 0.49$, followed by a chaotic régime in $0.41 < q < 0.49$. A two-frequency state returned for $0.38 < q < 0.41$, with all sideband energy transferred to $2q$ and $4q$ wavenumber modes: this reflects the fact that the $\pm 2q$ sidebands have greatest linear amplification rate at $q = 0.41$. A further bifurcation set in at $q = 0.38$, with modes q and $3q$ returning to prominence; chaotic behaviour resumed in the range $0.20 < q < 0.377$.

Analogous computations, for the conservative nonlinear Schrödinger equation ($c_0 = 0$) and Zakharov equation for deep-water waves, and for non-conservative mode interactions in Taylor–Couette flow and Bénard convection, are discussed in §§ 23.5, 24.4 and 25.2: rather similar transitions were found.

(iii) Bursting solutions

When $\delta_r > 0$ in (21.9), solutions $A(\xi, \tau)$ may, or may not, develop singularities at a single value of ξ after a finite time. Such singularities are no longer entirely due to focusing of a fixed amount of available energy, as for (19.3): now, the available energy in the disturbance also increases with time.

By judicious choice of similarity variables, Hocking & Stewartson (1971, 1972) investigated the structure of such localized bursting solutions, which they found to be of two distinct types. One or both types may occur throughout most of the a_i, δ_i parameter space; but there are two regions – broadly, with $|\delta_i|$ and $|a_i^{-1}|$ both large enough – where bursting is impossible. Haberman (1973a, 1977) has considered how bursting evolves from initial linear disturbances of the form (7.7). His analytical results mostly agree with computations, for similar initial packets, by Hocking, Stewartson & Stuart (1972) and Hocking & Stewartson (1972); but he found only one type of singularity.

For linearly-damped cases, d_1/d_{1r} in (21.8) is replaced by a constant with negative real part. Then, with sufficiently large initial amplitudes and $k > 0$, Hocking *et al.* (1972) found that bursts also occur. The three-dimensional counterpart

$$\frac{\partial A}{\partial \tau} - a_2 \frac{\partial^2 A}{\partial \xi^2} - b_2 \frac{\partial^2 A}{\partial \eta^2} = \frac{d_1}{d_{1r}} A + k |A|^2 A \tag{21.13}$$

of (21.8) was also investigated by Hocking *et al.* (1972) and Hocking & Stewartson (1971), who found some three-dimensional bursting solutions.

Marginally unstable disturbances in plane Poiseuille flow satisfy (20.7) and (20.8), not (21.13); but plane modulations of the form $A(\xi_1, \tau)$, $\xi_1 = \xi \cos\phi + \eta \sin\phi$, satisfy an equation like (21.8). Davey, Hocking & Stewartson (1974), correcting Hocking & Stewartson (1972), show that modulations skewed at angles $|\phi| > 57.3°$ to the flow direction may burst but that those with $|\phi| < 57.3°$ cannot. It is worth mentioning that, though these envelopes are skewed at a considerable angle, the actual waves must remain virtually straight-crested for the governing equations to hold. Little is known yet about three-dimensional solutions of (20.7)–(20.8) with complex coefficients, but bursting is sure to occur over much of the parameter range.

Despite claims to the contrary, such bursting is not directly connected with the development of turbulent spots in unstable shear flows. The theory outlined above is confined to disturbances centred on a single wavenumber and frequency, whereas turbulent spots are characterized by wavenumbers and frequencies much greater than those of the fundamental wave. Of course, since the 'burst' arises when the fundamental wave is modified so rapidly that weakly-nonlinear theory breaks down, subsequent events such as local secondary instabilities are liable to proceed rapidly.

21.4 *Korteweg–de Vries equation and its relatives*

Comprehensive accounts of the derivation and solution of the Korteweg–de Vries equation (18.9) and of Benjamin's (1967) analogous long-wave approximation for internal waves may be found elsewhere, particularly Miles (1980), Ablowitz & Segur (1981) and Dodd *et al.* (1982). The powerful method of inverse scattering, first developed for the Korteweg–de Vries equation, has greatly advanced the understanding of nonlinear wavemotion. It would be superfluous to add another account of the method here: instead, only a few remarks are made with the aim of relating the long-wave Korteweg–de Vries approximation to general water-wave theory.

Packets of long surface gravity waves in water of finite depth h may be characterized by two dimensionless parameters: a depth-related amplitude a/h and wavenumber kh. The limits $a/h \to 0$ and $kh \to 0$ are non-uniform, the result depending on the order in which these limits are taken. But Freeman & Davey (1975) showed that introduction of $\Delta \equiv (a/h)(kh)^{-2}$ in place of a/h leads to a *uniform* double limit $\Delta \to 0$, $kh \to 0$. They derive a generalization of the Korteweg–de Vries equation, with spatial variation in both ξ and η, which is valid as $kh \to 0$ with Δ finite. This is

$$
\left.
\begin{aligned}
2\frac{\partial \zeta}{\partial T} + 3\zeta \frac{\partial \zeta}{\partial \xi} + \frac{1}{3\Delta}\frac{\partial^3 \zeta}{\partial \xi^3} &= -\frac{\partial^2 \Phi}{\partial \eta^2}, \\
\zeta &= \partial \Phi / \partial \xi,
\end{aligned}
\right\}
\tag{21.14a, b}
$$

as earlier proposed by Kadomtsev & Petviashvilli (1970). Here, ζ is the (dimensionless) surface elevation and Φ the leading-order approximation to the velocity potential. If, for small Δ, one sets

$$
\left.
\begin{aligned}
\zeta &= AE + O(\Delta), \\
\Phi &= \Phi_0 + \Phi_1 E + O(\Delta),
\end{aligned}
\right\} \quad
E \equiv \exp i(\xi + \tfrac{1}{6}\Delta^{-1}T),
$$

where A, Φ_0 depend on scaled variables

$$
\xi_1 = \Delta(\xi + \tfrac{1}{2}\Delta^{-1}T), \quad \eta_1 = \Delta^{\frac{1}{2}}\eta, \quad \tau = \Delta T,
$$

the long-wave limit (21.6) of the Davey–Stewartson equations (19.1) is recovered at $O(\Delta)$ after a further slight rescaling (cf. Freeman & Davey).

Solutions of (21.14a, b) which represent obliquely interacting solitons have been found (Miles 1977a, b, 1980). Resonant interaction of three such solitons, reminiscent of three-wave resonance, occurs when conditions

$$
\mathbf{k}_3 = \mathbf{k}_2 \pm \mathbf{k}_1, \quad \omega_3 = \omega_2 \pm \omega_1
$$

are met. However, $\mathbf{k}_j$ and ω_j no longer represent wavenumbers and frequencies. Instead, $|\mathbf{k}_j|$ represents the spatial 'length' of the jth soliton, with the direction of $\mathbf{k}_j$ normal to the wave crest; and ω_j equals $|\mathbf{k}_j| c_j$ where c_j is the speed of the jth soliton. The existence of such resonant triads of solitons has important implications for reflection of solitary waves at a rigid wall. At sufficiently small angles of incidence, regular reflection is replaced by Mach reflection similar to that of shock waves, with three resonant solitons meeting at a single point (Miles 1977a, b; Melville 1980). Higher-order resonances are discussed by Anker & Freeman (1978b).

The range of validity of the Korteweg–de Vries equation, for wave propagation in one space dimension, was recently clarified by Fenton & Rienecker (1982). They obtained computer solutions of the full water-wave equations, by means of truncated Fourier series, both for solitary waves which overtake one another and for colliding solitary waves travelling in opposite directions. For the former, some deviations from the Korteweg–de Vries solutions were noted, but agreement was quite good. For the latter, the Korteweg–de Vries equation is inadequate since the direction of propagation is assumed *a priori* in its derivation.

Experimental confirmation of the Korteweg–de Vries model and demonstration of the properties of solitary wave interactions are given by various workers, notably Hammack & Segur (1974, 1978), Weidman & Maxworthy (1978), Koop & Butler (1981) and Segur & Hammack (1982).

The more general equation (18.10), which admits linear instability of a range of wavenumbers, has been investigated analytically and numerically by Cohen *et al.* (1976), Kawahara (1983) and Hooper & Grimshaw (1985). A modal decomposition

$$A = \sum_{-\infty}^{\infty} A_n(\tau) \exp(ink\xi), \quad A_{-n} = A_n^*$$

is appropriate for spatially-periodic solutions and A_0 may be taken to be zero without loss of generality. The various A_n then satisfy

$$dA_n/d\tau = a_n A_n - ink\sigma \left(\sum_{r=1}^{n} A_r^* A_{r+n} + \tfrac{1}{2} \sum_{r=1}^{n-1} A_r A_{n-r} \right),$$

$$a_n = \alpha(nk)^2 + i\beta(nk)^3 + \gamma(nk)^4.$$

With only a few linearly-unstable modes available, this set of equations may justifiably be truncated at some n. In particular, if only $n = \pm 1$ are linearly amplified, a reasonable approximation may be the two-mode equations

$$dA_1/d\tau = a_1 A_1 - ik\sigma A_2 A_1^*, \quad \operatorname{Re} a_1 = a_{1r} > 0,$$

$$dA_2/d\tau = a_2 A_2 - ik\sigma A_1^2, \quad \operatorname{Re} a_2 = a_{2r} < 0.$$

This has the equilibrium solution

$$|A_1|^2 = \frac{-a_{1r}|a_2|^2}{k^2\sigma^2 a_{2r}}, \quad A_2 = \frac{ik\sigma}{a_2} A_1^2$$

for real σ. A three-mode model ($n = \pm 1, \pm 2, \pm 3$) was considered by Cohen *et al.*; Lin & Krishna (1977) investigated similar interactions of two- and three-dimensional wave-modes in thin viscous films, using a generalization of (18.10).

If k is reduced so that the $n = \pm 2$ modes enter the unstable waveband, another two-mode equilibrium state becomes available, with A_2 and A_4 as dominant components. Obviously, further reduction of k admits still higher modes and more possible equilibrium states. Solutions may equilibrate to one such state or may exhibit continual modulations, regular or irregular, with energy exchange among several modes. Kawahara's numerical solutions for $\beta = 0$ developed no regular pattern, but non-zero dispersion led to formation of rows of 'solitary-wave'-like pulses of equal amplitude. No general analytic solution of equation (18.10) is available. Consideration of (18.10) as a perturbation of the Korteweg–de Vries equation, or of Burger's equation, might prove fruitful.

Though interesting in its own right, little weight should be attached to (18.10) as a model of observable waves in thin viscous layers: these usually have amplitudes a comparable with the mean depth h whereas (18.10) is valid only for $a/h \ll 1$.

22   Waves of larger amplitude

22.1   *Large-amplitude surface waves*

Nonlinear theory must be pursued to higher order if it is to describe steep waves. For a uniform train of progressive, inviscid, surface gravity waves of a given wavenumber k and amplitude a, Stokes (1847) developed a perturbation expansion to third order in the wave-slope parameter ka. The vertical surface displacement has sharper peaks and broader troughs than a sine-wave, and the frequency exceeds that of linear theory (see e.g. Lamb 1932; Kinsman 1969). Stokes (1880b) also introduced a conformal mapping which transformed the free boundary problem to that of finding the space co-ordinates as functions of velocity potential and streamfunction. Wilton's (1915) extension of Stokes' analysis to capillary–gravity waves encountered singularities now known to be due to resonance of the wave with its harmonic (see § 14 above).

Both the perturbation-expansion and conformal-transformation methods have recently been allied with powerful computation to yield solutions of high accuracy. A useful recent review is given by Schwartz & Fenton (1982). Schwartz (1974) and Cokelet (1977) carried the perturbation series to around a hundred terms and used Padé approximants to sum the series. The surface displacement and phase speed, the mass, momentum, energy and their fluxes were all found for waves of various heights in water of various depths. Particular attention was paid to determining the properties of the highest and nearly-highest waves. The highest has a pointed crest, with 120° included angle, as originally conjectured by Stokes (1880a) and calculated by Michell (1893). But Longuet-Higgins (1975) and Longuet-Higgins & Cokelet (1976) found that this wave does *not* have greatest energy per wavelength, the wave of maximum energy having a slightly lower, rounded, crest. Williams (1981) obtained similar results, via an integral equation method, and showed that in the shallow-water limit

$kh \to 0$ his results agree with the solitary-wave solutions of the Korteweg–de Vries equation; so also do those of Fenton & Rienecker (1982). The nature of wave breaking was clarified by Longuet-Higgins & Cokelet (1978) and Longuet-Higgins (1981). Large-amplitude waves are subject to normal-mode instabilities which destroy the upstream–downstream symmetry and lead to plunging or spilling breakers (see below). The 120° limiting solution is normally bypassed and this solution would be very difficult to realize experimentally.

Exact solutions for finite-amplitude capillary waves were obtained by Crapper (1957), Kinnersley (1976) and Vanden-Broeck & Keller (1980). The wave troughs are then sharper than the peaks and the solution of largest amplitude has 'bubbles' trapped in the troughs. The near-surface particle displacements and associated mean drift velocity turn out to be surprisingly large (Hogan 1984a). Schwartz & Vanden-Broeck (1979), Rottman & Olfe (1979), Hogan (1980, 1981) and Chen & Saffman (1980b) give numerical solutions for capillary–gravity waves.

For progressive gravity waves on an interface between fluids of differing densities, ρ_u and ρ_1, a sharp-crested wave cannot occur since this would necessitate infinite velocity in the upper field. Interfacial waves were examined by Holyer (1979), using similar numerical techniques, for fluids of infinite depth. The highest wave has points on the interface with longitudinal velocity equal to the phase speed. For free-surface waves, these points occur at the peak of the 120° crests; but, as the density ratio ρ_u/ρ_1 increases from zero, they move away from the top of the crests and approach the half-way points between crests and troughs as ρ_u/ρ_1 tends towards $1-$. For air over water, the maximum phase speed, momentum and energy still occur at amplitudes below the highest possible; but for $\rho_u/\rho_1 \geqslant 0.1$ these properties were all found to increase monotonically with amplitude. The maximum possible wave height increases with ρ_u/ρ_1: that for air–water is just 2% greater than for a free surface, but it reaches $2\frac{1}{2}$ times the free-surface value as $\rho_u/\rho_1 \to 1-$. Experiments of Thorpe (1968a) show reasonable agreement with the results of third-order theory but the observations are confined to relatively small amplitudes. Solutions found by Meiron & Saffman (1983), which represent overhanging large-amplitude interfacial waves, are certainly strongly unstable and so physically unrealizable.

Finite-amplitude waves on the (stable) interface between fluids of differing densities and velocities – i.e. the Kelvin–Helmholtz configuration of §2.1 – were recently studied both analytically and computationally by Saffman & Yuen (1982).

For standing waves on the free surface of infinitely deep liquid, Schwartz & Whitney (1981) carried out an expansion to 25th order. Earlier, Penney & Price (1952) had suggested that the free-surface standing wave of maximum height experiences a downwards acceleration of g at the crest, when this is at its highest point; in which case, the crest should form a 90° corner. Schwartz & Whitney's results support this conclusion, as do those of Rottman (1982), but Saffman & Yuen (1979) disagree. Meiron, Saffman & Yuen (1982) note a lack of uniqueness inherent in all such solutions for standing waves, which makes their physical significance uncertain. Nevertheless, experiments of Taylor (1953) and Edge & Walters (1964) show that the highest wave certainly attains a crest angle close to 90°, for air–water interfaces. For standing waves in water of finite depth, expansions to third order are given by Tadjbaksh & Keller (1960) and Chabert-d'Hieres (1960).

Standing gravity waves at an interface between infinite fluids are discussed by Rottman (1982), who carried the series expansion to 21st order, with Padé summation. A corner cannot occur at the highest crest unless the density ratio ρ_u/ρ_l is precisely zero. For other values, the wave of maximum amplitude has a vertical tangent at some point: higher waves would break by overturning. Though experiments by Thorpe (1968b) show breaking, this takes place at smaller amplitudes and is thought to be due to an instability associated with the shear layers induced by viscosity near the interface.

Formal convergence of Stokes' expansion was proved by Levi-Civita (1925) for progressive free-surface waves of sufficiently small amplitude. Corresponding proofs for interfacial and for standing waves are lacking. Convergence at large amplitudes has not been rigorously established for any configuration, but the circumstantial evidence of the high-order approximations appears strong. Proofs of the existence of large-amplitude solutions are given by Keady & Norbury (1978) and Toland (1978). But we shall see below that such uniform wave-trains are invariably unstable!

Other steady solutions, representing non-uniform and three-dimensional wave-trains, are given by Chen & Saffman (1980a), Olfe & Rottman (1980), Meiron, Saffman & Yuen (1982) and Roberts (1983). The first two papers concern two-dimensional modulated wave patterns with successive crests of differing heights, such that every alternate, or every third, crest is the same. Meiron *et al.* give corresponding three-dimensional solutions, both symmetric and skew-symmetric, which agree quite well with observations of Su (1982), made in a wide outdoor basin (see Figure 6.8).

22.2   Higher-order instability of wave-trains

Longuet-Higgins (1978a, b), McLean *et al.* (1981) and McLean (1982a, b) have reported a new type of instability for finite-amplitude gravity waves, the existence of which was conjectured by Zakharov (1968). Its discovery by Longuet-Higgins came from computations based on Stokes' (1880b) conformal mapping representation of two-dimensional gravity waves in deep water. McLean *et al.* and McLean (1982a, b) considered a wider class of three-dimensional disturbances, for gravity waves in both infinite and finite depths. Their computational method, based on the exact (inviscid) equations, employed Stokes' perturbation expansion.

Thus, the surface displacement of a finite-amplitude wave-train of wavelength $2\pi/k$ is represented as

$$\zeta = \sum_{n=0}^{\infty} A_n \cos n(x - Ct)$$

with normalized x, t co-ordinates and with known Fourier coefficients A_n and phase speed C depending on the wave-slope parameter ka. This wave is perturbed by an infinitesimal three-dimensional disturbance

$$\zeta' = \exp i[p(x - Ct) + qy - \Omega t] \sum_{n=-\infty}^{\infty} a_n \exp in(x - Ct) + \text{c.c.}$$

where $\Omega = \Omega(p, q, H)$ is an eigenvalue to be found. Instability arises when $\operatorname{Im}\Omega \neq 0$, with roots Ω occurring in complex-conjugate pairs.

McLean *et al.*'s computations were accomplished by truncating the expansions at high order, $|n| \leq 20$ or 50. They found two distinct regions of instability in the p–q plane for various values of ka. Examples are shown in Figure 6.6. The region nearer the origin (Type I instability) reduces to that for modulational instability of the three-dimensional Schrödinger equation (19.3) when $|p|, |q|$ and ka are sufficiently small. It also coincides, at larger p, q but small ka, with Phillips' (1960) four-wave resonance (see §23.1, Figure 7.2). Yuen (1983) also reports on Type I and II instabilities of linearly-stable waves in Kelvin–Helmholtz flow.

The outer region of (Type II) instability may be interpreted, at small ka, as a degenerate five-wave resonance of the form

$$\left.\begin{aligned}
\mathbf{k}_1 + \mathbf{k}_2 &= \mathbf{k}_3 + \mathbf{k}_4 + \mathbf{k}_5, \\
\omega_1 + \omega_2 &= \omega_3 + \omega_4 + \omega_5
\end{aligned}\right\} \tag{22.1a, b}$$

where, in normalized form,

$$\mathbf{k}_3 = \mathbf{k}_4 = \mathbf{k}_5 = (1, 0), \quad \omega_3 = \omega_4 = \omega_5 = C = 1$$

represent the fundamental wave-train and

$$\mathbf{k}_1 = (1+p, q), \quad \mathbf{k}_2 = (2-p, -q)$$

the perturbation. The linear dispersion relation

$$\omega = g^{\frac{1}{2}}(k^2 + l^2)^{\frac{1}{4}}, \quad \mathbf{k} = (k, l)$$

and (22.1b) yield

$$[(p+1)^2 + q^2]^{\frac{1}{4}} + [(2-p)^2 + q^2]^{\frac{1}{4}} = 3.$$

The highest growth rate occurs with $p = \frac{1}{2}$, when $\mathbf{k}_1$ and $\mathbf{k}_2$ have the same k-component, $k = \frac{3}{2}$: their x-phase velocity then coincides with C. At other values of p, there are two distinct pairs $\mathbf{k}_1$, $\mathbf{k}_2$ corresponding to positive and negative roots q.

Clearly, an acceptable description of the latter instability for small ka could be given by weakly-nonlinear analysis of three discrete wave-modes, pursued to fourth-order in the amplitudes. But this hardly seems worthwhile in view of the available computational results for the exact equations.

These two types of instability are probably the first two members of an

Figure 6.6. Type I and Type II instability of finite-amplitude gravity waves in deep water (from McLean *et al.* 1981). Cases (*a*)–(*d*) correspond to $ka/\pi = 0.064, 0.095, 0.111, 0.127$ respectively.

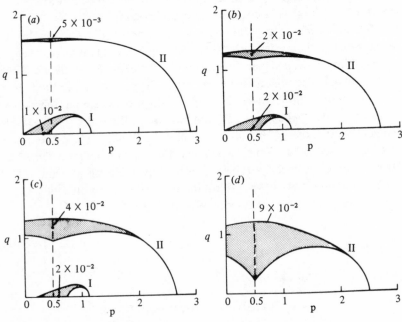

infinite class of higher-order resonances among three discrete (linear) wave-modes, as suggested by Zakharov (1968). These satisfy

$$\mathbf{k}_1 + \mathbf{k}_2 = N\mathbf{k}_0, \quad \omega_1 + \omega_2 = N\omega_0$$

where $\mathbf{k}_0 = (1, 0)$, $\omega_0 = 1$ denote the fundamental wave-train and N is any positive integer greater than 1. Since the characteristic growth rates of $\mathbf{k}_1$ and $\mathbf{k}_2$ are $O[(ka)^N]$ and $ka < 0.443$ even for the highest gravity waves, instability at sufficiently large N is likely to be suppressed by viscosity.

Longuet-Higgins' (1978a, b) earlier investigations (also Hasselmann, 1979) were confined to two-dimensional sub- and super-harmonic modes independent of y. His results agree with those of McLean *et al.* for $q = 0$. For $q = 0$ and at small ka there is no Type I instability of the discrete wavenumbers $p = M/8$ ($M = 1, 2, ..., 16$) and $p = \pm M$ ($M = 1, 2, ..., 7$) studied by Longuet-Higgins (though there *is* such instability at other values of p, notably Benjamin–Feir instability as $p \to 0$). But, as ka was increased, Longuet-Higgins found Type I instability among wavenumbers k_1, k_2 satisfying

$$k_1 + k_2 = 2$$

the fundamental wavenumber and frequency being normalized to unity. The corresponding frequency condition for four-wave resonance is

$$\omega_1 + \omega_2 = 2$$

but this cannot be satisfied as $ka \to 0$ for Longuet-Higgins' chosen wavenumbers. Onset of Type I instability, as ka increases, is due to nonlinear modification of the frequencies ω_1, ω_2 which causes the resonance condition to be nearly satisfied over a finite range of ka. For large and small ka, the waves are 'detuned' and no instability takes place. The domain of Type I instability of two-dimensional disturbances is shown in Figure 6.7. The most unstable Type I instability, at given ka, is two-dimensional.

At large values of ka, not far short of that for the highest wave, Longuet-Higgins encountered Type II instability of the mode with $k_1 = \frac{3}{2}$. Owing to nonlinearity, this attains the same phase speed as the fundamental (i.e. $2k_1 = 3$, $2\omega_1 = 3$) when $ka \approx 0.41$: this is the degenerate five-wave resonance condition. These results clearly demonstrate the rôle of non-linearity in tuning and detuning the resonance. But restriction to discrete two-dimensional modes conceals McLean *et al.*'s wider class of instabilities.

The onset of Type II instability of *two-dimensional* modes almost coincides with the attainment of the greatest possible amplitude of the first Fourier component of the fundamental wave (*not* quite the highest wave). The view that this instability is directly associated with wave breaking is

Figure 6.7. Stability boundary for growth of two-dimensional perturbations of uniform finite-amplitude gravity waves: ○, results of Longuet-Higgins (1978); the solid line is approximation (23.16) derived from Zakharov's equation (from Crawford *et al.* 1981).

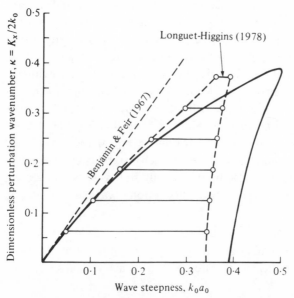

Figure 6.8. Three-dimensional wave configuration resulting from oblique Type II instability, for $ak = 0.33$ (from Su 1982).

supported by computations of Longuet-Higgins & Cokelet (1978). However, Type II instability of *oblique* modes is present at *all* fundamental wave amplitudes. It may be that breaking is associated with the near-coalescence of many members of the class of higher-order resonances noted above.

Experimental evidence of three-dimensional Type I and Type II instabilities is reported by Melville (1982), Su *et al.* (1982) and Su (1982). At relatively small wave-slopes, $ak \leqslant 0.3$, the Type I instability is stronger than Type II and the most rapid growth is associated with two-dimensional subharmonics of the fundamental. But oblique Type II instability is dominant for $0.3 < ak < 0.44$. The latter leads to three-dimensional waves: see Figure 6.8. Steady-state solutions corresponding to these are given by Ma (1982) and Meiron *et al.* (1982).

22.3   *Numerical work on shear-flow instability*

Truncation of the amplitude equation to exclude powers of $|A|$ higher than the third is justifiable only for rather small amplitudes. A higher-order expansion scheme was developed by Herbert (1980), who calculated the first seven coefficients a_j of a series expansion

$$\mathrm{d}A/\mathrm{d}t = A \sum_{j=0}^{\infty} a_j |A|^{2j}$$

for temporal evolution in plane Poiseuille flow, both at the critical point of linear theory and at $R = 5000$, $\alpha = 1.12$. He found that the a_j increase rapidly in magnitude with j rather faster than R^j, and that the signs of Re a_j alternate. The radius of convergence is therefore sure to be small. Calculation of the first three a_j by Gertsenshtein & Shtempler (1977) yielded similarly large values. Since the disturbance is considered to evolve under a constant mean pressure gradient, $\mathrm{d}P/\mathrm{d}x = -G$ say, the Reynolds number is best re-expressed as $R = R_p \equiv \frac{1}{2}Gh^3/\rho\nu^2$ where $2h$ is the channel width. Herbert (1982) gives a useful review of theory and experiment relating to the stability of plane Poiseuille flow.

Herbert (1983a) and Sen & Venkateswarlu (1983) re-examined the question of convergence and provided the expansion procedure with a more rational basis. The latter authors computed up to nineteen coefficients of the series expansion and used Shanks' method to sum the infinite series even when it diverges. Thereby, they calculated equilibrium amplitudes at various α and R, employing extensions of both Reynolds & Potter's (1967) and Watson's (1960) formulations. Convergence of the former was the more satisfactory: at $\alpha = 1.15$ and $R = 5000$, 4000 and 3500 it successfully yielded the equilibrium amplitudes, but below $R = 3500$ (at this α) a singularity of the series prevented this. These and other results suggest that the $|a_j|$ increase rapidly with j, roughly in proportion to $(3\alpha R)^j$ for the range of α, R considered, with the usual normalization that $\psi_1(0) = 2$.

Alternative methods, employing truncated Fourier series representations of a constant-amplitude disturbance, were employed by George, Hellums & Martin (1974), Zahn, Toomre, Spiegel & Gough (1974) and Herbert (1977) to seek equilibrium solutions of the Navier–Stokes equations. Figure 6.9 shows Herbert's results, with rather severe truncation after just two harmonics, of the neutral surface in $(R-\alpha-E)$ space, where E is a measure of the kinetic energy of a two-dimensional wavelike disturbance. This surface intersects the plane $E = 0$ in the linear neutral curve, with critical point at $R = 5772$, $\alpha = 1.02$. A minimum critical Reynolds number $R_e \approx 2702$ for finite-amplitude equilibrium is attained at $\alpha_e = 1.313$, a result which agrees closely with that of Zahn *et al.* A constant pressure gradient was imposed in both sets of computations. Retention of more Fourier modes causes some modification of the neutral surface: with four, R_e increases to 2935 at the wavenumber $\alpha_e = 1.323$.

Note that, for a range of wavenumbers α at each R, there are two equilibrium values of E. The smaller value denotes the threshold amplitude for subcritical nonlinear instability at prescribed α, R and this corresponds to *unstable* equilibrium: a small increase in amplitude leads to growth and a small decrease leads to decay. In contrast, the larger equilibrium value of E is stable to small amplitude changes; but this is a very restricted type

Figure 6.9. Herbert's (1977) results for the neutral surface in $R-\alpha-E$ space, with truncation after two harmonics.

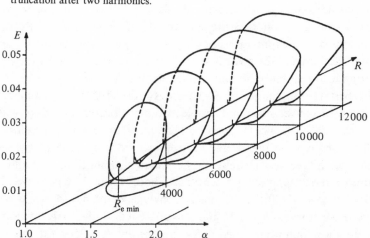

of stability which excludes disturbances of different wavenumber from that associated with E. It seems likely that instability of real flows to disturbances of the latter sort will prevent the attainment of these large amplitude two-dimensional periodic states (see §26).

With even more powerful computational facilities, Orszag & Kells (1980) and Orszag & Patera (1981) could retain up to 16 Fourier modes in x, and 16 in the spanwise direction y, while representing z-variations in terms of 33 Chebyshev polynomials. Thereby, they studied the temporal evolution of various initial disturbances, reporting satisfactory convergence of the truncated series and quite good agreement with Herbert's most accurate equilibrium solution for two-dimensional waves. They concluded that the threshold for finite-amplitude temporal instability of plane Poiseuille flow is near $R_e = 2800$, for initial disturbances with the form of a single Orr–Sommerfeld eigenmode. But these, and Herbert's conclusions are at variance with recent results of Rozhdestvensky & Simakin (1984), which at first agree with those of Orszag & Kells but are continued to much greater times.

Rozhdestvensky & Simakin carefully distinguish the Reynolds numbers R_p based on pressure gradient, defined above, and R_q based on mass flux Q, which equals $8Q/3\mu h$. In the absence of any disturbance, these both equal Vh/ν where V is the centreline velocity, but they take differing values when a disturbance is present. Rozhdestvensky & Simakin claim that all two-dimensional wavelike initial disturbances with $\alpha \geq 1$ eventually decay to zero if $R_p < 3250$. They found an equilibrium solution at $R_p = 3250$ ($R_q = 2855$) for $\alpha = 1.25$ and various other equilibria at higher R_p and other values of α. It seems that Orszag & Kells allowed insufficient time for true equilibration of their solutions: Rozhdestvensky & Simakin's computations continued for dimensionless times $T = O(10^3)$, which correspond to hundreds of wave periods, as compared with Orszag & Kells' $T = 150$.

Rozhdestvensky & Simakin also report two-dimensional equilibrium states with longer spatial perodicities, $\alpha \lesssim 0.34$ at Reynolds numbers R_p as low as 2700. However, these states comprise more than one dominant wave-mode, typically those with wavenumbers 3α and 4α which correspond to rather lightly-damped linear modes.

Orszag & Kells' and Rozhdestvensky & Simakin's results for finite three-dimensional and doubly-periodic disturbances are discussed in §26 below: meantime, we mention that such disturbances may grow at much lower values of R than those for amplification of a single two-dimensional wave-mode. In particular, for plane Couette flow, Orszag & Kells could find no undamped two-dimensional mode, but three-dimensional disturbances grew for $R \approx 1000$.

It appears that convergence of the Fourier-truncation method is more rapid than for the amplitude expansion technique (Herbert 1980). This may be due mainly to the requirement of a critical layer dominated by viscosity in the latter, which needs $|A| R^{\frac{2}{3}} \ll 1$ as shown in §22.4. Though the ranges of validity of both methods require further clarification, they seem destined to play an important rôle in future developments.

Related experiments mostly concern the downstream evolution of disturbances introduced by a vibrating ribbon and such configurations have also been treated numerically. Fasel (1976), Fasel, Bestek & Schefenacker (1977) and Fasel & Bestek (1980) employed a finite-difference scheme to solve the full Navier–Stokes equations with dependence on x,

Figure 6.10. Instantaneous disturbance velocities u', w' and spanwise vorticity Ω' for plane Poiseuille flow at $R = 10000$ and frequency $\omega = 0.2375$, for (a) 3% and (b) 5% input amplitudes. Direct numerical computations of Fasel & Bestek (1980).

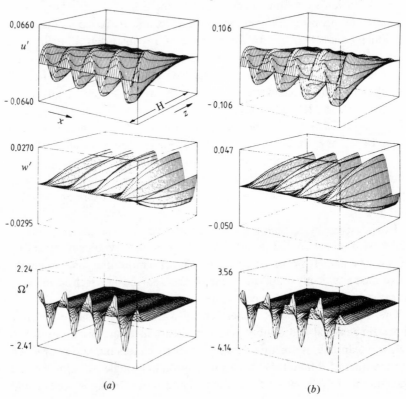

z and t. They studied finite disturbances in both plane Poiseuille flow and the Blasius boundary layer on a flat plate. Their domain of integration extended downstream for several wavelengths (typically five), the input disturbance was chosen as a Tollmien–Schlichting wave described by the linear Orr–Sommerfeld equation and the downstream boundary conditions permitted 'free passage' of the disturbance. At sufficiently small input amplitudes, good agreement was obtained with linear stability theory; at larger amplitudes, nonlinearity normally had a destabilizing influence. For Blasius flow, rather similar results were obtained by Murdock (1977) and Murdock & Taylor (1977) using a spectral method which employs Chebyshev polynomials.

Figures 6.10 and 6.11 display some of Fasel & Bestek's results. These show growing disturbances in plane Poiseuille flow at $R = 10\,000$ and frequency $\omega = 0.2375$, the input disturbances having a maximum streamwise velocity fluctuation which is 3% and 5% of the centre channel velocity. Figure 6.10 shows the disturbance velocity components u', w' and spanwise vorticity Ω'. Considerable distortion from the virtually sinusoidal x-variation of linear theory is apparent. Fourier decomposition of the signal is shown in Figure 6.11: the second harmonic and mean flow distortion are substantial but higher harmonics are still insignificant.

Nishioka, Iida & Ichikawa (1975) attempted to measure the threshold

Figure 6.11. Fourier components of u' in Figure 6.9, for 3% disturbance amplitude; from Fasel & Bestek (1980). 0, mean flow distortion; 1, fundamental; 2, second harmonic; 3, third harmonic.

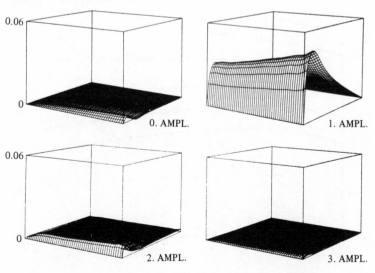

amplitude for subcritical instability of plane Poiseuille flow. Disturbances of prescribed amplitude, introduced by a vibrating ribbon, either grow or decay sufficiently far downstream (Figure 6.12). Measurements of the threshold amplitude at $R = 4000$, 5000 and 6000 are shown in Figure 6.13(a). The surprising decrease in threshold amplitude observed at larger ω is believed to be due to three-dimensional effects. Herbert's theoretical estimates, on retaining just two harmonics, are given in Figure 6.13(b) for comparison. However, in view of the large time, or the correspondingly large downstream distance, which is apparently necessary for true equilibrium states to emerge, it is arguable that accurate experimental determination of these two-dimensional threshold equilibria is unattainable in any apparatus yet constructed.

An earlier numerical study by Zabusky & Deem (1971) concerns rapidly growing disturbances to velocity profiles representative of symmetric

Figure 6.12. Amplitude variations, with downstream distance $x - x_0$, in plane Poiseuille flow at $R = 5000$ and $\omega = 0.323$. Experimental data of Nishioka *et al.* (1975). Note eventual decay of small initial amplitudes and subcritical instability of larger ones. ----- denotes Herbert's (1977) theoretical threshold amplitude. λ_{TS} is the wavelength (from Herbert 1982).

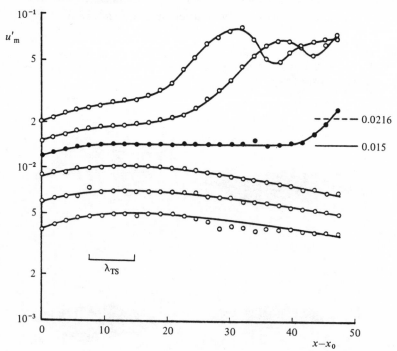

wakes and jets. They studied temporal evolution of spatially-periodic two-dimensional disturbances. Many features were found to be in agreement with laboratory experiments of Sato & Kuriki (1961) on the wake behind a thin flat plate, even though the latter concerned spatial, not temporal, growth. Such features include the development of a double vortex street of counter-rotating elliptical vortices which themselves nutate to produce a low-frequency oscillation. In contrast, unstable shear-layer profiles such as (20.3) yield a single street of like-rotating vortices, which subsequently

Figure 6.13. (a) Predicted threshold amplitudes at various frequencies ω, in plane Poiseuille flow at $R = 4000$, 5000 and 6000, given by Herbert's (1977) $N = 2$ truncation (cf. Figure 6.4). (b) Measured threshold amplitudes versus ω, at the same values of R, found by Nishioka *et al.* (1975).

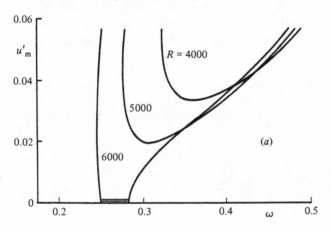

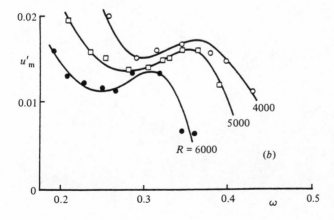

interact in pairs (Winant & Browand 1974). Similar structures have been reproduced in the numerical studies of Patnaik, Sherman & Corcos (1976) and Peltier, Hallé & Clark (1978). An example is shown in Figure 6.14.

The successes of these various numerical studies give confidence in their capacity to represent finite disturbances accurately beyond the range of weakly nonlinear analysis. Nevertheless, weakly nonlinear theories will long have a part to play in aiding physical understanding of nonlinear processes, in supplying a check on numerical results and, hopefully, in providing a source of fresh ideas for future progress.

22.4   *The nonlinear critical layer*

Obviously, weakly nonlinear theory cannot cope with strong nonlinearities. For waves in shear flows, these first appear in the vicinity of the critical layer $z = z_c$. The linear theory for waves of constant amplitude, outlined in §3.2, led to a jump in Reynolds stress across this critical layer. A logarithmic singularity of inviscid theory was resolved by

Figure 6.14. Computed constant-density contours at various times of development of Kelvin–Helmholtz instability of a viscous stratified shear layer. Dimensionless times τ in (a)–(f) are 0, 0.5, 1.0, 1.5, 2.0 and 2.42 (from Patnaik, Sherman & Corcos 1976).

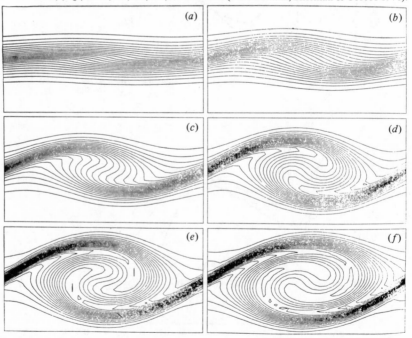

viscosity, at large R, a phase shift of π from the logarithmic term occurring across a viscous critical layer of width $O[(\alpha R)^{-\frac{1}{3}}]$.

The width of the Kelvin 'cat's eyes' within the critical layer is $O(\epsilon^{\frac{1}{2}})$ where ϵ is the wave amplitude at some distance from z_c (Drazin & Reid 1981, p. 141). Neglected nonlinear terms are unimportant only if the viscous critical layer thickness far exceeds the width of the cat's eyes: that is to say, $\lambda = R^{-1}\epsilon^{-\frac{3}{2}}$ must be large compared with unity for linear and weakly nonlinear theories to remain applicable.

The contrary assumption, that both $\lambda \leqslant O(1)$ and $\epsilon \ll 1$, underlies the theory of the nonlinear critical layer. This was first developed independently by Benney & Bergeron (1969) and Davis (1969), for $\lambda \to 0$, and subsequently improved and extended by Haberman (1972, 1976). A weakly nonlinear theory still suffices outside the critical layer region of thickness $O(\epsilon^{\frac{1}{2}})$ but strong nonlinearities are present within it. For $\lambda = 0$ the logarithmic phase shift ϕ, and consequently the jump in Reynolds stress, is zero, whereas the linear phase shift is π. Haberman's (1972) computations show how ϕ increases from 0 to π as λ increases from zero to large values. For large λ, Haberman (1976) has shown analytically that

$$\phi = -\pi[1 - \tfrac{1}{2}(\tfrac{3}{2})^{\frac{1}{3}}\,\Gamma(\tfrac{4}{3})\,\lambda^{-\frac{4}{3}} + O(\lambda^{-\frac{8}{3}})].$$

An analogous theory, differing in details, has been developed by Kelly & Maslowe (1970) and Haberman (1973b) for weakly stratified flows; while stronger stratification was investigated numerically by Maslowe (1972). Benney & Maslowe (1975) and Huerre & Scott (1980) incorporate nonlinear critical layers in their studies of marginally unstable waves in shear layers.

The critical-layer solutions exhibit closed-streamline 'cat's eyes', which, by virtue of the Prandtl–Batchelor theorem (Batchelor 1956), enclose regions of constant vorticity. Towards the edges of these cat's eyes, it is postulated that there are thin viscous layers across which the vorticity (and, in the stratified case, density) changes, in order to match with the flow in the outer regions. Though some doubts remained regarding the detailed flow within these layers, these studies made an important contribution to the analysis of nonlinear waves. A useful summary of the analytical details is given by Maslowe (1981).

Since then, Smith (1979b), Smith & Bodonyi (1982a) and Bodonyi, Smith & Gajjar (1983) have further elucidated the complex mathematical structure of finite neutrally-stable disturbances in boundary layers as $R \to \infty$. Separate detailed analyses are necessary in various parameter ranges; as the amplitude increases, the character of the critical layer changes from a nonlinear viscous one, like Haberman's (1972), to a strongly nonlinear, primarily inviscid, sort. Corresponding amplitude-dependent neutral disturbances in Poiseuille pipe flow were studied by Smith & Bodonyi (1982b).

Precisely how, and whether, a linearly unstable mode or a finite initial disturbance evolves in time towards such an equilibrium state is a matter of some complexity. It is unlikely that such equilibrium states could persist in real boundary-layer flows at the very large Reynolds numbers needed for the validity of the asymptotic theory. These solutions certainly *exist* and have an interest of their own; but their stability to other wave-modes – and so their practical significance – remains unknown.

There have been some related studies of the temporal evolution of finite-amplitude disturbances. For the shear-layer profile (20.3), Robinson (1974) developed a nonlinear inviscid analysis for rapidly growing waves which do not equilibrate. Later, Warn & Warn (1978), Stewartson (1978) and Brown & Stewartson (1978a, b) provided much insight into the development of marginally unstable Rossby waves and those in stratified shear layers. This work is conveniently reviewed by Stewartson (1981). A recent analysis by Gajjar & Smith (1985) considers unsteady nonlinear critical layers in channel and boundary-layer flows.

Brown & Stewartson (1978b) give an inviscid analysis of the marginally unstable stratified shear flow (5.1) with Richardson number J very close to $\frac{1}{4}$. They demonstrate how nonlinear critical-layer effects develop on the time scale $t = O(\epsilon^{-\frac{2}{3}})$, the wave amplitude $\tilde{A} = \epsilon A(t)$ then satisfying an evolution equation

$$\mathrm{d}\tilde{A}/\mathrm{d}t = (\tfrac{1}{4}-J)\,\tilde{A} + \alpha_4\,t^6\,|\,\tilde{A}\,|^4\,\tilde{A}$$

where α_4 is a small positive constant, approximately 10^{-6}. The unexpected appearance of the factor t^6 is at variance with Maslowe's (1977a) analysis for a viscous shear layer, which led to a more conventional Landau–Stuart equation of the form (8.8). This apparent discrepancy was reconciled by Brown, Rosen & Maslowe (1981). The details are complex, being sensitive to the relative strengths of viscous and thermal diffusion and of nonlinear effects, on various evolutionary time scales. Depending on circumstances, Maslowe's (1977a) cubic nonlinearity or Brown & Stewartson's quintic one may dominate in the viscous problem.

Numerical studies by Béland (1976, 1978) of forced nonlinear Rossby waves display their development and near-equilibration. The logarithmic phase shift ϕ is found to agree with Haberman's (1972) prediction: initial absorption of the incident wave at the critical layer is later replaced by reflection but virtually no transmission. The flow structure in the vicinity of the critical layer resembles that of Figure 6.14.

Similarly, with strong stratification, an internal gravity wave approaching the critical level where $U = c_r$ initially undergoes absorption, along with just a little reflection and transmission. The linear theory of Booker & Bretherton (1967) and others was described in § 5.3: their transmission and reflection coefficients are $O(e^{-\nu\pi})$, where $\nu = (J-\frac{1}{4})^{\frac{1}{2}}$, and these are small for all moderately large values of J. In the absence of dissipation, nonlinearity must eventually become important within the critical layer, since wave energy steadily accumulates in its vicinity.

Brown & Stewartson (1980, 1982a, b) and Burke (1983) have shown that the linear results remain valid on the time scale $1 \ll t \ll \epsilon^{-\frac{2}{3}}$; but that, when t is $O(\epsilon^{-\frac{2}{3}})$, nonlinear modification of the critical layer causes the reflection and transmission coefficients to develop with time. When $\nu \gg 1$, leading-order corrections are found: the transmission coefficient remains exponentially small, $O(e^{-\nu\pi})$, but the reflection coefficient increases to $O(\nu^{-1})$. Accordingly, the critical layer begins to restore wave energy to the outer flow by reflection, while continuing to act as an absorber and a very weak transmitter. These results are in qualitative agreement with the numerical studies of Breeding (1971, 1972) and of Jones & Houghton (1972). Experiments of Thorpe (1981) and Koop (1981) confirm that there is little transmission through the critical layer, but show no clear evidence of the above nonlinear effects.

The nonlinear critical theory has been developed only for a single dominant two-dimensional wave-mode. Though linear superposition of solutions remains permissible at leading order outside the critical layer region, this is not so within it. With two or more modes present, there must be strong nonlinear coupling within critical layers: as yet, no attempt has been made to construct a theory for such cases. Similarly, the stability of the nonlinear critical layer deserves attention: its similarity to a row of equally-spaced vortices suggests that instability and subsequent 'vortex pairing' may occur.

22.5   *Taylor–Couette flow and Rayleigh–Bénard convection*

These two are among the most fascinating of fluid flows, despite their deceptively simple geometry. In Taylor–Couette flow, the fluid is confined to the gap between concentric, differentially-rotating cylinders; in Bénard convection, fluid is confined between plane horizontal walls, and would be at rest were the temperature of the lower wall not higher than that of the upper.

In the simplest theoretical formulations, the cylinder lengths and the horizontal walls are regarded as infinite. If, also, the gap between the cylinders is small compared with the radii of the cylinders, the linear theory for small disturbances in the two flows takes identical form (see e.g. Chandrasekhar 1961; Drazin & Reid 1981). Instability gives rise to a stationary array of counter-rotating toroidal cells between cylinders, and to parallel roll cells between horizontal heated walls, the former driven by centrifugal forces and the latter by buoyancy. Since the roll cells of thermal convection have no preferred direction of orientation, composite cells comprising more than one Fourier mode in space commonly arise. The configuration of the side walls of the apparatus, together with weak nonlinearities, influences the type of convection cell, rectangular and hexagonal ones being most common.

For these flows, the prediction of onset of instability, by linear theory, agrees remarkably well with experiment. This is so largely because, near the critical point of linear theory, weak nonlinear terms have a stabilizing effect: the nonlinear equilibrium solution bifurcates supercritically from the linear critical point because the Landau constant λ_r (cf. §8) is negative.

A plethora of recent reviews describes the large quantity of analytical, computational and experimental work concerning these two flow configurations: see Palm (1975), Normand, Pomeau & Velarde (1977), Busse (1978, 1981a, b), DiPrima & Swinney (1981), DiPrima (1981) and Benjamin (1981). Further details are given in §§24–25 of the following chapter.

Chapter seven

CUBIC THREE- AND FOUR-WAVE INTERACTIONS

23 Conservative four-wave interactions

23.1 The resonance condition

For some systems, no three-wave resonance is possible and the strongest interactions take place among resonant quartets of waves. The dominant nonlinearities are then of cubic order in wave amplitude and so are much weaker than the quadratic nonlinearities of three-wave resonance. But even in systems which admit three-wave resonance, phenomena associated with four-wave resonances may be significant. For example, a single capillary–gravity wave resonating with its own third harmonic, mentioned in § 14.2, is a degenerate four-wave resonance. So, too, is the self-modulation of nearly uniform wave-trains examined in the previous chapter, for then the 'four' participating modes are all identical, or virtually so.

The general conditions for four-wave resonance are, of course,

$$\left.\begin{array}{l} \mathbf{k}_1 \pm \mathbf{k}_2 \pm \mathbf{k}_3 \pm \mathbf{k}_4 = 0, \\ \omega_1 \pm \omega_2 \pm \omega_3 \pm \omega_4 = 0, \end{array}\right\} \tag{23.1}$$

with corresponding signs, for participating modes with periodicities $\exp \pm i(\mathbf{k}_j \cdot \mathbf{x} - \omega_j t)$ $(j = 1, 2, 3, 4)$ and $\mathbf{x} = (x, y)$. The most studied resonance is that among deep-water gravity waves. Since these do not admit three-wave resonance, four-wave interactions provide the dominant coupling between modes. This plays an important rôle in establishing the spectrum of ocean waves, for direct wave-generation by wind operates effectively only at rather short wavelengths (Phillips 1977). Three- and four-wave resonance is also important in linking modes of different physical types, especially in plasma physics, where a great diversity of waves is available (see, e.g. Davidson 1972).

231

Phillips (1960, 1961) was the first to study four-wave resonance among gravity waves, and he has since given several accounts of the subject: see, especially, his monograph (Phillips 1977) and retrospective article (Phillips 1981a). Other early work was that of Longuet-Higgins (1962), Benney (1962) and Hasselman (1962, 1963).

The dispersion relation for deep-water gravity waves is $\omega = \pm (g\,|\,\mathbf{k}\,|)^{\frac{1}{2}}$, which defines an axisymmetric double trumpet-shaped surface in ω–k–l space, where $\mathbf{k} = (k, l)$. If a similar surface is constructed, with different origin O' as in Figure 7.1, they intersect in some curve. Any two points on this curve define four wavenumbers $\mathbf{k}_j$ (those shown in the figure having $l = 0$) which satisfy the resonance condition

$$\left.\begin{array}{l} \mathbf{k}_1 + \mathbf{k}_2 = \mathbf{k}_3 + \mathbf{k}_4, \\ \omega_1 + \omega_2 = \omega_3 + \omega_4. \end{array}\right\} \qquad (23.2a, b)$$

Movement of O' parallel to the ω-axis yields a family of such curves. Their

Figure 7.1. Diagram in ω–k plane, with $l = 0$, which determines a resonant quartet of surface gravity waves. The corresponding intersecting axisymmetric surfaces in ω–k–l space are readily envisaged (from Phillips 1974).

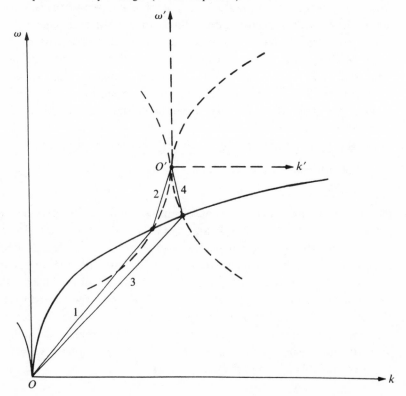

projection on the k–l plane is shown in Figure 7.2 and this denotes the complete family of resonance curves if its scale and orientation is regarded as arbitrary. This is because the dispersion relation for gravity waves has no preferred direction and ω is proportional to a single power of $|\mathbf{k}|$. For other dispersion relations, as for capillary–gravity waves or (stable) Kelvin–Helmholtz flow (see §2.1), variations with scale or direction occur and so all possible origins O' must be considered in analogous graphical constructions. The latter flows also admit stronger three-wave resonances.

23.2   The temporal evolution equations

For progressive gravity waves with wavenumbers satisfying condition (23.2a), let $A_j(t)$ ($j = 1, 2, 3, 4$) denote the complex amplitudes of the free-surface displacements associated with each wave. The evolution equations are found to have the form (Benney 1962; Bretherton 1964)

$$
\left.
\begin{aligned}
\mathrm{d}A_1/\mathrm{d}t &= \mathrm{i}A_1 \sum_{j=1}^{4} a_{1j} |A_j|^2 + \mathrm{i}h\omega_1 A_2^* A_3 A_4 \, \mathrm{e}^{-\mathrm{i}\,\Delta\omega t}, \\
\mathrm{d}A_2/\mathrm{d}t &= \mathrm{i}A_2 \sum_{j=1}^{4} a_{2j} |A_j|^2 + \mathrm{i}h\omega_2 A_1^* A_3 A_4 \, \mathrm{e}^{-\mathrm{i}\,\Delta\omega t}, \\
\mathrm{d}A_3/\mathrm{d}t &= \mathrm{i}A_3 \sum_{j=1}^{4} a_{3j} |A_j|^2 + \mathrm{i}h\omega_3 A_1 A_2 A_4^* \, \mathrm{e}^{\mathrm{i}\,\Delta\omega t}, \\
\mathrm{d}A_4/\mathrm{d}t &= \mathrm{i}A_4 \sum_{j=1}^{4} a_{4j} |A_j|^2 + \mathrm{i}h\omega_4 A_1 A_2 A_3^* \, \mathrm{e}^{\mathrm{i}\,\Delta\omega t}, \\
\Delta\omega &\equiv \omega_1 + \omega_2 - \omega_3 - \omega_4,
\end{aligned}
\right\}
\tag{23.3}
$$

Figure 7.2. Curves in k–l wavenumber plane given by resonance condition (23.2). Lines from A and B to any pair of points on one such curve determine a resonant quartet as shown (after Phillips 1977).

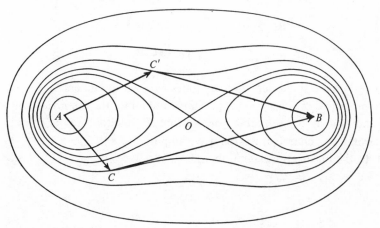

when interactions of higher than cubic order are ignored. If the resonance condition (23.2b) is not nearly met, the terms in h contain rapidly-oscillating exponential factors and these average to zero over the modulation timescale. Only if the mismatch $\Delta\omega$ from resonance is small need these terms be retained.

The sixteen elements of the matrix a_{ij} are real and so also is h. Their values depend on the wavenumber configuration and determination of these is algebraically tedious. The terms in a_{ij} modify the phases but not the magnitudes of the amplitudes A_i: that is, they provide a nonlinear modification to the phase velocities. The diagonal terms a_{jj} represent Stokes' (1847) frequency shift due to self-interaction. Only the h terms act to transfer energy between modes.

Benney's (1962) derivation employs techniques of perturbation theory which are now standard. The governing equations are as (14.2), but with surface tension γ set equal to zero. The two free-surface boundary conditions are just (11.13) and (11.14), which contain all nonlinearities. These may be replaced by Taylor expansions about the mean surface $z = 0$, namely,

$$\left.\begin{aligned}\zeta_t-\phi_z+(\zeta\phi_x)_x+(\zeta\phi_y)_y+\tfrac{1}{2}(\zeta^2\phi_{xz})_x+\tfrac{1}{2}(\zeta^2\phi_{yz})_y = 0,\\[2mm] g\zeta+\phi_t+\zeta\phi_{zt}+\tfrac{1}{2}|\nabla\phi|^2+\tfrac{1}{2}\zeta^2\phi_{zzt}+\tfrac{1}{2}\zeta(|\nabla\phi|^2)_z = 0,\end{aligned}\right\} \quad (z = 0) \quad (23.4\text{a, b})$$

correct to cubic order in wave amplitude. Elimination of ζ then yields a single boundary condition for ϕ.

A velocity potential, satisfying Laplace's equation and the boundary condition at $z = -\infty$,

$$\begin{aligned}\phi = \epsilon\, &\mathrm{Re}\sum_l P_l(t)\, \mathrm{e}^{\mathrm{i}\mathbf{k}_l\cdot\mathbf{r}}\, \mathrm{e}^{|\mathbf{k}_l|z}+\epsilon^2 P_{0,0}(t)\\ &+\epsilon^2\, \mathrm{Re}\sum_{l,m} P_{l,m}(t)\, \mathrm{e}^{\mathrm{i}(\mathbf{k}_l+\mathbf{k}_m)\cdot\mathbf{r}}\, \mathrm{e}^{|\mathbf{k}_l+\mathbf{k}_m|z}\\ &+\epsilon^3\, \mathrm{Re}\sum_{l,m,n} P_{l,m,n}(t)\, \mathrm{e}^{\mathrm{i}(\mathbf{k}_l+\mathbf{k}_m+\mathbf{k}_n)\cdot\mathbf{r}}\, \mathrm{e}^{|\mathbf{k}_l+\mathbf{k}_m+\mathbf{k}_n|z}+\dots,\end{aligned}$$

describes a discrete set of wave-modes with wavenumbers $\mathbf{k}_l$ ($l = 1, 2, \dots$) in propagation space $\mathbf{r} = (x, y)$. Substitution into the nonlinear boundary condition at $z = 0$ enables the various P-functions to be found. The parameter ϵ characterizes the wave amplitudes.

In the linear approximation, (23.4a, b) reduces to

$$\phi_{tt}+g\phi_z = 0 \quad (z = 0)$$

and so

$$\mathrm{d}^2 P_l/\mathrm{d}t^2+g\,|\mathbf{k}_l|\,P_l = 0,$$

which yields two primary-wave frequencies $\pm(g\,|\,\mathbf{k}_l\,|)^{\frac{1}{2}}$. For propagating wave-trains, one or other value is selected, say ω_l for each $\mathbf{k}_l$: for standing waves, both must be retained as in §14.2. At $O(\epsilon^2)$,

$$\frac{d^2P_{l,m}}{dt^2}+g\,|\,\mathbf{k}_l+\mathbf{k}_m\,|\,P_{l,m} = 2(\mathbf{k}_l\cdot\mathbf{k}_m-|\,\mathbf{k}_l\,|\,|\,\mathbf{k}_m\,|)\,\frac{d}{dt}\,(P_l\,P_m). \quad (23.5)$$

Here, both sums and differences of the primary wavenumbers are retained by allowing m to take both positive and negative values, with $\mathbf{k}_{-m}=-\mathbf{k}_m$, $\omega_{-m}=-\omega_m$ and $P_{-m}=P_m^*$, the complex conjugate of P_m. Since no three-wave resonance is possible, free modes with frequencies $\pm(g\,|\,\mathbf{k}_l+\mathbf{k}_m\,|)^{\frac{1}{2}}$, for which the left-hand side of (23.5) vanishes, may be disregarded. The remaining, forced, solution for $P_{l,m}$ has periodicity $\omega_l+\omega_m$ and is proportional to $P_l\,P_m$: this is readily found from (23.5).

On proceeding to $O(\epsilon^3)$, terms in $\exp(i\mathbf{k}_l\cdot\mathbf{r})$ may be extracted from the nonlinear boundary conditions. The corresponding equation for P_l has the form

$$d^2P_l/dt^2+\omega_l^2\,P_l = i\epsilon^2\,e^{-i\omega_l t}\,F_l \quad (l=1,2,\ldots) \quad (23.6)$$

for each primary wave amplitude. The nonlinear interaction terms represented by F_l derive from various sources. Those proportional to $|\,P_k\,|^2\,P_l$ and $P_{l,-k}\,P_k$ automatically have the correct periodicity; others, proportional to $P_r\,P_m\,P_n$ and $P_r\,P_{m,n}$, have periodicity

$$e^{i(\mathbf{k}_r+\mathbf{k}_m+\mathbf{k}_n-\mathbf{k}_l)\cdot\mathbf{r}}\,e^{-i(\omega_r+\omega_m+\omega_n-\omega_l)t}$$

and these terms match the periodicity of P_l only if the resonance conditions (23.1) are satisfied. With $\epsilon P_j(t)$ re-expressed as $(-i\omega_j\,|\,\mathbf{k}_j\,|^{-1})\,A_j(t)\,e^{-i\omega_j t}$, where the A_j denote slowly-varying complex amplitudes of the free-surface displacements, (23.6) yield

$$dA_j/dt = -\tfrac{1}{2}ig^{-1}F_j \quad (j=1,2,3,4).$$

With four wavenumbers satisfying (23.2a), these are just the evolution equations (23.3) above. The complicated expressions for the coefficients a_{ij} and h are given by Benney (1962).

23.3   *Properties of the evolution equations*
Equations (23.3) satisfy the conservation relations

$$\frac{d}{dt}\left(\frac{|\,A_1\,|^2}{\omega_1}\right)=\frac{d}{dt}\left(\frac{|\,A_2\,|^2}{\omega_2}\right)=\frac{d}{dt}\left(\frac{-|\,A_3\,|^2}{\omega_3}\right)=\frac{d}{dt}\left(\frac{-|\,A_4\,|^2}{\omega_4}\right) \quad (23.7)$$

which connect the wave actions of the modes. It follows that total wave action is conserved,

$$\frac{d}{dt}\left(\sum_{j=1}^{4}\frac{|\,A_j\,|^2}{\omega_j}\right)=0;$$

a result which is *not* true of three-wave resonance. Also, the energy equation is

$$\frac{\mathrm{d}E_{\text{tot}}}{\mathrm{d}t} = \frac{\mathrm{d}}{\mathrm{d}t}\left(\sum_{j=1}^{4} |A_j|^2\right) = \mathrm{i}h(\omega_1 + \omega_2 - \omega_3 - \omega_4)(A_1^* A_2^* A_3 A_4 - A_1 A_2 A_3^* A_4^*)$$

and this also vanishes by virtue of the resonance condition (23.2). As for the three-wave case, E_{tot} is not constant when there is a small detuning $\Delta\omega = \omega_1 + \omega_2 - \omega_3 - \omega_4$ from exact resonance, despite the fact that the system is non-dissipative.

Because of (23.7), the four evolution equations (23.3) may be reduced to an equation for a single variable $x(t)$ just as in §15.1. Complete analytical solutions may then be given in terms of Jacobian elliptic functions (Bretherton 1964; Turner & Boyd 1978).

The extension to spatial, as well as temporal, modulations is accomplished on replacing $\mathrm{d}A_j/\mathrm{d}t$ by $(\partial/\partial t + \mathbf{v}_j \cdot \nabla)A_j$ in (23.3), where $\mathbf{v}_j$ denote the respective group velocities $\partial\omega_j/\partial\mathbf{k}_j$: but this is justified only for waves in deep water. With finite depth, spatial modulations in amplitude drive an $O(\epsilon^2)$ mean flow as described in §11.2 and this must be taken into account when deriving the $O(\epsilon^3)$ interaction equations. Apparently, little work has been done on this problem. In the deep-water case, uniformly-propagating modulations which depend on a *single* variable, say $\tau \equiv t - ax - by$ with constant a and b, obey equations of similar form to (23.3) and so may be solved. Likewise, waves with equal linear damping rates may be treated by transforming to a new 'time' co-ordinate, as in §16.2, which recovers equations of undamped form (Bingham & Lashmore-Davies 1979).

Initial scepticism of the importance of resonances among gravity waves was dispelled by the experiments of Longuet-Higgins & Smith (1966) and McGoldrick, Phillips, Huang & Hodgson (1966). Two wave-trains, with mutually perpendicular wavenumbers $\mathbf{k}_1$, $\mathbf{k}_2$ and corresponding frequencies ω_1, ω_2 were generated mechanically from adjacent sides of a square wave tank. When these were chosen to satisfy the resonance conditions

$$2\mathbf{k}_1 = \mathbf{k}_2 + \mathbf{k}_3, \quad 2\omega_1 = \omega_2 + \omega_3,$$

the oblique mode $\mathbf{k}_3$ was seen to grow spontaneously with distance across the tank. Quite good agreement was obtained with the theoretical initial growth rate of $\mathbf{k}_3$ but the tank was not large enough to confirm the existence of periodic modulations. At other, non-resonant, choices of $\mathbf{k}_1$ and $\mathbf{k}_2$, no significant $\mathbf{k}_3$ component was generated. Figure 7.3, from McGoldrick *et al.*, shows measured frequency spectra for resonant and non-resonant cases.

Subsequent experiments on the modulational instabilities of initially uniform wave-trains, already described in §21, demonstrated the resonant growth of sidebands. At the other extreme, measurements of the evolution of ocean-wave frequency spectra (see e.g. Phillips 1977) support the view that resonance is responsible for broadening the spectrum to low frequencies which cannot be generated directly by wind.

23.4   *Zakharov's equation for gravity waves*

Hasselmann (1962, 1963) was first to consider resonant interactions within a continuous spectrum of gravity waves. Later, in an important but

Figure 7.3. Gravity-wave frequency spectra measured by McGoldrick *et al.* (1966), with two forced components f_1, f_2. Four-wave resonance occurs with $f_1/f_2 = 1.775$ at component with frequency $2f_1 - f_2$. This component is not prominent in the non-resonant case $f_1/f_2 = 1.600$.

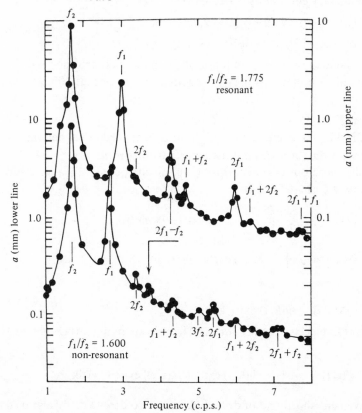

terse paper, Zakharov (1968) gave a rather different formulation. Only recently has Zakharov's work gained deserved recognition through its fuller rederivation by Crawford, Saffman & Yuen (1980), also described in Yuen & Lake (1982).

Instead of referring the surface boundary conditions to the mean level $z = 0$, Zakharov introduced the potential ϕ^s at the free surface as a dependent variable,

$$\phi^s(\mathbf{r}, t) = \phi(\mathbf{r}, \zeta(\mathbf{r}, t), t),$$

where $\mathbf{r}$ denotes (x, y), $z = \zeta(\mathbf{r}, t)$ is the free surface and $\phi(\mathbf{r}, z, t)$ the usual velocity potential. It follows that

$$\phi_t^s(\mathbf{r}, t) = \phi_t(\mathbf{r}, \zeta(\mathbf{r}, t), t) + \phi_z(\mathbf{r}, \zeta(\mathbf{r}, t), t)\,\zeta_t(\mathbf{r}, t),$$

$$\nabla_{\mathbf{r}}\phi^s(\mathbf{r}, t) = \nabla_{\mathbf{r}}\phi(\mathbf{r}, \zeta(\mathbf{r}, t), t) + \phi_z(\mathbf{r}, \zeta(\mathbf{r}, t), t)\,\nabla_{\mathbf{r}}\zeta(\mathbf{r}, t)$$

where $\nabla_{\mathbf{r}} \equiv (\partial/\partial x, \partial/\partial y)$.

The boundary conditions (14.2c, d) yield

$$\left.\begin{array}{l} \zeta_t + (\nabla_{\mathbf{r}}\phi^s)\cdot(\nabla_{\mathbf{r}}\zeta) - \phi_z[1 + (\nabla_{\mathbf{r}}\zeta)^2] = 0, \\[2mm] \phi_t^s + g\zeta + \tfrac{1}{2}(\nabla_{\mathbf{r}}\phi^s)^2 - \tfrac{1}{2}\phi_z^2[1 + (\nabla_{\mathbf{r}}\zeta)^2] = \gamma\kappa/\rho, \end{array}\right\} \qquad (23.8a, b)$$

and it remains to express ϕ_z in terms of ϕ^s and ζ. Zakharov showed that these boundary conditions may be rewritten in the elegant Hamiltonian form

$$\zeta_t = \delta E/\delta\phi^s, \quad \phi_t^s = -\delta E/\delta\zeta \qquad (23.9)$$

where E is the total energy and $\delta/\delta\phi^s$, $\delta/\delta\zeta$ denote variational derivatives; but this reformulation was not used by Crawford *et al.*

The (double) Fourier transforms of the velocity potential ϕ and surface displacement ζ are defined as

$$\hat{\phi}(\mathbf{k}, z, t) = \frac{1}{2\pi}\int_{-\infty}^{\infty}\phi(\mathbf{r}, z, t)\exp\left(-i\mathbf{k}\cdot\mathbf{r}\right)d\mathbf{r},$$

$$\hat{\zeta}(\mathbf{k}, t) = \frac{1}{2\pi}\int_{-\infty}^{\infty}\zeta(\mathbf{r}, t)\exp\left(-i\mathbf{k}\cdot\mathbf{r}\right)d\mathbf{r},$$

where

$$\hat{\phi}(\mathbf{k}, z, t) = \hat{\Phi}(\mathbf{k}, t)\exp\left(|\mathbf{k}|z\right)$$

by virtue of Laplace's equation (14.2a). The corresponding surface potential is

$$\phi^s(\mathbf{r}, t) = \frac{1}{2\pi}\int_{-\infty}^{\infty}\hat{\Phi}(\mathbf{k}, t)\exp\left[|\mathbf{k}|\zeta(\mathbf{r}, t)\right]\exp\left(i\mathbf{k}\cdot\mathbf{r}\right)d\mathbf{k}.$$

The convenient shorthand of $d\mathbf{r}$ and $d\mathbf{k}$ is used to denote $dx\,dy$ and $dk\,dl$ respectively, where $\mathbf{k} = (k, l)$. For sufficiently small surface displacements, the exponential in $|\mathbf{k}|\zeta$ may be expanded by Taylor series. With ζ expressed in terms of ξ, this yields

$$\phi^s(\mathbf{r}, t) = \frac{1}{2\pi} \int_{-\infty}^{\infty} \hat{\Phi}(\mathbf{k}) \exp(i\mathbf{k}\cdot\mathbf{r}) \, d\mathbf{k}$$

$$+ \frac{1}{(2\pi)^2} \iint_{-\infty}^{\infty} \hat{\Phi}(\mathbf{k}) \,|\mathbf{k}|\, \hat{\zeta}(\mathbf{k}_1) \exp[i(\mathbf{k}+\mathbf{k}_1)\cdot\mathbf{r}] \, d\mathbf{k} \, d\mathbf{k}_1$$

$$+ \frac{1}{(2\pi)^3} \iiint_{-\infty}^{\infty} \hat{\Phi}(\mathbf{k}) \tfrac{1}{2}|\mathbf{k}|^2 \hat{\zeta}(\mathbf{k}_1) \hat{\zeta}(\mathbf{k}_2) \exp[i(\mathbf{k}+\mathbf{k}_1+\mathbf{k}_2)\cdot\mathbf{r}] \, d\mathbf{k} \, d\mathbf{k}_1 \, d\mathbf{k}_2$$

$$+ \dots$$

with the time-dependence of $\hat{\Phi}$ and $\hat{\zeta}$ implied.

The Fourier transform $\hat{\phi}^s(\mathbf{k}, t)$ of $\phi^s(\mathbf{r}, t)$ is therefore found as a series of integrals involving $\hat{\Phi}$ and $\hat{\zeta}$. This may be inverted iteratively, giving $\hat{\Phi}$ in terms of $\hat{\phi}^s$ and $\hat{\zeta}$ as

$$\hat{\Phi}(\mathbf{k}) = \hat{\phi}^s(\mathbf{k}) - \frac{1}{2\pi} \iint_{-\infty}^{\infty} |\mathbf{k}_1| \, \hat{\phi}^s(\mathbf{k}_1) \, \hat{\zeta}(\mathbf{k}_2) \, \delta(\mathbf{k}-\mathbf{k}_1-\mathbf{k}_2) \, d\mathbf{k}_1 \, d\mathbf{k}_2$$

$$- \frac{1}{(2\pi)^2} \iiint_{-\infty}^{\infty} \mathbb{R} \, \hat{\phi}^s(\mathbf{k}_1) \, \hat{\zeta}(\mathbf{k}_2) \, \hat{\zeta}(\mathbf{k}_3) \, \delta(\mathbf{k}-\mathbf{k}_1-\mathbf{k}_2-\mathbf{k}_3) \, d\mathbf{k}_1 \, d\mathbf{k}_2 \, d\mathbf{k}_3 + \dots,$$

$$\mathbb{R} \equiv \tfrac{1}{4}|\mathbf{k}_1|(2|\mathbf{k}_1| - |\mathbf{k}-\mathbf{k}_1| - |\mathbf{k}-\mathbf{k}_2| - |\mathbf{k}_1+\mathbf{k}_2| - |\mathbf{k}_1+\mathbf{k}_3|),$$

where $\delta(\ )$ denotes the delta function

$$\delta(\mathbf{k}) = \frac{1}{(2\pi)^2} \int_{-\infty}^{\infty} \exp(i\mathbf{k}\cdot\mathbf{r}) \, d\mathbf{r}.$$

It follows that ϕ_z may be constructed as a convergent series in terms of $\hat{\phi}^s$ and $\hat{\zeta}$: accordingly, in the boundary conditions (23.8a, b), ϕ_z is now known in terms of $\phi^s(\mathbf{r}, t)$ and $\zeta(\mathbf{r}, t)$.

These boundary conditions may be combined into a single equation by defining the complex variable

$$b(\mathbf{k}, t) = (\tfrac{1}{2}\omega(\mathbf{k})|\mathbf{k}|^{-1})\, \hat{\zeta}(\mathbf{k}, t) + i(\tfrac{1}{2}|\mathbf{k}|\,\omega^{-1}(\mathbf{k}))\, \hat{\phi}^s(\mathbf{k}, t). \qquad (23.10)$$

The free surface ζ and potential ϕ are recovered as a sum and difference, respectively, of the inverse Fourier transforms of b and its conjugate b^*. The equation for $b(\mathbf{k}, t)$ is

$$b_t(\mathbf{k}) + i\omega(\mathbf{k})\, b(\mathbf{k}) + i \iint_{-\infty}^{\infty} V^{(1)}(\mathbf{k}, \mathbf{k}_1, \mathbf{k}_2)\, b(\mathbf{k}_1)\, b(\mathbf{k}_2)\, \delta(\mathbf{k}-\mathbf{k}_1-\mathbf{k}_2) \, d\mathbf{k}_1 \, d\mathbf{k}_2$$

$$+ i \iint_{-\infty}^{\infty} V^{(2)}(\mathbf{k}, \mathbf{k}_1, \mathbf{k}_2)\, b^*(\mathbf{k}_1)\, b(\mathbf{k}_2)\, \delta(\mathbf{k}+\mathbf{k}_1-\mathbf{k}_2) \, d\mathbf{k}_1 \, d\mathbf{k}_2$$

$$+ i \iint_{-\infty}^{\infty} V^{(3)}(\mathbf{k}, \mathbf{k}_1, \mathbf{k}_2)\, b^*(\mathbf{k}_1)\, b^*(\mathbf{k}_2)\, \delta(\mathbf{k}+\mathbf{k}_1+\mathbf{k}_2) \, d\mathbf{k}_1 \, d\mathbf{k}_2$$

(over)

$$+\mathrm{i}\iiint_{-\infty}^{\infty} W^{(1)}(\mathbf{k},\mathbf{k}_1,\mathbf{k}_2,\mathbf{k}_3)\,b(\mathbf{k}_1)\,b(\mathbf{k}_2)\,b(\mathbf{k}_3)\,\delta(\mathbf{k}-\mathbf{k}_1-\mathbf{k}_2-\mathbf{k}_3)\,\mathrm{d}\mathbf{k}_1\,\mathrm{d}\mathbf{k}_2\,\mathrm{d}\mathbf{k}_3$$

$$+\mathrm{i}\iiint_{-\infty}^{\infty} W^{(2)}(\mathbf{k},\mathbf{k}_1,\mathbf{k}_2,\mathbf{k}_3)\,b^*(\mathbf{k}_1)\,b(\mathbf{k}_2)\,b(\mathbf{k}_3)\,\delta(\mathbf{k}+\mathbf{k}_1-\mathbf{k}_2-\mathbf{k}_3)\,\mathrm{d}\mathbf{k}_1\,\mathrm{d}\mathbf{k}_2\,\mathrm{d}\mathbf{k}_3$$

$$+\mathrm{i}\iiint_{-\infty}^{\infty} W^{(3)}(\mathbf{k},\mathbf{k}_1,\mathbf{k}_2,\mathbf{k}_3)\,b^*(\mathbf{k}_1)\,b^*(\mathbf{k}_2)\,b(\mathbf{k}_3)\,\delta(\mathbf{k}+\mathbf{k}_1+\mathbf{k}_2-\mathbf{k}_3)\,\mathrm{d}\mathbf{k}_1\,\mathrm{d}\mathbf{k}_2\,\mathrm{d}\mathbf{k}_3$$

$$+\mathrm{i}\iiint_{-\infty}^{\infty} W^{(4)}(\mathbf{k},\mathbf{k}_1,\mathbf{k}_2,\mathbf{k}_3)\,b^*(\mathbf{k}_1)\,b^*(\mathbf{k}_2)\,b^*(\mathbf{k}_3)\,\delta(\mathbf{k}+\mathbf{k}_1+\mathbf{k}_2+\mathbf{k}_3)\,\mathrm{d}\mathbf{k}_1\,\mathrm{d}\mathbf{k}_2\,\mathrm{d}\mathbf{k}$$

$$+\ldots=0, \tag{23.11}$$

where the real interaction coefficients $V^{(j)}$, $W^{(j)}$ are known, but lengthy, functions of the wavenumbers $\mathbf{k}_j$ and frequencies $\omega_j(\mathbf{k}_i)$. These coefficients are stated by Crawford *et al.* (1980) (also Crawford *et al.* 1981 and Yuen & Lake 1982) who correct minor errors in Zakharov's account.

Zakharov (1968) derived this result rather differently. He first showed that Hamilton's equations (23.9) are equivalent to

$$\partial b/\partial t = -\mathrm{i}\,\delta E/\delta b^*,$$

then calculated the right-hand side from the energy functional E.

The leading-order approximation $b_t = -\mathrm{i}\omega b$ yields the linear result. The three double integrals in $V^{(j)}$ represent quadratic interactions among Fourier components. If we set

$$b(\mathbf{k}_j,t)=\epsilon B_j(\mathbf{k}_j,t)\exp[-\mathrm{i}\omega(\mathbf{k}_j)\,t]$$

where ϵ is a measure of the small wave-slope $|\mathbf{k}|\,\zeta$, (23.11) yields

$$\mathrm{i}\frac{\partial B}{\partial t}=\epsilon\left\{\iint_{-\infty}^{\infty} V^{(1)}B_1\,B_2\,\delta(\mathbf{k}-\mathbf{k}_1-\mathbf{k}_2)\exp[\mathrm{i}(\omega-\omega_1-\omega_2)\,t]\,\mathrm{d}\mathbf{k}_1\,\mathrm{d}\mathbf{k}_2\right.$$
$$\left.+I_2+I_3\right\}+O(\epsilon^2) \quad (23.12)$$

at quadratic order, where I_2 and I_3 denote corresponding integrals in $V^{(2)}$ and $V^{(3)}$. If no three-wave resonance is possible, the right-hand side is a rapidly-varying function of t, and B may be decomposed into a dominant $O(1)$ slowly-varying part with $O(\epsilon^2)$ time derivative and a small $O(\epsilon)$ rapidly-varying part, $\epsilon B'$ say. The latter may be found by integrating (23.12) directly with respect to t; for, only the dominant parts of B_1, B_2 need be retained in the integrals, and they are constant on the $O(1)$ timescale t. If three-wave resonance occurs, direct integration of (23.12) is impossible: the integrals have slowly-varying parts and B evolves on the timescale ϵt.

With no three-wave resonance, the slow evolution of B on the timescale $\epsilon^2 t$ is found by retaining the $O(\epsilon^3)$ terms of (23.11). Some of these derive from the $V^{(j)}$ integrals through products of the dominant $O(\epsilon)$ part of b and the known rapidly-varying part $\epsilon^2 B'$; others come from the $W^{(j)}$ integrals and are of cubic order in the dominant part of b. Again, many of these terms are rapidly-varying and may be 'filtered out' since they do not contribute to the slow modulation of B. In particular, the only resonances among deep-water gravity waves are those with $\omega + \omega_1 - \omega_2 - \omega_3 = 0$ (where the ω's are constrained to be positive without loss of generality) and so the integrals in $W^{(1)}$, $W^{(3)}$ and $W^{(4)}$ may be suppressed. The final result has the form

$$i\frac{\partial B}{\partial t} = \iiint_{-\infty}^{\infty} T(\mathbf{k}, \mathbf{k}_1, \mathbf{k}_2, \mathbf{k}_3)\, B_1^* \, B_2 \, B_3 \, \delta(\mathbf{k} + \mathbf{k}_1 - \mathbf{k}_2 - \mathbf{k}_3)$$

$$\times \exp\left[i(\omega + \omega_1 - \omega_2 - \omega_3)\, t\right] d\mathbf{k}_1 \, d\mathbf{k}_2 \, d\mathbf{k}_3, \quad (23.13)$$

where T is known in terms of the $V^{(j)}$ and $W^{(2)}$. This is Zakharov's integral equation for the slow evolution of a weakly nonlinear wave field. Only those modes for which $|\omega + \omega_1 - \omega_2 - \omega_3|$ is $O(\epsilon^2\omega)$ or less need be retained in the integral.

23.5   *Properties of Zakharov's equation*

Suppose that the Fourier spectrum $B(\mathbf{k}, t)$ comprises just four discrete resonant modes,

$$B(\mathbf{k}, t) = \sum_{n=1}^{4} B^{(n)}(t)\, \delta(\mathbf{k} - \boldsymbol{\kappa}_n)$$

with conditions (23.2) satisfied by the wavenumbers κ_n and frequencies ω_n. It is readily confirmed that (23.13) then yields results (23.3).

Alternatively, for a narrow spectrum closely centred around a single wavenumber $\mathbf{k}_0$ and frequency $\omega(\mathbf{k}_0)$, the dispersion relation is approximately

$$\omega(\mathbf{k}) = \omega(\mathbf{k}_0) + aK_1 + bK_1^2 + cK_2^2 + \ldots$$

where a, b, c are constants, K_1 is the component of $\mathbf{k} - \mathbf{k}_0$ parallel to $\mathbf{k}_0$ and K_2 that perpendicular to $\mathbf{k}_0$. Obviously, a is the group velocity of $\mathbf{k}_0$. In (23.13), $T(\mathbf{k}, \mathbf{k}_1, \mathbf{k}_2, \mathbf{k}_3)$ may be replaced by the constant, $T_0 = T(\mathbf{k}_0, \mathbf{k}_0, \mathbf{k}_0, \mathbf{k}_0)$, which equals $|\mathbf{k}_0|^3 / 4\pi^2$. Also, $B(\mathbf{k}, t)$ may be rewritten as

$$B(\mathbf{k}, t) = \hat{A}(\mathbf{k}, t)\, \exp\left[i(aK_1 + bK_1^2 + cK_2^2 + \ldots)\, t\right] \quad (23.14)$$

to incorporate the frequency shift. Equation (23.13) then gives

$$\partial \hat{A}/\partial t + i(aK_1 + bK_1^2 + cK_2^2 + \ldots)\, \hat{A}$$

$$= -iT_0 \iiint_{-\infty}^{\infty} \hat{A}_1^* \, \hat{A}_2 \, \hat{A}_3 \, \delta(\mathbf{k} + \mathbf{k}_1 - \mathbf{k}_2 - \mathbf{k}_3)\, d\mathbf{k}_1 \, d\mathbf{k}_2 \, d\mathbf{k}_3. \quad (23.15)$$

The inverse Fourier transform of $\hat{A}$ is

$$A(\mathbf{r}, t) = \frac{1}{2\pi} \int\int_{-\infty}^{\infty} \hat{A}(\mathbf{k}, t) \exp\left[i(K_1 x + K_2 y)\right] dK_1 dK_2$$

to leading order, with x measured along $\mathbf{k}_0$. Correspondingly, (23.15) gives

$$\frac{\partial A}{\partial t} + a \frac{\partial A}{\partial x} - \frac{i}{2}\left(b \frac{\partial^2 A}{\partial x^2} + c \frac{\partial^2 A}{\partial y^2}\right) = -iT_0 |A|^2 A,$$

which is just the nonlinear Schrödinger equation (19.3). This derivation was originally given by Zakharov.

Zakharov also examined the stability of periodic wave-trains (see §21.1) and this work was extended by Crawford *et al.* (1981). A solution of (23.13), which represents a uniform wave-train with wavenumber $\mathbf{k}_0 = (k_0, 0)$ is

$$B(\mathbf{k}, t) = \begin{cases} B_0 \exp\left(-iT_0 B_0^2 t\right), & \mathbf{k} = \mathbf{k}_0 \\ 0, & \mathbf{k} \neq \mathbf{k}_0. \end{cases}$$

The amplitude a_0 of the wave is given, from (23.10), by

$$B_0 = \pi a_0 (2\omega_0/k_0)^{\frac{1}{2}}.$$

If this wave-train is perturbed by two modes with wavenumbers $\mathbf{k}_0 \pm \mathbf{K}$ and small amplitudes $B_{\pm}(t)$, the equations for $B_{\pm}$ are found from (23.13) to be

$$i\,dB_+/dt = T_{1+} B_0^2 B_-^* \exp\left[-i(\Omega + 2T_0 B_0^2) t\right] + 2T_{2+} B_0^2 B_+,$$

$$i\,dB_-/dt = T_{1-} B_0^2 B_+^* \exp\left[-i(\Omega + 2T_0 B_0^2) t\right] + 2T_{2-} B_0^2 B_-,$$

$$T_{1\pm} = T(\mathbf{k}_0 \pm \mathbf{K}, \mathbf{k}_0 \mp \mathbf{K}, \mathbf{k}_0, \mathbf{k}_0), \quad T_{2\pm} = T(\mathbf{k}_0 \pm \mathbf{K}, \mathbf{k}_0, \mathbf{k}_0, \mathbf{k}_0 \pm \mathbf{K}),$$

$$\Omega = 2\omega(\mathbf{k}_0) - \omega(\mathbf{k}_0 + \mathbf{K}) - \omega(\mathbf{k}_0 - \mathbf{K}).$$

The substitution

$$B_+ = \hat{B}_+ \exp\left[-i(\tfrac{1}{2}\Omega + T_0 B_0^2) t - i\sigma t\right],$$

$$B_-^* = \hat{B}_-^* \exp\left[i(\tfrac{1}{2}\Omega + T_0 B_0^2) t - i\sigma t\right]$$

leads to an eigenvalue problem with eigenvalues σ given by

$$\sigma = (T_{2+} - T_{2-}) B_0^2 \pm \{-T_{1+} T_{1-} B_0^4 + [-\tfrac{1}{2}\Omega + B_0^2(T_{2+} + T_{2-} - T_0)]^2\}^{\frac{1}{2}}.$$
(23.16)

This result is correct to $O(k_0^2 a_0^2)$ for all wavenumbers $\mathbf{K}$ such that the frequency mismatch Ω from exact resonance is $O(T_0 B_0^2)$ or less. In contrast, the corresponding results (21.3) and (21.5), derived from the two- and three-dimensional nonlinear Schrödinger equations (21.1) and (19.3), are valid only for wavenumbers $\mathbf{K}$ small compared with $\mathbf{k}_0$: in particular, (21.5) incorrectly predicts instability for *all* L/K close to $\pm 2^{-\frac{1}{2}}$. Result (23.16) may also be derived from (23.3), with frequency mismatch $\Delta \omega = \Omega$, $A_1 = A_2 = B$ and $A_3, A_4 = B_+, B_-$.

Crawford *et al.* (1981), also Yuen & Lake (1982), compare the stability boundary $\text{Im}\{\sigma\} = 0$ with those of (21.3) and (21.5) and also with Longuet-Higgins' (1978a, b) computations from the full nonlinear equations (see §22.2). For two-dimensional disturbances, this comparison is shown in Figure 6.7: result (23.16) is a clear improvement on the Benjamin–Feir result (21.3) and gives reasonable agreement with the full nonlinear computations even at quite large wave-slopes.

For three-dimensional disturbances, the unstable region given by (23.16) is shown in Figure 7.4 for various wave-slopes: the fundamental wave-number is $\mathbf{k}_0 = (k_0, 0)$ and the perturbations have $\mathbf{k}_\pm = (k_0 \pm K_x, \pm K_y)$. As $k_0 a_0$ approaches zero, the region of instability is situated narrowly around Phillips' figure-of-eight curve for four-wave resonance with $2\mathbf{k}_0 - \mathbf{k}_+ - \mathbf{k}_- = 0$. For fairly small $k_0 a_0$, the unstable region resembles a pair of touching 'horseshoes' and, as $k_0 a_0$ further increases, the 'horseshoes' move apart due to stabilization of modes with small $|\mathbf{K}|$. The unstable region becomes increasingly concentrated near $\mathbf{K} = \pm 0.78\, \mathbf{k}_0$ and finally disappears when $k_0 a_0 \approx 0.5$. This reproduces, at least qualitatively, the stabilization of Type I instability at large wave-slopes found by McLean *et al.* (1981) and already discussed in §22.2. However, the Zakharov equation (23.13) is no longer a rational approximation at such large wave-slopes. Moreover, the Zakharov equation is incapable of predicting the higher-order Type II instability which corresponds to five-wave resonance. However, Zakharov's analysis has recently been extended by Stiassnie & Shemer (1984) to incorporate the next-order five-wave

Figure 7.4. Regions of instability in K_x–K_y space as given by (23.16), for various values of wave-slope $k_0 a_0$: ---- $k_0 a_0 = 0$, —— $k_0 a_0 = 0.01$, — — — $k_0 a_0 = 0.1$, —·— $k_0 a_0 = 0.4$, ····· $k_0 a_0 = 0.48$ (from Crawford *et al.* 1981).

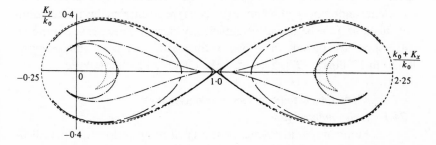

resonances in water of both infinite and finite depth: their extended equation represents Type II instability fairly satisfactorily.

In principle, the evolution of any finite number of discrete modes may be studied numerically, using Zakharov's equation. Caponi, Saffman & Yuen (1982) – repeated in Yuen & Lake (1982) – have done so for a seven mode system of two-dimensional waves, comprising a fundamental $\mathbf{k} = (k_0, 0)$ and six sidebands $\mathbf{k} = (k_n^\pm, 0)$, $k_n^\pm = k_0 \pm n \Delta k$ ($n = 1, 2, 3$): $\Delta k / k_0$ was chosen to be 0.2. Initially, a uniform wave-train with wavenumber k_0 and amplitude a_0 was prescribed, together with small $O(10^{-6} a_0)$ perturbations at the other wavenumbers. At sufficiently small $k_0 a_0$ ($= 0.05$), the motion was periodic and the sidebands remained small; but at $k_0 a_0 = 0.1$ quasi-periodic (Fermi–Pasta–Ulam) modulations occurred as a consequence of Benjamin–Feir instability of the $k_1^\pm$ sidebands. At $k_0 a_0 = 0.2$, both $k_1^\pm$ and $k_2^\pm$ lie in the unstable domain of Figure 6.7 and the modulations were found to be chaotically irregular. At $k_0 a_0 = 0.45$, the sidebands $k_1^\pm$, $k_2^\pm$ are stable but $k_3^\pm$ now lies in the unstable domain: then, seemingly periodic modulations at first occurred, but the $k_1^\pm$ and $k_2^\pm$ modes later grew to prominence (because of their resonance with k_0 and $k_3^\pm$) and chaotic behaviour ensued. At $k_0 a_0 = 0.5$, all the sidebands are stable and the periodic wave-train remained unmodulated – but Type II instabilities may then be expected. In accord with expectations, corresponding calculations based on the nonlinear Schrödinger equation (21.1) showed similar quasi-periodic and chaotic modulations, but no restabilization at large $k_0 a_0$ where (21.1) is not a good approximation.

Of much oceanographic interest is the evolution of a field of *random* gravity waves. Following on from Hasselmann's (1962, 1963) work, Longuet-Higgins (1976) and Alber (1978) considered the limiting case of a narrow-band spectrum, for which the nonlinear Schrödinger equation provides an acceptable starting point. In contrast, Crawford *et al.* (1980) employed Zakharov's equation to investigate a narrow-band random spectrum. All agree that Benjamin–Feir instability is suppressed by sufficient randomness, but a homogeneous spectrum remains modulationally unstable if the bandwidth is sufficiently narrow. Discussion of random wave spectra is outside the scope of this work: recent accounts may be found in Phillips (1977), Yuen & Lake (1982) and West (1981).

24   Mode interactions in Taylor–Couette flow

24.1   *Axisymmetric flow*

Finite-amplitude axisymmetric Taylor-vortex flow was first calculated by Davey (1962), by an amplitude-expansion method similar to that of Stuart (1960) and Watson (1960). Velocity perturbations have the form

$$\mathbf{u}(r, z, t) = \mathrm{Re}\{A(t)\,\mathbf{u}_1(r)\exp(\mathrm{i}\alpha z)\} + \mathrm{h.o.t.}$$

where higher-order terms include harmonics, mean-flow distortion and modifications of the leading-order eigenmode. Here, z and r denote distances along and perpendicular to the axis of the cylinders, which (meantime) are taken to have infinite length.

The amplitude equation has the typical Landau form (cf. §22.3)

$$\mathrm{d}A/\mathrm{d}t = \sigma A - a_1 A\,|A|^2 + a_2 A\,|A|^4 + \dots,$$

where $\sigma(T, \alpha)$ is the real eigenvalue of the corresponding linear stability problem. The Taylor number T is a dimensionless measure of the strength of the rotation, proportional to the square of a Reynolds number $R = \Omega_1\,r_1\,d/\nu$. Other parameters are the gap to inner radius ratio d/r_1 and the ratio of outer- to inner-cylinder rotation speeds Ω_2/Ω_1, which may be regarded as constants: the case $\Omega_2/\Omega_1 = 0$ and the small-gap approximation $d/r_1 \to 0$ have received most attention. The Taylor number is usually defined as

$$T = \frac{4\Omega_1\,d^4}{\nu^2}\left(\frac{r_1^2\Omega_1 - r_2^2\Omega_2}{r_2^2 - r_1^2}\right) = 4R^2\left(\frac{d}{r_1 + r_2}\right)\left(1 - \frac{\Omega_2\,r_2^2}{\Omega_1\,r_1^2}\right)$$

where $r_2 = r_1 + d$ is the outer radius.

At the critical Taylor number T_0 for onset of linear instability, σ changes sign; also, Davey (1962) found a_1 to be real and positive near T_0. Accordingly, there is a super-critical finite-amplitude equilibrium state

$$|A|^2 \approx \sigma/a_1 \quad (T \geqslant T_0)$$

with axial wavenumber $\alpha = \alpha_0$ equal to that of the least stable linear mode. Subsequently, DiPrima & Eagles (1977) calculated a_2, and several higher-order coefficients have been computed by Herbert (1981).

At each $T > T_0$, there is a finite bandwidth of linearly unstable wavenumbers α, centred on α_0, each of which may yield a finite-amplitude equilibrium state. For T only slightly greater than T_0, when the bandwidth is narrow, the stability of these states to 'sideband' disturbances with wavenumbers $\alpha \pm \delta$ has been investigated. This was first done by Kogelman & DiPrima (1970), following an earlier general approach of Eckhaus (1963, 1965). Stuart & DiPrima (1978) have noted the similarity of Eckhaus' mechanism to that of Benjamin & Feir (1967) already discussed in §21.1. It was found that wavenumbers sufficiently close to α_0 are stable equilibrium states, but that those outside a band of width $3^{-\frac{1}{2}}$ times that of the linearly-unstable bandwidth are not. A later analysis by Nakaya (1974), taken to higher order, found a slightly narrower bandwidth of stable vortex flows. The nonlinear coupling between the modes α, $\alpha+\delta$ and $\alpha-\delta$ is essentially a four-wave resonance with $\mathbf{k}_1 = \mathbf{k}_2 = \alpha$ and all four frequencies zero.

24.2 *Periodic wavy vortices*

The transition from axisymmetric Taylor vortices to wavy vortex flow (see Figure 7.5) is also a four-wave interaction phenomenon. Davey, DiPrima & Stuart (1968) considered the interaction of axisymmetric modes with azimuthal velocity components

$$v = v_{c0}(r, t) \cos \alpha z, \quad v_{s0}(r, t) \sin \alpha z$$

and two non-axisymmetric modes with

$$v = \mathrm{Re}\,\{v_{c1}(r, t) \cos \alpha z\, e^{in\phi}\}, \quad \mathrm{Re}\,\{v_{s1}(r, t) \sin \alpha z\, e^{in\phi}\}.$$

Here, v_{c0}, v_{s0} are real and v_{c1}, v_{s1} are complex functions, n is an integer and ϕ is the azimuthal angle. To leading order, the v-functions may be written as

$$v_{c0} \approx A_c(t) f_0(r), \quad v_{s0} \approx A_s(t) f_0(r),$$
$$v_{c1} \approx B_c(t) h_0(r), \quad v_{s1} \approx B_s(t) h_0(r)$$

where A_c, A_s are real and B_c, B_s are complex functions with complex conjugates B_c^*, B_s^*. Higher-order expansions may then be developed in ascending powers and products of the A's, B's and B^*'s. In this way, the third-order amplitude equations were found to be

$$
\left.
\begin{aligned}
\mathrm{d}A_c/\mathrm{d}t &= a_0 A_c + a_1 A_c(A_c^2 + A_s^2) + A_c(a_3 \,|\, B_c \,|^2 + a_4 \,|\, B_s \,|^2) \\
&\quad + \mathrm{Re}\,\{2a_5 A_s B_c B_s^*\}, \\
\mathrm{d}A_s/\mathrm{d}t &= a_0 A_s + a_1 A_s(A_c^2 + A_s^2) + A_s(a_3 \,|\, B_s \,|^2 + a_4 \,|\, B_c \,|^2) \\
&\quad + \mathrm{Re}\,\{2a_5 A_c B_s B_c^*\}, \\
\mathrm{d}B_c/\mathrm{d}t &= b_0 B_c + B_c(b_1 \,|\, B_c \,|^2 + b_2 \,|\, B_s \,|^2 + b_3 A_c^2 + b_4 A_s^2) \\
&\quad + (b_3 - b_4) B_s A_c A_s + (b_1 - b_2) B_c^* B_s^2, \\
\mathrm{d}B_s/\mathrm{d}t &= b_0 B_s + B_s(b_1 \,|\, B_s \,|^2 + b_2 \,|\, B_c \,|^2 + b_3 A_s^2 + b_4 A_c^2) \\
&\quad + (b_3 - b_4) B_c A_s A_c + (b_1 - b_2) B_s^* B_c^2,
\end{aligned}
\right\}
\qquad (24.1a\text{–}d)
$$

with known real a- and complex b-coefficients. Real and imaginary parts of the latter will be identified by subscripts r and i respectively.

The real A_c and A_s equations may be combined as a single complex equation for $A \equiv A_c + iA_s$. Also, $B_+ \equiv B_c - iB_s$ and $B_- \equiv B_c^* - iB_s^*$ may be identified as the respective complex amplitudes of modes with periodicities exp $(i\alpha z + in\phi)$, exp $(i\alpha z - in\phi)$. Equations (24.1) might therefore be recast as

$$
\left.
\begin{aligned}
dA/dt &= A(a_0 + a_1\,|\,A\,|^2 + c_2\,|\,B_+\,|^2 + c_3\,|\,B_-\,|^2) + c_4\,B_+\,B_-\,A^*,\\
dB_+/dt &= B_+(b_0 + d_1\,|\,A\,|^2 + d_2\,|\,B_+\,|^2 + d_3\,|\,B_-\,|^2) + d_4\,A^2B_-^*,\\
dB_-/dt &= B_-(b_0^* + d_1^*\,|\,A\,|^2 + d_3^*\,|\,B_+\,|^2 + d_2^*\,|\,B_-\,|^2) + d_5\,A^2B_+^*
\end{aligned}
\right\}
\qquad (24.1)'
$$

with the c- and d-constants known in terms of the a's and b's.

Figure 7.5. Taylor-vortex flows at various values of R/R_c where R_c is the linear critical Reynolds number. (a) $R/R_c = 1.1$, steady axisymmetric Taylor vortices; (b) $R/R_c = 6.0$ and (c) $R/R_c = 16.0$, wavy vortices; (d) $R/R_c = 23.5$, waves have disappeared (from Fenstermacher *et al.* 1979).

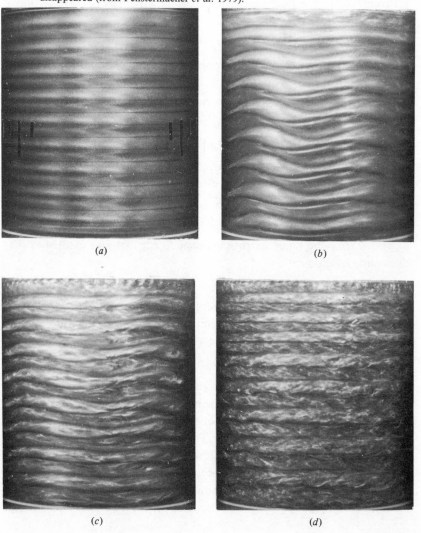

(a)

(b)

(c)

(d)

This demonstrates the connection with the four-wave resonance equations (23.3) with two modes identical; but now the coupling coefficients are no longer real and the modes experience linear growth or damping. Note that the linear frequencies of the modes A, B_+ and B_- are 0, b_{0i}, $-b_{0i}$ respectively and so satisfy (23.2b). A similar analysis by Nakaya (1975) allowed the axial wavenumber of the A- and B-modes to differ: the interaction criterion

$$2(\alpha, 0) - (\alpha + \delta, n) - (\alpha - \delta, -n) = 0$$

is met and this includes both the axisymmetric sideband instability ($n = 0$, $\delta \neq 0$) and wavy-vortex instability ($\delta = 0$, $n \neq 0$).

From (24.1), it is easily seen that axisymmetric Taylor vortices have $B_c = B_s = 0$, $A_s = CA_c$ and

$$A_c^2 + A_s^2 = K a_0\, e^{2a_0 t}(1 - K a_1\, e^{2a_0 t})^{-1}$$

for arbitrary constants C, K. The constant a_1 is known to be negative and there is a supercritical equilibrium state for $a_0 > 0$ ($T > T_0$). Similarly, a simple non-axisymmetric mode has $A_c = A_s = 0$ and

$$B_c = \beta_{ce}\, e^{i\omega(t - t_0)}, \quad B_s = \beta_{se}\, e^{i\omega(t - t_0)},$$

$$\beta_{ce}^2 + \beta_{se}^2 = \beta_e^2, \quad \beta_e = (-b_{0r}/b_{1r})^{\frac{1}{2}}, \quad \omega = b_{0i} - (b_{1i}\, b_{0r}/b_{1r}).$$

The stability of the axisymmetric equilibrium state to infinitesimal non-axisymmetric disturbances is established by linearizing (24.1) with respect to B_+ and B_-. Since A_c and A_s are then constant, with

$$A_s = CA_c, \quad A_c^2 + A_s^2 = -a_0/a_1 > 0,$$

B_c and B_s may be taken as proportional to $\exp(\lambda t)$ and λ satisfies the eigenvalue relationship

$$\lambda = b_0 - (a_0/2a_1)[b_3 + b_4 \pm (b_4 - b_3)(1 + C^4)^{\frac{1}{2}}(1 + C^2)^{-1}],$$

provided (24.1) may be truncated after the cubic terms. The largest growth rate λ_r occurs with C zero or infinity; that is, when $A_s = B_c = 0$ or $A_c = B_s = 0$. For these cases, the non-axisymmetric mode grows whenever

$$(-a_0/a_1) \sup(b_{3r}, b_{4r}) > -b_{0r}. \tag{24.2}$$

The right-hand side is the linear damping rate of this mode, and is positive near T_0. The coefficients b_0, b_3, b_4 depend on the azimuthal mode number n. As T increases above T_0, the equilibrium amplitude $(-a_0/a_1)^{\frac{1}{2}}$ of the Taylor vortices increases from zero. The criterion (24.2) therefore yields a set of critical Taylor numbers T_n ($n = 1, 2, 3, \ldots$) for onset of instability with azimuthal periodicity $\exp(\pm in\phi)$.

In the small gap approximation, the azimuthal wavenumber is n/r_1 since azimuthal arc length equals $r_1\phi$. Also, the critical axial wavenumber α_0

equals $3.127/d$ and this is much larger than n/r_1 for many integers n. Consequently, the wavenumber vectors $(\alpha_0, \pm n/r_1)$ of the lowest non-axisymmetric modes remain close to that of the fundamental and it comes as no surprise that the first few critical Taylor numbers T_n lie close to T_0.

Improved estimates of the T_n, based on retaining quintic terms in (24.1) and relaxing the small gap approximation, were given by Eagles (1971): his first few values, with Davey *et al.*'s (1968) added in parentheses, are $T_0 = 3506$ (3390), $T_1 = 3892$ (3670), $T_2 = 3902$ (3676), $T_4 = 3962$ (3710), for $\Omega_2/\Omega_1 = 0$, $d/r_1 = 0.05$ and $\alpha = \alpha_0 = 3.127/d$. Herbert (1981) later developed a computational amplitude-expansion approach (similar to that described in §22.3) which retained terms of up to 15th order for the small-gap limit. Jones (1981) employed an amplitude and Galerkin expansion to compute T_n for various gap ratios d/r_1. Their results are in reasonable agreement with those just quoted and are probably more accurate.

Transition to the lowest $n = 1$ mode was observed by Schwartz, Springett & Donnelly (1964), but the first unstable non-axisymmetric mode found by Coles (1965) and others was that with $n = 4$: the apparent preference for the $n = 4$ mode is not understood. Measurements by King *et al.* (1984) of the azimuthal wave speeds of non-axisymmetric modes agree well with theory.

24.3   *Effects of finite length*

Experiments are necessarily conducted with finite, not infinite, cylinders; but when the ratio of length to radius is sufficiently large, reasonable agreement is normally obtained with theoretical results for the infinite case. Coles (1965) studied the formation and destruction of stable non-axisymmetric flows by very gradually increasing and decreasing the rotation speed of his inner cylinder with the outer cylinder held fixed. A repeatable sequence of events was recorded, with sharp transitions between various states having p Taylor vortices and n azimuthal waves, each stable for some range of T. The transitions displayed marked hysteresis, according to whether the rotation was increasing or decreasing: different (p, n) states could be maintained indefinitely at the same rotation speed, that present being dependent on the history of the motion. For no other flow configuration has the non-uniqueness of stable solutions been demonstrated so dramatically: all who solve boundary-value problems by computational methods should ponder this work.

An even greater variety of flows was recently observed by Andereck, Dickman & Swinney (1983), using an apparatus in which *both* cylinders rotated. They describe 'twisted' and 'braided' Taylor vortices and vortices with wavy inflow and/or outflow boundaries, as well as the more familiar wavy-vortex flows. These complex flows have yet to be studied theoretically.

Considerable attention has recently focused on end effects. Experiments by Benjamin & Mullin (1981, 1982) and Mullin (1982) with rather short cylinders have revealed that there exist many distinct stable axisymmetric flows which satisfy the same boundary conditions. Transition between modes with differing numbers of cells exhibits marked hysteresis as the rotation speed and effective cylinder length are independently varied. Transition to time-periodic flows associated with the appearance of azimuthal waves was studied by Cole (1976), Mullin & Benjamin (1980) and Mullin *et al.* (1981) for short cylinders. Figure 7.6, from Benjamin's (1981) review, shows some of their results. The axes are $R = \Omega_1 r_1 d/\nu$ and $\Gamma = l/d$ where l is the annulus length. The curves denote the onset (and hysteretic disappearance) of periodic flow for various axisymmetric cellular modes. The sharpness and heights of the 'peaks' are remarkable: the maximum values of R are an order of magnitude greater than estimates for similar cells between *infinite* cylinders. In Benjamin's words, 'the prospects for a definite theoretical account still appear remote', but

Figure 7.6. Results of Mullin & Benjamin (1980) on stability to azimuthal waves of Taylor-vortex flows between short cylinders. Flows are normal two-cell mode (2), three-cell mode (3), abnormal four-cell mode (A4) and normal four-cell mode (4). Instability results on crossing the solid curves from below; instability disappears on crossing dashed curves from above (from Benjamin 1981).

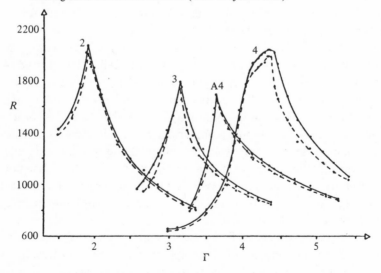

qualitative insights are gained by approaches employing abstract mathematics (e.g. Benjamin 1978).

End effects are more easily modelled for *long*, but finite, cylinders with ends at $z = \pm l$. This was done by Hall (1980a, b, 1982) and Stuart & DiPrima (1980). Axisymmetric modes denoted by $A_c \equiv A$ and $A_s \equiv B$ are even and odd, respectively, in z and may then satisfy

$$\left.\begin{aligned}
\mathrm{d}A/\mathrm{d}t &= \sigma_A A + a_{20} A^2 + a_{02} B^2 + a_{30} A^3 + a_{12} AB^2 + \ldots, \\
\mathrm{d}B/\mathrm{d}t &= \sigma_B B + a_{11} AB + a_{03} B^3 + a_{21} A^2 B + \ldots,
\end{aligned}\right\} \quad (24.3)$$

far from the ends. As $l \to \infty$, $\sigma_A/\sigma_B \to 1$, a_{20}, a_{02} and a_{11} approach zero and a_{30}, a_{12}, a_{03}, a_{21} become equal: these equations then resemble (24.1a, b) without the B_s and B_c modes.

Small, but non-zero, quadratic coefficients a_{20}, a_{02} and a_{11}, along with slightly different linear growth rates and cubic coefficients, destroy symmetry and indicate a preference for one or other mode. Odd equilibrium solutions, in addition to the trivial $B = 0$, are

$$B \approx \pm(-\sigma_B/a_{03})^{\frac{1}{2}}, \quad A = 0 \quad (T \geqslant T_B),$$

which resemble the supercritical equilibrium states described above. As σ_B increases from zero at $T = T_B$ with a_{03} negative, these give the 'pitchfork' bifurcation from $B = 0$ shown in Figure 7.7(a). Even solutions are in equilibrium when

$$\sigma_A A + a_{20} A^2 + a_{30} A^3 + O(A^4) = 0, \quad B = 0,$$

where a_{20} is numerically small; the two non-trivial solutions show asymmetric 'transcritical' bifurcation from $A = 0$ as in Figure 7.7(b). One solution branch now extends to *subcritical* Taylor numbers.

The supercritical, dashed, portion of the curve is known to be unstable with respect to small changes in A (see e.g. Joseph 1976). The solution with $A < 0$ is stable to small changes in A; but it would be unstable to phase

Figure 7.7. Typical (a) 'pitchfork' and (b) 'transcritical' bifurcations. Solid portions denote equilibrium solutions stable to amplitude perturbations, dashed portions denote unstable solutions.

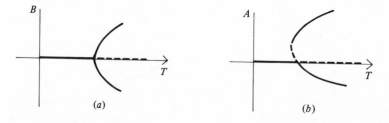

(a)   (b)

variations, and so evolve towards the solution with $A > 0$, if such phase variations were compatible with the end-wall boundary conditions.

But equations (24.3) are appropriate only for end-wall boundary conditions at $z = \pm l$ which admit a purely azimuthal primary flow. In fact, realistic end conditions drive flows with some axial and radial dependence at *all* rotation speeds of the inner cylinder (cf. Benjamin & Mullin 1981). Fourier decomposition shows that additional constant terms e_A, e_B must then be added to the right-hand sides of (24.3): these constants approach zero as $l \to \infty$, and e_A or e_B is zero if the boundary conditions are respectively symmetric or antisymmetric. With non-zero values e_A, e_B, the bifurcations are 'imperfect' and single-mode equilibrium solutions are as shown schematically in Figure 7.8(a), (b), (c).

The dashed portions of the curves are unstable to small amplitude changes. Some curves, but not others, admit a smooth transition between the primary flow and Taylor-vortex flow as T is increased. Imperfect bifurcations of similar kind were first analysed, for thermal convection in finite containers, by Kelly & Pal (1976, 1978), Hall & Walton (1977) and Daniels (1977). Mixed modes, with both A and B present, are discussed by Hall (1980b, 1982) for Taylor-vortex flow and by Hall & Walton (1979) for Bénard convection. Benjamin (1978), Schaeffer (1980) and Joseph (1981) have demonstrated the generic nature of such solutions and clarified their relationship to Thom's (1975) catastrophe theory.

Figure 7.8. Equilibrium solutions with imperfect bifurcation. (a) $B(T)$ for $e_B > 0$ (inverted for $e_B < 0$); (b) $A(T)$ for $e_A > 0$; (c) $A(T)$ for $e_A < 0$.

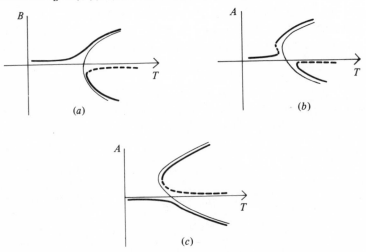

24.4   Doubly-periodic and 'chaotic' flow

Wavy-vortex flow remains strictly periodic, with a dominant frequency ω_1 and its harmonics, until the Reynolds number R is an order of magnitude larger than its value R_c at onset of Taylor instability. Fenstermacher, Swinney & Gollub (1979), employing a laser-Doppler technique to measure the flow at a single point, found rather sudden onset of a second, incommensurate frequency ω_3† at $R/R_c = 10.1$: see Figure 7.9. (Strictly, 'incommensurate' means that ω_3/ω_1 is irrational; but, for present purposes, it is taken to mean that no *small* integers m, n, less than 10 say, exist with $m/n = \omega_3/\omega_1$). Gorman & Swinney (1982) identified ω_3 with a modulation of the wavy vortices, which was usually spatial as well as temporal. Their visual observations showed a periodic 'flattening' of one or more wavelengths around the annulus. The evolution with azimuthal angle ϕ and time t took various forms, characterized by a pair of integers $(n \,|\, k)$: here, n denotes the azimuthal mode number of the waves and k that of the modulation. Gorman & Swinney's schematic representation of several of these forms is reproduced in Figure 7.10.

The purely temporal modulations (cases $n \,|\, 0$) might be thought to correspond to periodic solutions of (24.1), with continuous energy exchange among the modes A, B_c and B_s: after all, the rather similar four-wave resonance equations (23.3) display periodic modulations. DiPrima (1981) has noted that Eagles' (1971) calculated coefficients do not admit this possibility, but these results relate to fairly small values of $(R - R_c)/R_c$. Also, these truncated equations are formally valid only as the linear growth rates a_0, b_0 simultaneously approach zero: and this is *never* so for $n \neq 0$! The practical range of approximate validity of (24.1) is unknown, but is unlikely to extend to $R/R_c \gtrsim 10$.

Gorman & Swinney found their results to be qualitatively consistent with a rather abstract but general theory of Rand (1981). They also seem to be consistent with the growth of modes with azimuthal mode numbers differing from that of the fundamental wavy vortices. In particular, azimuthal modes $n \pm 1, n \pm 2, \ldots$ interact resonantly with the n-mode (as well as with the primary Taylor vortices) since their azimuthal phase speeds are known to be the same: the growth of such modes might well induce azimuthal modulations like those of Figure 7.10. Modes with slightly different axial wavenumbers α could also play a part. Further computational work may be expected to clarify the issue.

Shortly after the appearance of two incommensurate frequencies,

---

† Subscript 3 is used to accord with Fenstermacher *et al.*'s notation.

Figure 7.9. Measured time-dependence of radial velocity component and corresponding frequency power spectra at various R/R_c (from Fenstermacher *et al.* 1979). Values of R/R_c in (a)–(h) are 9.6, 10.1, 11.0, 15.1, 18.9, 21.7, 23.0, 43.9 respectively.

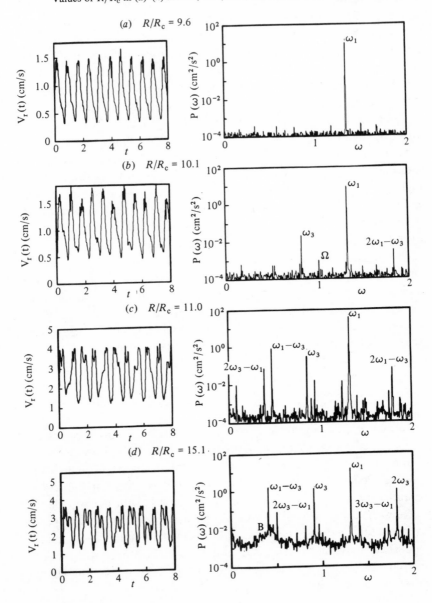

Fenstermacher *et al.* (1979) observed the development, at about $R/R_c = 12$, of a weak broadband contribution to the frequency spectrum measured at a single point. As R/R_c further increased, so too did the intensity of this broadband 'peak' and also the background noise. The sharp modulation-frequency peak $\omega \approx \omega_3$ disappeared at $R/R_c = 19.3$ and

Figure 7.9(*e–h*). For legend see opposite.

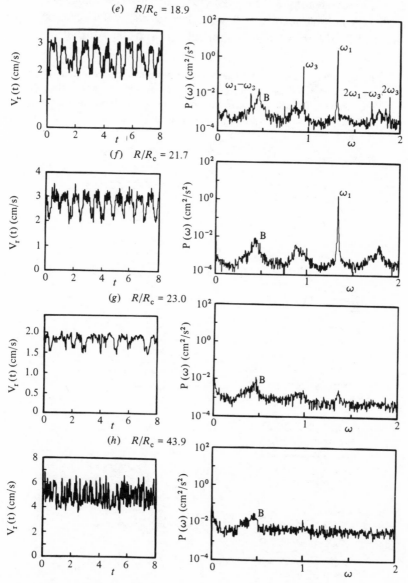

(*e*) $R/R_c = 18.9$

(*f*) $R/R_c = 21.7$

(*g*) $R/R_c = 23.0$

(*h*) $R/R_c = 43.9$

256   *Cubic three- and four-wave interactions*

the wavy-vortex frequency peak itself vanished at $R/R_c = 21.9$. The spectrum then appeared to be continuous and so suggestive of fully-developed turbulence. However, after similar events in a different apparatus, Walden & Donnelly (1979) witnessed the surprising re-emergence of a sharp spectral peak for $28 < R/R_c < 36$.

Koschmieder (1980), DiPrima (1981), DiPrima & Swinney (1981) and L'vov, Predtechensky & Chernykh (1983) review these and other related studies. A notable feature of flows at large R/R_c is the persistence of a strong Taylor-vortex configuration along with the 'turbulence'. The

Figure 7.10. Gorman & Swinney's (1982) schematic representation of periodic temporal modulations, in reference frame rotating with the waves. Each diagram represents evolution in time of a *single* vortex outflow boundary (*not* a stack of vortices at fixed time). Heavy lines denote greatest 'flattening'. Corresponding values of $(n|k)$ are noted.

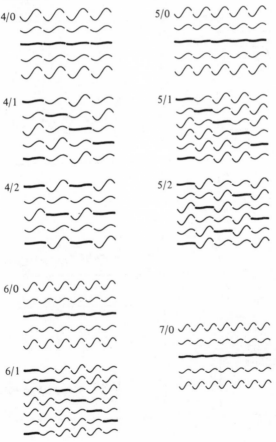

robustness of these axisymmetric vortices is remarkable. So, too, is the fact that the observed number of cells, even in 'turbulent' flow, is determined by the past history of the flow and *not* just the operational value of R. Such a 'memory' is inconsistent with most notions of *strong* turbulence.

Much attention has recently focused on theoretical analyses of model systems of low dimension which display 'strange attractor' behaviour: see, e.g. Swinney & Gollub (1981). One such model, for three-wave interactions, was described in § 16.2. The first, and most celebrated, was that of Lorenz (1963), whose three-variable analogue of Rayleigh–Bénard convection exhibited apparently chaotic behaviour. Ruelle & Takens (1971) later proposed that a generic feature of low-dimensional strange-attractor models is the emergence of two incommensurate frequencies followed by 'chaotic' motion with a broadband frequency spectrum. With this work and the experiments of Fenstermacher *et al.* (1979) in mind, Sherman & McLaughlin (1978) investigated a set of five amplitude equations with quadratic coupling. These exhibited transition from sharply-peaked to broadband frequency spectra as a growth-rate parameter was varied, but their model is not strictly applicable to the experiments. It may be conjectured that equations (24.1) would display qualitatively similar properties for some ranges of the parameters. There are obvious similarities, too, with the computations of Yuen & Lake (1982) described above, on sideband modulations of gravity waves.

More elaborate computational models have been constructed by Yahata (1981, 1983a), who employed a Galerkin representation of axisymmetric and $n = 4$ modes comprising 56 separate functions. Essentially, his .configuration represents four-wave interaction among modes $(\alpha, 0)$, $(2\alpha, 0)$, $(\alpha, 4)$, $(\alpha, -4)$ and mean flow $(0, 0)$. The results exhibit the appearance and then disappearance of two and three dominant frequencies (plus sum and difference harmonics) and the onset of chaotic motion. These may have a bearing on Gorman & Swinney's modulated wavy $(n \,|\, k)$-vortices with $k = 0$, but the computations do not allow for azimuthal modulations. Finally, it should be remembered that these chaotic motions take place in a low-dimensional system comprising just a few spatially-periodic modes (C'est chaotique, mais ce n'est pas la turbulence!). In contrast, much of the 'noise' which comes to dominate turbulent flow at large R seems to be associated with small spatial scales: these are evident in Figure 7.5 even at the fairly low values $R/R_c = 16$ and 23.5. Recent overviews of 'order in chaos' are given in Campbell & Rose (1983) and Barenblatt, Iooss & Joseph (1983).

25 Rayleigh–Bénard convection

25.1 Introduction

Thermally-driven convective instabilities in horizontal fluid layers heated uniformly from below have many similarities with Taylor–Couette flow. But there are also important differences: while the latter are strongly anisotropic, with well-defined axial, azimuthal and radial dependence, convective flows are usually much less so. Convection cells in layers of infinite lateral extent have *no* preferred horizontal orientation and horizontal anisotropy is weak in layers of large aspect ratio where the horizontal dimensions greatly exceed the depth.

Convective instabilities may lead to stable rolls, with well-defined wavenumber **k**, which are very similar to steady axisymmetric Taylor vortices, both mathematically and physically. Governing dimensionless parameters are the Rayleigh number R and the Prandtl number P, defined as

$$R = \frac{\gamma_1 g \, \Delta T \, d^3}{\nu \kappa}, \quad P = \frac{\nu}{\kappa}.$$

(Note that, in this section, R no longer denotes Reynolds number.) Here, ν and κ are kinematic viscosity and thermal diffusivity, d the layer depth, g gravitational acceleration, γ_1 the coefficient of thermal expansion and ΔT the temperature difference between bottom and top boundaries.

Instability first sets in at a particular value k_c of $|\mathbf{k}|$ when the Rayleigh number R (a dimensionless measure of the importance of buoyancy over diffusive processes) exceeds a critical value R_c. The value R_c depends on the boundary conditions but not on the Prandtl number P of the fluid. With rigid horizontal boundaries, $R_c = 1708$ for layers of infinite lateral extent and the orientation of **k** is immaterial. In practice, however, lateral boundaries play an important part in selecting a preferred mode, especially when the aspect ratio is not large. For instance, rectangular side walls usually favour rolls aligned parallel to the shorter side and circular walls may give rise to concentric circular cells. Nevertheless, a marked lack of uniqueness remains and many stable convective flows may satisfy identical boundary conditions.

Cellular convection was first systematically examined by Bénard (1900, 1901) in a shallow liquid layer with free upper surface, but there are many earlier reports of the phenomenon (see Normand, Pomeau & Velarde 1977). The hexagonal cells observed by Bénard correspond to three superposed roll patterns, with wavenumbers $\mathbf{k}_j$ ($j = 1, 2, 3$) of equal magnitude but differing in orientation by 120°. Motion may be either upwards or downwards at the centres of cells depending on small vertical asymmetries, such as the temperature dependence of fluid viscosity, and on surface tension effects. In shallow layers like Bénard's, horizontal temperature-induced gradients in surface tension are more powerful than buoyancy in driving the flow (Pearson 1958). Here, attention is mainly focused on buoyancy-driven convection in layers confined by rigid top and bottom boundaries. Previous reviews of theory and experiment include those of Palm (1975), Normand, Pomeau & Velarde (1977) and Busse (1978, 1981a, b). A rigorous approach to theoretical aspects is given by Joseph (1976).

25.2   Instabilities of rolls

A weakly nonlinear amplitude-expansion analysis by Schlüter, Lortz & Busse (1965) extended earlier work of Gor'kov (1957) and Malkus & Veronis (1958). This elucidated the properties of steady finite-amplitude convection in horizontally-infinite layers between rigid parallel boundaries at small positive values of $R - R_c$. They found that the only stable flows are longitudinal rolls with a range of $|\mathbf{k}|$ centred on k_c that is narrower than the bandwidth for linear instabilities. (In contrast, configurations with vertical asymmetry are subcritically unstable ($R < R_c$) to finite-amplitude disturbances and hexagonal cells are preferred near R_c: see below.)

Near R_c, the instability of rolls with wavenumbers outwith this range is due to growth of modes with wavenumbers closer to k_c. For instance, when $\mathbf{k} = (k, 0)$ and k is too small, an instability bends the roll axis into a sinusoid, thereby exciting (resonant) modes $(k, \pm l)$ with greater wavenumber $(k^2 + l^2)^{\frac{1}{2}}$: both conditions (23.1) are satisfied since the rolls have zero frequency. This is the 'zig–zag instability' observed by Busse & Whitehead (1971) with angles of about 40° and 140° between the new and original rolls. In contrast, rolls with k too far above k_c – but still within the linearly-unstable bandwidth – are normally replaced, through 'cross-roll instability', by rolls at right angles with wavenumber close to $(0, k_c)$. Both types of instability are shown in Figure 7.11 taken from Busse & Whitehead (1971). Also possible is a two-dimensional Eckhaus instability (Stuart & DiPrima 1978) of sidebands $(k \pm \delta, 0)$ with the same orientation as the original rolls.

Newell & Whitehead (1969) analysed roll cells with wavenumber $\mathbf{k} = (k_c, 0)$ and slowly-varying amplitude $\epsilon A(x, y, t)$ at R just above R_c. With suitably-scaled variables, the amplitude equation may be written in the form

$$\frac{\partial A}{\partial T} - \left(\frac{\partial}{\partial X} - \frac{\mathrm{i}}{2}\frac{\partial^2}{\partial Y^2}\right)^2 A = (1 - |A|^2)\, A.$$

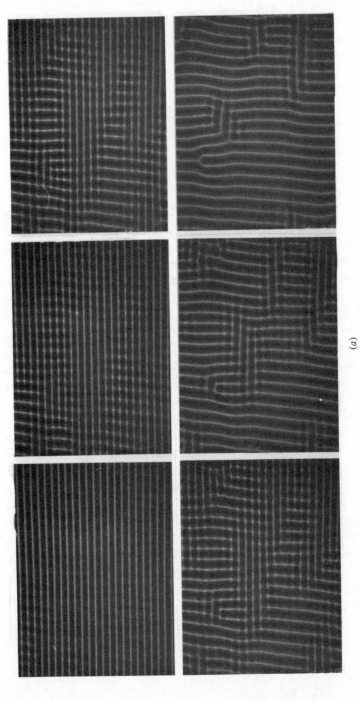

(a)

Figure 7.11. Temporal development of (a) 'cross-roll' instability and (b) 'zig-zag' instability of convection rolls (from Busse & Whitehead 1971). In (a), $R = 3000$, depth = 5 mm, $k = 2\pi/1.64$; times between successive photographs are 10, 4, 3, 7, 24 min respectively. In (b), $R = 3600$, depth = 5 mm, $k = 2\pi/2.8$; times between photographs are 9, 10, 10, 26, 72 min.

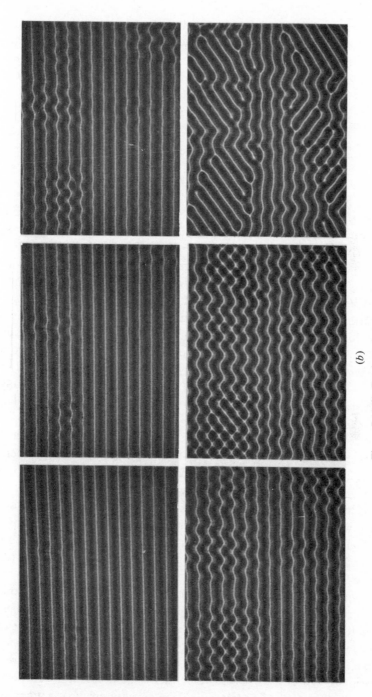

(b)

Figure 7.11(b). For legend see opposite.

262   Cubic three- and four-wave interactions

The Eckhaus instability, and similar sideband instability of oblique modes $\mathbf{k} = (k_c \pm \delta_1, \pm \delta_2)$ with $|\delta_j/k_c| \ll 1$, may be deduced from this equation much as outlined in §21.1 for gravity waves. The form of the linear operator on the left-hand side is a consequence of the linear growth-rate relation,

$$\sigma(R, k) \approx \left(\frac{\partial \sigma}{\partial R}\right)_c (R - R_c) - \left(\frac{\partial^2 \sigma}{\partial k^2}\right)_c (k - k_c)^2, \quad k = |\mathbf{k}|,$$

near (R_c, k_c). Setting $\mathbf{k} = (k_c + K, L)$ where K, L are small, gives

$$\sigma \approx \left(\frac{\partial \sigma}{\partial R}\right)_c (R - R_c) - \left(\frac{\partial^2 \sigma}{\partial k^2}\right)_c (K + \tfrac{1}{2} L^2 k_c^{-1})^2$$

and transformations $K \to -i\, \partial/\partial X$, $L \to -ik_c^{\frac{1}{2}}\, \partial/\partial Y$ yield the required form. (In contrast, the nonlinear Schrödinger equation results from dispersion relations with a leading-order L-term proportional to L^2.)

At larger R, a 'skewed varicose instability' affects even those rolls with $|\mathbf{k}| \approx k_c$. This is apparently associated with the growth of 'subharmonic' modes $(\tfrac{1}{2}k, \pm l)$ and so may be an instance of three-wave resonance.

A further type of instability differs from the above-mentioned (quasi-) steady flows in being time-periodic. This oscillatory instability of steady rolls sets in close to R_c if the Prandtl number P is very small and if both upper and lower boundaries are stress-free: but transition is delayed to larger R for rigid boundaries. For $P \lesssim 1$, the rolls are sinusoidally distorted as in the 'zig–zag' instability, but they also oscillate in time, either as travelling or standing waves.

For onset of oscillatory instability, there is qualitative agreement between calculations (Clever & Busse 1974; Lipps 1976) and experiment for $P \lesssim 2$ (Krishnamurti 1970; Busse & Whitehead 1974), but observed and theoretical frequencies differ somewhat. The similarity with onset of wavy vortices in Taylor–Couette flow is striking, though oscillatory standing waves have not been observed in the latter flow. At larger P, onset of oscillatory instability is delayed to higher values of R, and other types of instability (notably cross rolls) may appear first. The observed period decreases as R increases.

Krishnamurti (1973) describes the periodicity, for large P and R, as due to the orbiting of local hot or cold spots (with associated high shear) around the roll-cell, rather than distortions of the cell axis. Behaviour of this sort was reproduced in computations of Moore & Weiss (1973), for free boundaries and with the motion artificially restricted to be two-dimensional: with $P = 6.8$, as for water, their oscillations set in at around $R/R_c = 50$. Lipps (1976) undertook a three-dimensional direct numerical simulation for air ($P = 0.7$) between rigid upper and lower boundaries and stress-free 'sidewalls' which admitted spatially-periodic solutions. Previously two-dimensional rolls became wavy and time-periodic at $R \approx 6500$ ($R/R_c \approx 3.8$); more complex spatial and time dependence set in for $R \geqslant 9000$. Though the sidewall conditions admit only discrete cell sizes, these results are convincing.

The current theoretical situation is more fully set out by Busse (1981a, b). His map of the predicted region of stable finite-amplitude two-dimensional rolls in k–R–P parameter space is reproduced in Figure 7.12. This region is bounded by onset of various kinds of instability and summarizes results

Figure 7.12. Region of stable two-dimensional convection rolls in k–R–P space. This is bounded by various types of instability: oscillatory (OS), skewed varicose (SV), cross-roll (CR), knot (KN) and zig–zag (ZZ) (Busse 1981a, b).

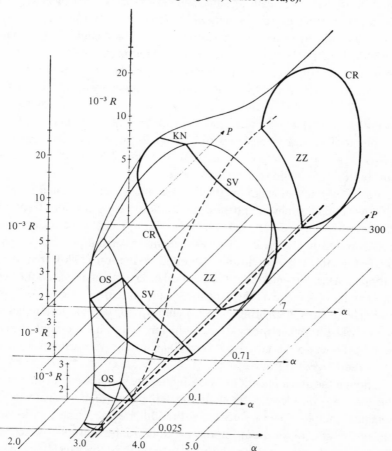

of several earlier numerical studies by Busse and Clever. These studies all employed Galerkin representations of the rolls, as truncated series of orthogonal functions; the imposition of small disturbances of the required form led to linear eigenvalue problems which determined their stability. Though these results are probably quite accurate, the range of validity of such series-truncation methods is difficult to establish, other than empirically.

25.3   Rolls in finite containers

The influence of aspect ratio modifes the picture. With large aspect ratios, the multi-roll pattern which first emerges from random background disturbances is patchy, with each patch containing rolls of differing orientation. Dislocations at patch boundaries, where different cell structures meet, move with time, causing some patches to shrink and others to grow. An experimental study of the merging, or 'pinching', of artificially created roll cells with the same orientation but differing wavenumber was made by Whitehead (1976). Gollub, McCarriar & Steinman (1982) have recently examined evolving convection patterns in layers of large aspect ratio.

In contrast, the theoretical configurations analysed by Kelly & Pal (1976, 1978), Daniels (1977) and Hall & Walton (1977, 1979) are strictly two-dimensional, consisting of parallel rolls with fixed orientation. These accounts of secondary and imperfect bifurcations, already mentioned in §24.3 above, apply only rather close to the critical Rayleigh number R_c for onset of convection.

For convection in a box of small aspect ratio, only a limited number of flow configurations are possible and rather sharp transitions from one to the other may be observed. For this reason, recent experiments have concentrated on high-precision measurements in small-aspect-ratio apparatus of small scale. For instance, Gollub & Benson (1980) used laser-Doppler measurement to study convection in a rectangular box about 1 cm deep and less than 3 cm long, which typically contained only two or three roll cells. Ahlers & Behringer (1978) employed cryogenic techniques to study heat transfer in (normal) liquid helium ^{4}He. This was contained in small cylinders of about 1 or 2.5 cm diameter and various aspect ratios ($\Gamma = \text{radius}/\text{depth}$) from 2 to 57.

Though the above-mentioned instabilities are suppressed or delayed by the proximity of sidewalls, the evolution of the flow as R is gradually raised remains complex. The influence of Prandtl number, aspect ratio and sidewall configuration are all important. Gollub & Benson (1980) described four quite separate 'routes to turbulent convection' (by which they mean the development of temporally-periodic flows into flows with broadband frequency spectrum).

In the first, a two-cell configuration at $P = 5.0$ became oscillatory, with some frequency ω_1, at $R/R_c = 27.2$ (R_c being the value of R for onset of convection in an infinite layer, *not* in their apparatus); a second apparently unrelated frequency ω_2 appeared at $R/R_c = 32$; then, at about $R/R_c = 44.4$, these frequencies 'phase-locked' at $\omega_1/\omega_2 \approx 7/3$; next, substantial broadband noise developed at the slightly higher $R/R_c = 46.0$ and this increased while ω_1/ω_2 decreased as R/R_c was made larger; ultimately, the peaks at ω_1 and ω_2 became submerged in the broadband spectrum.

In the second, a similar two-cell configuration at the lower Prandtl number $P = 2.5$ became oscillatory at $R/R_c = 17$; sharp spectral peaks at ω_1 and its harmonics were observed until, at $R/R_c = 21.5$, a period-doubling subharmonic component $\frac{1}{2}\omega_1$ appeared; by $R/R_c = 26.5$, the $\frac{1}{2}\omega_1$ component had dominated the fundamental, and a second subharmonic bifurcation, to $\frac{1}{4}\omega_1$, set in; the flow became non-periodic at about $R/R_c = 28$ and the peaks of the frequency spectrum merged with the broadband contribution at about $R/R_c = 40$.

Three-cell configurations behaved very differently. At $P = 2.5$, for example, period flow set in at $R/R_c \approx 30$ and apparently independent frequencies ω_2, ω_3 appeared at 39.5 and 41.5; broadband noise began to grow at $R/R_c = 43$. At higher P, similar behaviour was found, but onset of ω_2 and ω_3 was delayed to larger R/R_c.

Finally, instances were found where ω_2/ω_1 was rather small, such that the velocity records showed slow periodic modulations in the amplitude of ω_1-oscillations. In a three-cell case with $P = 5.0$, the ω_2 modulations appeared at $R/R_c = 95$ and *intermittent* noise of irregular duration set in at $R/R_c = 102$; at greater R/R_c, a featureless broadband spectrum was obtained.

Ahlers & Behringer (1978) reported onset of aperiodic 'turbulent' flows at R/R_c close to 1 for their largest aspect ratio of $\Gamma = 57$; the value of R/R_c at onset increased as Γ was reduced, reaching about 11 when $\Gamma = 2.08$. Rather similar results were found by Libchaber & Maurer (1978). The additional effect of a magnetic field on convection in mercury was explored by Libchaber, Fauve & Laroche (1983).

Gollub & Benson could discern 'no simple rules for predicting which sequence will occur for a given aspect ratio, Prandtl number, and mean flow' but each of their observed sequences is characterized by mode interactions. The route through two incommensurate frequencies to broadband spectrum parallels the behaviour of Taylor vortices. The subharmonic bifurcations are reminiscent of quadratic three-wave resonance, but lack of information about the spatial structure precludes firm identification. For the same reason, emergence of three distinct frequencies defies precise interpretation: no direct connection with the theory of Ruelle & Takens (1971) has yet been established, though qualitative similarities exist. As to intermittent noise, it is tempting to draw parallels with the onset of intermittent, spatially-localized, patches of turbulence in parallel shear flows (see §27.2) but here, again, data on the spatial structure is lacking.

A 14-mode Lorenz-attractor model investigated by Curry (1978) and a system of several modes solved numerically by Yahata (1982, 1983b) reproduce various of the observed features, including period-doubling bifurcations and transition to 'chaotic' time-dependence. However, these mode-truncated models should not be regarded as precise representations of the flow. An interesting example of an experiment which apparently *does* correspond exactly to a known strange-attractor system of ordinary differential equations is that described by Roux, Simoyi & Swinney (1983): this concerns the remarkable Belousov–Zhabotinskii reaction of chemical kinetics.

In a numerical study of doubly-diffusive thermosolutal convection, Huppert & Moore (1976) solved the governing *partial* differential equations by finite differences and found transitions from time-periodic to chaotic motion. Further work by Da Costa, Knobloch & Weiss (1981) and by Moore *et al.* (1983) has revealed further details of the solution structure, for two-dimensional convection with fixed roll spacing and stress-free horizontal boundaries. As the thermal Rayleigh number R_T was increased, with other flow parameters assigned constant values, finite-amplitude oscillatory rolls gave way to chaotic motion through a sequence of period-doubling bifurcations. As R_T was further increased, a narrow 'window' was found which yielded oscillatory solutions with three times the original dimensionless period $\tau = 1$; these then passed through more period-doubling bifurcations till chaotic motion was re-established. Solutions at still larger R_T yielded 'an inverse cascade of period-halving bifurcations leading to solutions with period 1, followed by another transition back to chaos through a period-doubling sequence'. Figure 7.13(*a*)–(*h*) shows examples of phase-plane trajectories and frequency power spectra of the total kinetic energy. Cases (*a, b*) and (*g, h*) are the 2τ and 3τ periodic limit cycles while (*c, d*) and (*e, f*) describe 'semi-periodic' chaotic motion.

A recent review by Swinney (1983), of order and chaos in nonlinear

Figure 7.13. Phase-plane trajectories of kinetic energy E versus its time-derivative $\dot{E}$ and corresponding frequency power spectra $\phi(\omega)$ of E (on logarithmic scale), found by Moore *et al.* (1983). (a, b) $R_T = 10450$, period-doubling $\tau = 2$ limit cycle with basic frequency $\omega_0 \approx 30$ $(\tau = 1)$; (c, d) $R_T = 10475$, 'semi-periodic' trajectories with noisy but sharply-peaked spectrum; (e, f) $R_T = 10500$, chaotic solutions with peaks only at ω_0 and harmonics; (g, h) $R_T = 10510$, $\tau = 3$ limit cycle.

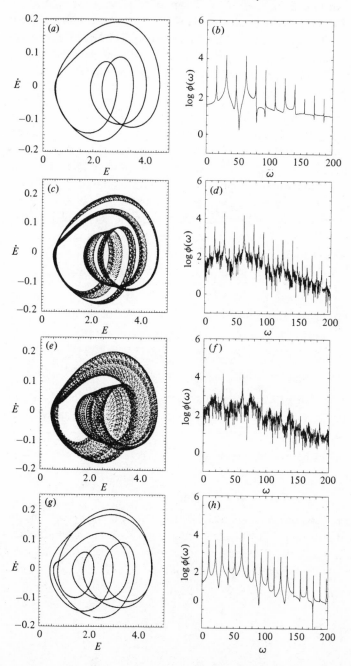

systems, mentions many other related experiments and recent theoretical developments which cannot be discussed here: the richness of structure revealed in recent years is both remarkable and bewildering.

25.4   *Three-roll interactions*

An interesting case of three-mode interaction for convection in infinite rotating layers was studied by Küppers & Lortz (1969) and Busse & Clever (1979) (also reported by Busse 1981a). The latter's perturbation analysis relates to three sets of rolls, each with the same wavenumber modulus $|\mathbf{k}|$, but orientations differing by 120°. Accordingly,

$$\mathbf{k}_1 + \mathbf{k}_2 + \mathbf{k}_3 = 0$$

and, since the corresponding frequencies are zero, quadratic resonance might be expected. In layers of fluid otherwise at rest, each set of rolls is stable provided R/R_c is not too far above unity, and one possibility is finite-amplitude hexagonal cells. However, the mode-interactions are *not* here of quadratic order: though the governing evolution equations have the form of (16.7), the quadratic coefficients are identically zero (by reason of vertical symmetry) and the dominant interactions are cubic. Also, the linear-growth rates $-\sigma_j$ are equal and the cubic coupling coefficients α_{ij} are found to be imaginary with $i\alpha_{jk} > 0$.

By means of symmetry arguments and the transformations

$$G_j = |\alpha_{jj}/\sigma_j| |A_j|^2, \quad \tau = -\sigma_j t,$$

the governing equations may be recast as

$$\left.\begin{aligned}
\dot{G}_1 &= G_1(1 - G_1 - \alpha G_2 - \beta G_3), \\
\dot{G}_2 &= G_2(1 - G_2 - \alpha G_3 - \beta G_1), \\
\dot{G}_3 &= G_3(1 - G_3 - \alpha G_1 - \beta G_2),
\end{aligned}\right\} \tag{25.1}$$

where α and β are positive constants. For convection in layers at rest, $\alpha = \beta$ and possible finite-amplitude equilibrium states are the individual-roll configurations

$$G_i = 1, \quad G_j = 0 \quad (j \neq i = 1, 2 \text{ or } 3),$$

and the hexagons

$$G_i = (1 + 2\alpha)^{-1} \quad (i = 1, 2 \text{ and } 3).$$

Differing positive values of α and β arise when the layer is rotating about a vertical axis: then, individual rolls remain possible equilibrium solutions, but these are unstable whenever one of α, β is less than unity. The equilibrium combined-mode solution is

$$G_i = (1 + \alpha + \beta)^{-1} \quad (i = 1, 2, 3)$$

and this is stable to small perturbations of the G_i if and only if $\alpha + \beta < 2$. Accordingly, in the two regions of parameter space

$$R_1 = \{\alpha, \beta : \ 0 < \alpha < 1, \alpha + \beta > 2\}, \quad R_2 = \{\alpha, \beta : \ 0 < \beta < 1, \alpha + \beta > 2\},$$

no equilibrium solution is stable.

Surprisingly, this absence of stable equilibrium solutions does *not* signal transition to time-periodic flow: such oscillatory solutions cannot exist, at least near R_c, when $P > 1$. Instead, solutions of (25.1) rapidly approach the plane $G_1 + G_2 + G_3 = 1$ and then wander around seeking, but never attaining, one or other of the single-roll fixed points $G_i = 1$ ($i = 1, 2$ or 3). The solutions are neither periodic nor 'chaotic' but spend ever-increasing times in the vicinity of each fixed point. This behaviour was first described by May & Leonard (1975), who investigated (25.1) as a model of three-species competition in biological population dynamics. A similar 'period lengthening' was noted in §16.2 for damped three-wave interactions.

Both May & Leonard (1975) and Busse (1981a) reject continued period lengthening as unphysical, since the minimum value of each G_j for each cycle is ever-decreasing. In practice, random low-level disturbances inevitably affect the temporal and spatial evolution, as apparently found in experiments by Busse & Heikes (1980): the 'period' is then an irregular, stochastic function.

Segel (1962) examined the interaction of two sets of roll cells with parallel wavenumbers k_1, k_2 of differing magnitude. The governing equations, with vertical symmetry, are then as (25.1), but with one mode, say G_3, equal to zero. When rolls with wavenumbers $k_1 \pm k_2$ are not too strongly damped, one would expect these to be driven unstable by the cubic interaction.

In earlier theoretical investigations of hexagonal convection, Palm (1960), Segel & Stuart (1962) and Segel (1965) incorporated vertical asymmetry due to weak temperature dependence of the fluid viscosity;

$$\nu(T) = \nu_0 + \Delta\nu \cos\left[\pi(T - T_0)/(T_1 - T_0)\right], \quad \Delta\nu/\nu_0 \ll 1,$$

say, where T_1, T_0 are the temperatures at stress-free lower and upper boundaries respectively. At leading order, perturbation quantities may be written in the form

$$W = \phi(x, y, t)\, g(z),$$

$$\phi = A_1 \cos(\alpha y + \theta_1) + A_2 \cos\left[\tfrac{1}{2}\alpha(y - 3^{\frac{1}{2}}x) - \theta_2\right] + A_3 \cos\left[\tfrac{1}{2}\alpha(y + 3^{\frac{1}{2}}x) - \theta_3\right].$$

Here, the A_j and θ_j depend on t only and are the (real) amplitudes and phases of three roles inclined at angles of 60° to one another. A perturbation analysis, taken to cubic order in A_j, leads to equations of the form

$$\left.\begin{aligned}
\dot{A}_1 &= \sigma A_1 - a A_2 A_3 \cos(\theta_1+\theta_2+\theta_3) - A_1(rA_1^2+pA_2^2+pA_3^2), \\
\dot{A}_2 &= \sigma A_2 - a A_1 A_3 \cos(\theta_1+\theta_2+\theta_3) - A_2(pA_1^2+rA_2^2+pA_3^2), \\
\dot{A}_3 &= \sigma A_3 - a A_1 A_2 \cos(\theta_1+\theta_2+\theta_3) - A_3(pA_1^2+pA_2^2+rA_3^2), \\
A_1^2\dot{\theta}_1 &= A_2^2\dot{\theta}_2 = A_3^2\dot{\theta}_3 = a A_1 A_2 A_3 \sin(\theta_1+\theta_2+\theta_3),
\end{aligned}\right\} \tag{25.2}$$

with terms of higher than cubic order omitted. These are equivalent to the three complex amplitude equations (16.7) with real coupling coefficients. Segel & Stuart (1962) retain terms in the second time derivatives, $\ddot{A}_j$, but these are negligible when A_j are small. The constant coefficients σ, p and r are just those obtained by neglecting the temperature variations of ν, but a is proportional to $\Delta\nu$. In other words, vertical asymmetry is responsible for quadratic coupling which is absent in the symmetric case.

Below R_c, $\sigma < 0$ and there are no non-trivial equilibrium states if $a = 0$, since p and r are positive. But quadratic coupling admits a class of subcritical hexagonal cells,

$$A_1 = A_2 = A_3 \equiv A, \quad \theta_1+\theta_2+\theta_3 = \pi,$$

with

$$A = \tfrac{1}{2}(r+2p)^{-1}\{a\pm[a^2+4\sigma(r+2p)]^{\frac{1}{2}}\}, \quad a > 0,$$

Figure 7.14. Schematic representation of finite equilibrium states corresponding to hexagons and rolls. A denotes amplitude, R the Rayleigh number; linear instability ($\sigma > 0$) occurs for $R > R_c$. Solid curves are stable, dashed curves unstable, to amplitude perturbations.

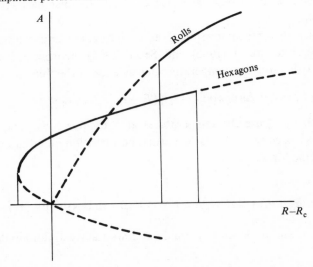

Figure 7.15. Hexagonal convection with a free surface: (a) $R < R_c$, (b) $R > R_c$, from Koschmieder (1967).

whenever $\sigma > -\frac{1}{4}a^2(r+2p)^{-1}$. These are indicated schematically in Figure 7.14: the lower branch, with negative sign, is unstable to perturbations of A and the upper branch is stable. Corresponding solutions, with the sign of A reversed, are obtained for $\theta_1+\theta_2+\theta_3 = 0$ and also for $a < 0$.

Above R_c, equilibrium solutions include the hexagons just described and also individual rolls. Other equilibrium solutions of 'mixed' type also exist, but these are always unstable (Segel 1965). The regions of stability of hexagons and rolls are indicated qualitatively in Figure 7.14. The hexagons shown with $A > 0$ have upwards motion at the cell centres when $\Delta\nu < 0$ and downwards motion when $\Delta\nu > 0$; those shown with $A < 0$ (or equivalently, $A > 0$ with $\theta_1+\theta_2+\theta_3 = 0$) have opposite directions of motion and are unstable to phase perturbations. It must be remembered that the truncated equations (25.2) give valid approximations for the equilbrium solutions only when σ, a (and so A) are sufficiently small to permit the omission of higher-order terms.

Palm, Ellingsen & Gjevik (1967) derived corresponding results for more realistic rigid–rigid and rigid–free boundary conditions at the bottom and top. Busse (1967) included temperature variations of thermal diffusivity κ and the coefficient of expansion γ_1. Busse's general formulation allowed for interactions among a large number of discrete modes and bears a close resemblance to the discretized form of Zakharov's formulation (23.11). Scanlon & Segel (1967) discussed surface-tension driven hexagonal convection. In all cases, stable hexagons were found in some parameter range near R_c. Such hexagons were realized experimentally by Koschmieder (1967) and others, for both $R < R_c$ and $R > R_c$: see Figure 7.15. Time-dependent interaction of differing hexagonal configurations has been modelled numerically, at rather large R, by Toomre, Gough & Spiegel (1982). Mode interactions of hexagonal configuration in baroclinic instability were studied by Newell (1972) and, more generally, by Loesch & Domaracki (1977).

26   Wave interactions in planar shear flows

26.1   Three dominant waves

Cubic nonlinearities influencing three-wave resonant interactions in parallel shear flows were already mentioned in §16.3: for temporal modulations, the governing equations have the form (16.7). Non-resonant three-wave interactions among arbitrary wavenumbers $\mathbf{k}_i$ give rise to similar cubic terms, without significant quadratic coupling; but then further 'sum and difference' wavenumber components may grow to prominence unless these experience stronger linear damping than the three waves chosen.

Both resonant and non-resonant cases were considered by Usher & Craik (1975), who suggested, on the basis of asymptotic estimates for large R, that particularly large cubic interaction coefficients may usually influence the growth of symmetric oblique-mode configurations like those of §17.2. These coefficients may be largest when interaction is resonant, for all three modes then have the same critical layer.

The parametric excitation of resonant oblique-wave modes by a two-dimensional wave was briefly discussed in §14.3, where only quadratic nonlinearities were retained. With the configuration of §17.2, the resonance condition selects a particular value of the spanwise wavenumber β which might be expected to lie close to that of the most strongly excited mode. But cubic (and higher) nonlinearities also act to modify the frequencies and to enhance or suppress oblique modes: accordingly, these may shift the preferred value of β away from that at resonance. Moreover, in the special case of plane Poiseuille flow, the quadratic coupling coefficients are zero by reason of symmetry, for interactions among lowest, even, Orr–Sommerfeld eigenmodes: obviously, there is then less reason for the 'resonant' value of β to be strongly excited.

Experiments of Saric & Thomas (1984) and Kachanov & Levchenko (1984) on the Blasius boundary layer all found evidence of subharmonic oblique-wave growth roughly as predicted by Craik (1971): but observed spanwise wavenumbers did not always agree closely with those predicted by the resonance criterion. Also, subharmonic oblique waves in plane Poiseuille flow are reported by Ramazanov (1985).

In an attempt to clarify the situation, Herbert (1983b, c; 1984) recently made several computational studies. All of these concern the temporal eigenvalues of infinitesimal, symmetric, oblique-wave pairs in the presence of a finite-amplitude two-dimensional wave. His results of (1983b) for plane Poiseuille flow appear to be inaccurate, though they, like those of (1984), show the expected parametric growth of subharmonic oblique waves at subcritical Reynolds numbers and a preferred β with maximum growth rate. Results of (1983c) for the Blasius boundary layer (neglecting non-parallelism) display some interesting features which may be interpreted as follows.

The evolution equations of near-resonant, infinitesimal, oblique modes A_1, A_2 in the presence of a finite two-dimensional wave with supposed constant amplitude A_0 have the form (cf. 16.7 and Craik 1975)

$$\left.\begin{array}{l} \dot{A}_1 = \sigma A_1 + \mu A_0 A_2^* + \lambda \,|\, A_0 \,|^2 A_1, \\ \dot{A}_2 = \sigma A_2 + \mu A_0 A_1^* + \lambda \,|\, A_0 \,|^2 A_2, \end{array}\right\} \tag{26.1}$$

with truncation after cubic terms. In a reference frame moving with the phase speed ω/α of the two-dimensional wave, σ (and so $A_{1,2}$) incorporates the effect of detuning from resonance. Setting $A_1 \propto \exp st$, $A_2 \propto \exp s^*t$ yields eigenvalues

$$s = \sigma_r + \lambda_r |A_0|^2 \pm \{-(\sigma_i + \lambda_i |A_0|^2)^2 + |\mu A_0|^2\}^{\frac{1}{2}}.$$

When $|A_0|$ is sufficiently small, these are complex conjugates.

As $|A_0|$ is increased, the two roots become real whenever $|\mu|^2 > 4\sigma_i \lambda_i$: modes A_1 and A_2 are then phase-locked to the fundamental wave A_0. At still larger $|A_0|$, the roots become complex conjugates again. The waves are driven unstable if, for some $|A_0|$, $\mathrm{Re}\, s > 0$. Behaviour of this sort was found by Herbert (1983c), but it must be noted that his series truncation, like (26.1), is invalid for large $|A_0|$.

Herbert (1983c) also discovered that higher eigenstates with the same x–y periodicity as the Orr–Sommerfeld eigenmodes A_1, A_2 may grow to prominence. These new modes originate from the least damped modes of the 'Squire equation' (7.4). If these are represented by A_3 and A_4, the linearized system (26.1) must be replaced by the four-mode equations

$$
\left.
\begin{aligned}
\dot{A}_1 &= \sigma A_1 + A_0(\mu A_2^* + \nu A_4^*) + |A_0|^2 (\lambda A_1 + \kappa A_3), \\
\dot{A}_2 &= \sigma A_2 + A_0(\mu A_1^* + \nu A_3^*) + |A_0|^2 (\lambda A_2 + \kappa A_4), \\
\dot{A}_3 &= \sigma' A_3 + A_0(\mu' A_2^* + \nu' A_4^*) + |A_0|^2 (\lambda' A_1 + \kappa' A_3), \\
\dot{A}_4 &= \sigma' A_4 + A_0(\mu' A_1^* + \nu' A_3^*) + |A_0|^2 (\lambda' A_2 + \kappa' A_4).
\end{aligned}
\right\}
\tag{26.2}
$$

Here, σ' is the least-damped eigenvalue $\frac{1}{2}i\alpha(c_r - \tilde{c})$ of the homogeneous form of (17.3), in a reference frame moving with the phase speed c_r of the wave A_0, and several additional quadratic and cubic coupling coefficients arise. With A_1, $A_3 \propto \exp st$ and A_2, $A_4 \propto \exp s^*t$, the eigenvalues s are found as roots of a fourth-degree polynomial. Two of Herbert's examples show that as $|A_0|$ increases, one of the two new roots coalesces with the smaller of two real roots associated with A_1 and A_2: the resultant complex-conjugate pair may, or may not, become unstable ($\mathrm{Re}\, s > 0$) depending on the chosen α, β and R. In another example, the A_1 and A_2 modes are sufficiently detuned from resonance that their roots remain complex conjugates as $|A_0|$ is increased and a real root originating as a 'Squire mode' is the first to become unstable. Saric & Thomas (1984) have invoked 'Squire modes' to explain some of their experimental observations. But this seems unconvincing since true 'Squire modes' have no velocity component normal to the wall: it must be remembered that modes can exchange their physical identities when dispersion curves pass close together (cf. §§2.2, 6.1).

The experiments of Kachanov & Levchenko (1984) merit further description. By a vibrating ribbon, they excited modes with two frequencies, ω_1 and $\omega_2 = \frac{1}{2}\omega_1 + \Delta\omega$, the latter at first having much smaller amplitudes than a two-dimensional mode with frequency ω_1 and wavenumber $(\alpha_1, 0)$. Because of the frequency mismatch $\Delta\omega$, symmetric wavenumber pairs $(\frac{1}{2}\alpha_1, \pm\beta)$ with frequency ω_2 are not exactly resonant with the mode $(\alpha_1, 0)$. However, there are modes $(\frac{1}{2}\alpha_1 + \delta, \beta)$, $(\frac{1}{2}\alpha_1 - \delta, -\beta)$ for appropriate δ and β with respective frequencies $\frac{1}{2}\omega_1 + \Delta\omega$ and $\frac{1}{2}\omega_1 - \Delta\omega$ and these form asymmetric resonant triads with $(\alpha_1, 0)$. The results of Kachanov & Levchenko show the emergence of a frequency peak at $\frac{1}{2}\omega_1 - \Delta\omega$ and amplitude modulations consistent with the growth of asymmetric resonant modes (Figure 5.12). In contrast with these quantitative experiments, those of Saric & Thomas (1984) are flow-visualization studies, using an ingenious smoke–wire technique. These show a characteristic 'staggered peak–valley structure' with Λ-shaped corrugations of streaklines when subharmonic modes are prominent; otherwise, a 'regular peak–valley structure' emerges, as in the earlier experiments of Klebanoff, Tidstrom & Sargent (1962). Examples of these distinctive patterns are reproduced in Figure 5.11.

For plane Poiseuille flow, recent experiments of Kozlov & Ramazanov (1984) and Nishioka & Asai (1985) concern the growth of three-dimensional modes in the presence of a finite-amplitude two-dimensional Tollmien–Schlichting wave: The latter find agreement with some of Herbert's results.

26.2   *Analysis of four-wave interactions*

The experiments of Klebanoff *et al.* showed clear evidence of spanwise-periodic longitudinal vortices maintained by the wave motion. This led Benney & Lin (1960) and Benney (1961, 1964) to examine the quadratic interaction of a two-dimensional wave with periodicity $A \exp \mathrm{i}(\alpha x - \omega t)$ and an oblique-wave combination of the form $B \cos \beta y \exp \mathrm{i}(\alpha x - \omega t)$. As already described in §13.1, this gave rise to longitudinal vortices like those observed: their assumption that two- and three-dimensional modes had equal x-wavenumbers *and* frequencies is unjustified, but this was later relaxed by Antar & Collins (1975). Benney & Lin did not continue their analysis to calculate the effect of cubic nonlinearities on the waves and gave no means of estimating the preferred spanwise wavenumber β.

Stuart (1962b) first set out the analytical formalism leading to equations for the temporal evolution of the complex amplitudes A and B, namely

$$\left.\begin{aligned}
\mathrm{d}A/\mathrm{d}t &= A(a_0 + a_1 |A|^2 + a_2 |B|^2 + \ldots) + a_3 A^* B^2 + \ldots \\
\mathrm{d}B/\mathrm{d}t &= B(b_0 + b_1 |A|^2 + b_2 |B|^2 + \ldots) + b_3 B^* A^2 + \ldots
\end{aligned}\right\} \quad (26.3)$$

Omitted terms are of higher than cubic order and hopefully negligible for small but finite amplitudes. The coefficients a_0, b_0 arise from linear theory. If ω is chosen to be the real frequency of the mode with wavenumber $(\alpha, 0)$, a_0 is real and denotes the linear growth rate αc_i; b_0 is then complex, with real part equal to the growth rate of oblique modes $(\alpha, \pm\beta)$ and imaginary part given by the frequency difference between these and the two-dimensional mode. The connection between (26.3) and the four-wave interaction equations (24.1)' with $|B_+| = |B_-|$ is clear. The wavenumber configuration satisfies

$$2(\alpha, 0) - (\alpha, \beta) - (\alpha, -\beta) = 0$$

but the frequencies are somewhat detuned from resonance.

The main task lies in determining the interaction coefficients a_j, b_j ($j = 1, 2, 3$) by the method of amplitude expansion. Second-order terms in A^2, AB and B^2 have x–t periodicity $\exp 2i(\alpha x - \omega t)$, the AB and some of the B^2 terms also being y-dependent. Others in $|A|^2$, AB^* and $|B|^2$ represent mean-flow modification and include Lin–Benney longitudinal vortices with spanwise wavenumbers β and 2β. These quadratic terms in turn interact with the A and B waves (and conjugates) to yield cubic nonlinearities, some with the same periodicity as the original waves. Much as for the single-wave equation of §20, compatibility conditions for the existence of solutions at this order yield the temporal evolution equations (26.3). The various interaction coefficients are found as integrals over the flow domain involving eigenfunctions of the linear and second-order problems. Clearly, a_1 is the Landau constant for the two-dimensional mode, but numerical determination of the other coefficients was not carried out until Itoh (1980b) did so for plane Poiseuille flow. As in the single-wave case, treatment of the growth-rate terms poses a dilemma (cf. §20.3): Itoh supposed that mean flow and harmonic terms are in equilibrium.

Itoh chose to re-express (26.3) in the real form

$$\begin{aligned}
&\tfrac{1}{2}|A|^{-2}(\mathrm{d}|A|^2/\mathrm{d}t) = a_0 + a_{1r}|A|^2 + [a_{2r} + |a_3|\sin(\theta + \theta_1)]|B|^2, \\
&\tfrac{1}{2}|B|^{-2}(\mathrm{d}|B|^2/\mathrm{d}t) = b_{0r} + [b_{1r} + |b_3|\sin(\theta + \theta_2)]|A|^2 + b_{2r}|B|^2, \\
&\tfrac{1}{2}\mathrm{d}\theta/\mathrm{d}t = b_{0i} + [b_{1i} - a_{1i} - |b_3|\cos(\theta + \theta_2)]|A|^2 \\
&\qquad\qquad\quad + [-a_{2i} + b_{2i} + |a_3|\cos(\theta + \theta_1)]|B|^2,
\end{aligned}\right\}$$

$$\text{(26.4a, b, c)}$$

$\theta \equiv 2\,\mathrm{ph}\,B - 2\,\mathrm{ph}\,A$, $\quad \theta_1 \equiv \mathrm{ph}\,(ia_3)$, $\quad \theta_2 \equiv \mathrm{ph}\,(ib_3)$.

Neither for plane Poiseuille nor Blasius flow can exact resonance ($b_{0i} = 0$) occur with the chosen wavenumber configuration. When detuning is sufficiently large that $|b_{0i}|$ is by far the largest term on the right of (26.4c), $\theta \approx 2b_{0i}t + \theta(0)$. Then, the θ-terms may be eliminated from (26.4a, b) by averaging over the period π/b_{0i} of 'fast' oscillations. Equilibrium solutions are then possible when

$$|A|^2\{a_0 + a_{1r}|A|^2 + a_{2r}|B|^2\} = |B|^2\{b_{0r} + b_{1r}|A|^2 + b_{2r}|B|^2\} = 0.$$

Corresponding equilibrium states in boundary-layer flows were earlier considered by Volodin & Zel'man (1977).

If $|B|$ is zero, the solution $|A|^2 = -a_0/a_{1r}$ denotes subcritical equilibrium (cf. §8.3). As R is reduced below the critical Reynolds number for the chosen α (where $a_0 = 0$ but $b_0 < 0$), the equilibrium value of $|A|$ increases from zero and infinitesimal three-dimensional modes B are driven unstable when $|A|^2$ exceeds $-b_{0r}/b_{1r} \equiv A_T^2$. Itoh has calculated this threshold amplitude A_T for onset of three-dimensional growth over a range of spanwise wavenumbers β at $R = 5500$, 5000 and 4500 and $\alpha = 1$. The lowest threshold is attained at $\beta \approx 0.3$ in each case. Itoh also found the threshold amplitudes $|A|$ for *temporary* growth of infinitesimal modes $|B|$ by considering $\sin(\theta + \theta_2)$ to equal unity in (26.4b): these are of course smaller than A_T and would replace the former time-average threshold if b_{0i} were sufficiently small to permit phase-locking of A and B modes with $\theta + \theta_2 = \frac{1}{2}\pi$.

A more general phase-locked criterion for sustained growth of (infinitesimal) $|B|$ is

$$\cos(\theta + \theta_2) = (b_{0i}|A|^{-2} + b_{1i} - a_{1i})|b_3|^{-1},$$

$$|b_0|^2 + 2|A|^2[b_{0r}b_{1r} + b_{0i}(b_{1i} + a_{1i})] + |A|^4[b_{1r}^2 - |a_3|^2 + (b_{1i} + a_{1i})^2] < 0.$$

Equations (26.4a, b, c) may also be employed to examine finite-amplitude solutions with $|B| = O(|A|)$, but the equations are certainly invalid when amplitudes are large. An erroneous symmetry argument advanced by Orszag & Patera (1983) purports to show that two- and three-dimensional modes are *necessarily* locked in phase: this physically absurd result is due to an incorrect Fourier representation of the disturbance (which fortunately does not invalidate their numerical results).

When $\alpha = 1$, the amplitude $|A|$ denotes the magnitude of velocity fluctuations normal to the wall at the channel centre, with the free-stream velocity taken as unity. The least value, $|A| = A_T$, which can support sustained three-dimensional growth is very small: Itoh's values are about 0.0031, 0.0047 and 0.0063 for $\alpha = 1$ and $R = 5500$, 5000 and 4500 respectively. These fall well below the corresponding equilibrium values $|A| = (-a_0/a_{1r})^{\frac{1}{2}}$ of the two-dimensional mode. But whether neglected higher-order terms are in fact negligible, even at such small amplitudes, remains an open question; this could, and should, be answered by extending to three-dimensional modes the methods employed by Sen & Venkateswarlu (1983).

Hall & Smith (1984) recently carried out an asymptotic analysis, valid as $R \to \infty$, of the weakly-nonlinear interaction of two or more non-resonant oblique modes, each having wavenumbers close to the lower branch of the neutral curve for linear stability. Of course, the choice of such wavenumbers, when other linearly-unstable modes are available, is artificial; but this is imposed by the requirements of rational asymptotic development in powers of the small parameter $R^{-\frac{1}{8}}$. For two modes, Hall & Smith's equations for downstream spatial evolution take the form

$$dA/dX = q_1(X - X_1) A + (a_1 |A|^2 + a_2 |B|^2) A$$
$$dB/dX = q_2(X - X_2) B + (b_1 |A|^2 + b_2 |B|^2) B,$$

where X_1, X_2 are real and q_j, a_j, b_j complex constants. Dependence of the respective linear-growth terms on scaled distance X derives from non-parallelism of their boundary-layer flow. Hall & Smith emphasize that non-parallelism can play a crucial rôle in determining the ultimate state achieved from given initial (i.e. upstream) conditions: solutions may reach a stable finite amplitude or terminate in a singularity. The calculated constants yield ratios $|a_{2r}/a_{1r}|$ and $|b_{1r}/b_{2r}|$ (where r denotes real part) which are typically rather large: that is to say, the mutual interaction of even non-resonant modes is usually stronger than their self-interaction, as Usher & Craik (1975) earlier inferred in less rigorous fashion.

Benney & Gustavsson (1981) have examined the nonlinear interaction of an oblique Orr–Sommerfeld mode (A) and an eigenmode (B) of the vertical-vorticity equation (7.4). As described in Gustavsson & Hultgren's (1980) linear analysis (see §7.1), these two modes are envisaged to be close to direct (linear) resonance and so have nearly the same periodicity $\exp[i(\alpha x + \beta y - \bar{\omega}t)]$. Weakly-nonlinear evolution of the amplitudes A and B then satisfies equations like (26.3), with $a_2 = a_3 = b_2 = 0$ but with an additional term $c_0 A$ added to the expression for dB/dt. The imaginary parts of a_0 and b_0 describe the slight detuning of either mode from the frequency $\bar{\omega}$. Solutions may be examined much as for (26.3). Benney & Gustavsson also consider interaction of several such modes. The practical importance, or otherwise, of vertical-vorticity modes in hydrodynamic stability needs further clarification.

26.3 Direct computational approach

Direct numerical solution of the three-dimensional, time-dependent Navier–Stokes equations has been accomplished by Orszag & Kells (1980), Orszag & Patera (1983) and Rozhdestvensky & Simakin (1984). Their spectral method was briefly discussed in §22.3, along with the two-dimensional solutions. A two-dimensional disturbance consists of the fundamental Fourier wavenumber component $(\alpha, 0)$, higher harmonics $(n\alpha, 0)$ and mean-flow distortion; three-dimensional Fourier components are chosen as $(n\alpha, m\beta)$ with integer n, m and some fixed β. Accordingly, subharmonic disturbances like those discussed in §17 and §26.1 are excluded. Earlier, rougher, computations by Maseev (1968) examined similar configurations.

A particular solution, for plane Poiseuille flow at $R = 2935$ with $\alpha = 1.3231$ and $\beta = 1$, is described by Orszag & Kells. A two-dimensional wave was imposed with quite large initial amplitude close to that for subcritical equilibrium; three-dimensional modes were then found to grow exponentially with time from very small amplitudes, in accord with expectations from (26.4). Figure 7.16, from Orszag & Patera, shows a series of similar calculations with $\alpha = \beta = 1.25$ and various subcritical Reynolds

Figure 7.16. Evolution of three-dimensional modes in presence of an initially dominant two-dimensional wave, for plane Poiseuille flow at $\alpha = \beta = 1.25$ and various Reynolds numbers R; from Orszag & Patera (1983).

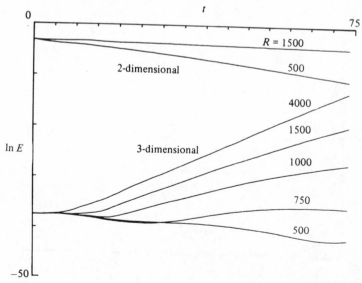

numbers. The chosen two-dimensional wave $(\alpha, 0)$ is initially large compared with the imposed three-dimensional modes $(\alpha, \pm\beta)$ but the Reynolds numbers are too low to permit self-supporting finite-amplitude two-dimensional disturbances. Though the two-dimensional wave decays, the small-amplitude oblique modes grow rapidly for all $R \geqslant 1000$. Of course, truly infinitesimal oblique modes must ultimately decay again to zero, since the wave which forces them itself decays: but initially-small, finite, oblique modes might attain sufficient amplitude to support self-sustaining disturbances. Orszag & Kells give such examples, the one reproduced in Figure 7.17 showing apparently chaotic modulations at $R = R_\mathrm{p} = 1250$, $\alpha = \beta = 1$. (See p. 221 for definitions of R_p, R_q).

The same authors have carried out corresponding computations for plane Couette flow and pipe Poiseuille flow, neither of which sustains linear instability. They found that 'with moderate two-dimensional amplitudes, the critical Reynolds numbers for substantial three-dimensional growth are about 1000 in plane Poiseuille and Couette flows and several thousand in pipe Poiseuille flow'. They go on to conjecture that turbulence, once established, might be sustained at somewhat lower Reynolds numbers than this.

Rozhdestvensky & Simakin (1984) carried out similar computations of

Figure 7.17. Temporal evolution of mixed two- and three-dimensional disturbances at $R = 1250$, $\alpha = \beta = 1$ in plane Poiseuille flow; from Orszag & Kells (1980).

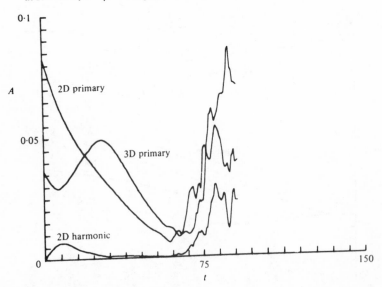

plane channel flows for much longer times. At $R_p = 1250$, $\alpha = \beta = 1$, they found that their disturbance eventually decayed to zero; but they found self-sustaining disturbances for $R_p \geqslant 2100$ ($R_q \geqslant 1313$), the lowest values of R_p and R_q corresponding to $\alpha = 1.25$ and $\beta = 2.0$. The value of R_q is in broad agreement with various experiments (e.g. Patel & Head 1969) which found transition to turbulence in plane channels for $R_q \gtrsim 1000$. More surprising is the quite good agreement between the observed (turbulent) mean-velocity profiles and those computed. Is it too much to hope that this approach may lead to a satisfactory representation of turbulent shear flows, despite the restriction to just a few three-dimensional Fourier modes? Other numerical studies specifically related to turbulence modelling are reviewed by Deissler (1984).

Orszag & Kells emphasize the computational difficulties of correctly describing these flows. 'One disturbing feature is the high resolution in both space and time that seems to be necessary'; 'low horizontal resolution can give spurious predictions of transition [which] must be considered carefully in future work...'

The importance of three-dimensional disturbances in promoting sub-critical instability and transition to turbulence in shear flows is clear: a two-dimensional viewpoint is seldom tenable. Three-dimensionality may manifest itself by the four-wave interactions (with higher-order modifications) here discussed or a subharmonic three-wave mechanism of §§ 17.2 and 26.1. Possible alternative routes are interaction with pre-existing spanwise variations of the mean flow or boundary walls (Komoda 1967; Nayfeh 1981; Dhanak 1983) and spanwise-vortex instability of the mean flow in the presence of two-dimensional waves (Craik 1982d; Herbert & Morkovin 1980; cf. § 13.2). The rôles of weak free-stream turbulence and surface roughness in promoting subcritical instability in boundary layers are well known but difficult to model mathematically (cf. Reshotko 1976; Tani 1969, 1981). Other, more exotic, mechanisms are doubtless possible.

Further theoretical and experimental work on three- and four-wave interactions is currently in progress: accounts presented at the Second IUTAM Symposium on Laminar–Turbulent Transition (Novosibirsk, 1984) have just been published (Kozlov 1985). A long-awaited monograph by Morkovin (1986?) will also discuss experiments on instability and transition to turbulence.

Chapter eight

STRONG INTERACTIONS, LOCAL
INSTABILITIES AND TURBULENCE:
A POSTSCRIPT

27   Strong interactions, local instabilities and turbulence: a postscript

27.1   *Short waves and long waves*

Small-wavelength disturbances may ride on large-amplitude long gravity waves. The orbital velocities of fluid particles due to the long waves provide a variable surface current through which the short waves propagate. When this current is comparable with the propagation velocity of the short waves relative to the long ones, their interaction is no longer weak. Nevertheless, the characteristics of the short waves may still be described, at least in part, by Whitham's theory of slowly-varying wave-trains in an inhomogeneous medium (see §11.3). Phillips (1981b) deduced from wave action conservation that capillary waves are likely to be 'blocked' by steep gravity waves. In much the same way, Gargett & Hughes (1972) earlier showed that short gravity waves may be trapped by long internal waves, so leading to caustic formation and local wave breaking. Untrapped modes also undergo amplitude modulations by the straining of the dominant wave field.

Computations of Longuet-Higgins (1978a, b) and McLean *et al.* (1981), already described in §22.2, display generation of short waves by high-order instability of steep gravity waves. In addition, finite-amplitude wave-trains necessarily contain bound harmonics, which travel with the fundamental Fourier component. Weakly-nonlinear interaction of neighbouring frequency components may also give rise to phase-locking of modes.

For these reasons, and doubtless others besides, measurements of the phase speeds of Fourier components of wave fields often reveal significant departures from the linear dispersion relation, even after allowance is made for wave-induced mean currents. Ramamonjiarisoa & Coantic (1976), Ramamonjiarisoa & Giovanangeli (1978) and others have observed a strong tendency for short waves to move with the phase speed of dominant longer waves (see also Yuen & Lake 1982).

Waves on a sharp density interface produce an oscillatory shear layer since the (inviscid) tangential velocity fluctuations on either side of the interface are of opposing sign. When the wave amplitude is large, this shear layer may exhibit Kelvin–Helmholtz-like instability, with billows of much smaller scale than the fundamental wavelength. Thorpe (1968b) observed such local instability induced by standing internal waves. This instability causes mixing of denser and lighter fluid and may contribute significantly, though intermittently, to thermocline erosion in sea and lakes.

Airflow over large-amplitude water waves is known to separate from near the crests, with consequent eddy formation. If a 'trapped' eddy forms on the downwind side of the crest, there is enhanced energy transfer from wind to wave. Jeffreys' (1925) old 'sheltering hypothesis', originally advanced to explain the generation of infinitesimal waves, is then more relevant. However, the mutual interaction of finite-amplitude water waves and a turbulent wind remains the most intractable of problems.

27.2   *Local transition in shear flows*

Growing disturbances in unstable shear flows induce increasingly strong distortions of the instantaneous velocity profile. Local, instantaneous profiles are likely to develop one or more inflection points and so may support a secondary inviscid instability like that of §3.1. When the primary disturbance contains appreciable oblique-wave components, the strongest inflectional, enhanced-shear profiles arise at 'peak' locations where the waves are largest. The subsequent instability may be regarded as a short-wave 'wrinkling' of this shear layer. Klebanoff, Tidstrom & Sargent (1962) detected this, for the flat-plate boundary layer, first as a rapid one-, two- or three-spike fluctuation in the hot-wire signal recording downstream velocity at 'peak' positions; further downstream, more spikes appeared and the small-scale disturbance soon evolved into a localized, but growing, turbulent spot (Figure 8.1).

Instantaneous velocity profiles were measured by Kovasznay, Komoda & Vasudeva (1962) for boundary layers and by Nishioka, Asai & Iida (1980) for plane Poiseuille flow. Their results are qualitatively very similar: viewed in a reference frame travelling downstream with the wave speed, their sequences of velocity profiles at peak locations display a region of enhanced shear inclined at an angle to the wall (Figure 8.2). On either side of peak positions, the maximum intensity of shear diminishes.

The enhanced shear layer was for long attributed mainly to distortion

Figure 8.1. Oscillograms of Klebanoff *et al.* (1962), of downstream velocity fluctuations, showing progress of 'breakdown' at various downstream distances x_1 from the vibrating ribbon ($U_\infty/\nu = 3.1 \times 10^5$ ft^{-1}, frequency 145 c s^{-1}). The lower trace is a sinusoidal reference signal.

$z = 0.12$ in. $y = -0.2$ in.

$x_1 = 8.00$ in.   9.00 in.   9.25 in.   9.50 in.

9.75 in.   10.00 in.   10.25 in.   10.50 in.

11.00 in.   12.00 in.   13.00 in.   15.00 in.

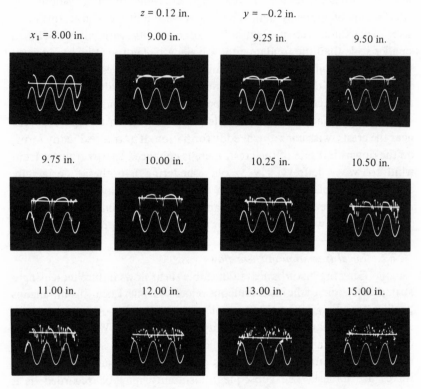

Figure 8.2. Instantaneous downstream velocity profiles measured during passage of 9.4% disturbance in plane Poiseuille flow (i.e. just before formation of the first 'spike'). Note inclination to wall of enhanced shear region. From Nishioka *et al.* (1980).

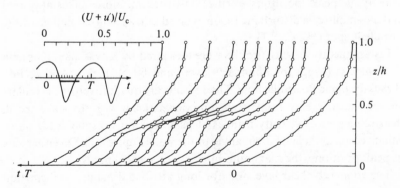

of the primary velocity profile by the spanwise-periodic Lin–Benney vortices with accompanying oblique waves (cf. Tani, 1969). However, instantaneous profiles are most strongly influenced by the waves themselves, oblique waves being particularly effective because of the contribution of the velocity component parallel to wave crests. It is the latter contribution which explains the inclination to the wall of the high-shear layer (cf. Craik 1980).

Nishioka, Asai & Iida (1980) investigated the local secondary instability of the high-shear layer by introducing artificial small-scale disturbances and measuring their spatial growth rates for a range of imposed frequencies much higher than that of the primary waves. Growth rates much larger than that of Tollmien–Schlichting waves were recorded.

The secondary instability is strongly three-dimensional, the 'wrinkle' of the high-shear layer being rapidly swept back on either side of the peak location. Klebanoff *et al.* interpreted this as the origin of the discrete 'hairpin-eddy' structure known to develop further downstream. A similar secondary instability, occurring within a localized packet of primary waves, was described by Gaster (1978): this appears as a smaller-scale 'packet within a packet' which later manifests a tertiary instability to even smaller scales. Flow-visualization experiments of Wortmann (1977) give yet another view of these complex events. It seems that the secondary instability at first comprises a small number of swept-back short waves; but these quickly roll up, like Kelvin–Helmholtz billows, into discrete vortices with strongly three-dimensional structure.

Attempts to develop a theoretical description of the secondary transition have had only limited success. Greenspan & Benney (1963) tried to model it by considering the linear stability of a set of two-dimensional inviscid parallel flows. With a profile chosen to represent that just prior to breakdown, they found rapid instability with a preferred wavenumber about five times that of the primary wave: an encouraging result from an over-crude model.

Later, Landahl (1972) and Itoh (1981) attempted to extend Whitham's kinematic-wave theory to dissipative waves in a slowly-varying shear flow. Landahl's formulation proved controversial and was strongly criticized by Stewartson (1975), Tani (1980, 1981) and others on grounds of mathematical inconsistency. Itoh's attempt to resolve these was only partly successful and the matter was re-examined by Landahl (1982). Though a totally convincing theory is still lacking, the underlying idea, that short secondary waves grow not only through local instability but also by focusing of wave energy at particular locations, remains an attractive one. Such focusing can certainly take place when capillary waves ride on longer gravity waves, for instance; however, secondary transition in shear flows is not only dissipative but strongly three-dimensional. The Landahl–Itoh theory confronted, without entirely resolving, the difficulties of handling dissipation, but three-dimensionality was ignored.

The strength, and distance from the wall, of this high-shear layer depends both on downstream distance $x' = x - c_r t$ and spanwise distance z. Furthermore, its orientation does not normally coincide with that of streamlines (viewed relative to the wave). Accordingly, it is doubtful whether a quasi-parallel approximation is justified, even though the wavelength of the secondary instability is rather small. Craik (1980) conjectured that secondary instability may be most likely when the primary waves are large enough to nearly 'match' streamlines and high-shear layer over a substantial part of the primary wavelength: only in this event do individual fluid particles 'see' a slowly-varying inflectional profile. Perhaps the best hope of success lies in refining the resolution of full-scale numerical models, like those developed by Orszag and Herbert for the growth of the primary waves: at present, these cannot cope with the small spatial scales of secondary transition.

The secondary transition is an important precursor of the development of turbulence: the local 'spikes' or 'wrinkles' of the high-shear layer in turn lead to local turbulent spots which grow and eventually merge to yield a fully-turbulent state. Conversely, an artificially-introduced turbulent spot generates oblique Tollmien–Schlichting waves somewhat analogous to ship waves (Wygnanski, Haritonidis & Kaplan 1979). For reports of other current work, see Kozlov (1985): especially the view of Kachanov *et al.* (1985) that high-frequency 'spikes' indicate rapid deterministic growth of high harmonics of oblique modes, rather than local secondary instability.

27.3   *Some thoughts on transition and turbulence*

Much attention and effort has recently centred on the investigation of 'coherent structures' in fully-developed turbulent flows. These are wave-like or billow-like features which resemble laminar-flow phenomena: they may persist amid the 'incoherent' turbulence for some time and similar structures continually appear and then decay. Hussain (1983) has recently reviewed much of the experimental literature and current quasi-theoretical speculations in this area.

It has been repeatedly conjectured that these large-scale coherent motions within turbulent flows may originate from linear instability of the mean velocity profile. Though the mean velocity profile doubtless plays a part in the development of such disturbances, it is inconceivable to the present writer that strong, large-scale turbulent fluctuations should not also do so: to restrict attention to linear instability of the mean-flow profile seems naïve.

Apparently even more naïve is the customary 'engineering approximation' for prediction of transition to turbulence in boundary-layer flows. According to conventional lore, disturbances introduced at an upstream location (typically the leading edge of an aircraft wing or turbine blade), trigger transition to turbulence when their amplitude is increased in accordance with linear theory by a factor of e^N: the value of N is usually, but not always, taken to be 9. This criterion pays no heed to the size of the initial disturbance nor to any nonlinear mechanism. Nevertheless, it is a firmly-established, simple, empirical criterion which would not have survived in use were it totally unreliable (see e.g. Mack 1977). It may be defended on two main grounds: firstly, the level of background disturbances is often much the same for a variety of flow conditions; and secondly, the time (and distance) of linear amplification of Tollmien–Schlichting waves is much longer than the interval between onset of significant nonlinearity and breakdown to turbulence. The latter view is supported by the known rapid growth of weakly nonlinear three-dimensional disturbances and the short timescale of local secondary transition. Ironically, it is the very strength and rapidity of nonlinear growth mechanisms which enables them to be neglected in the empirical e^N criterion!

A recent growth area is the development of computer codes for describing fully turbulent flows, usually offered as a service to industry. Almost all involve empirical modelling of small-scale 'sub-grid' motions and so lack secure theoretical support: claims to accuracy, in a competitive commercial market, may owe more to salesmanship than scholarship. Notable and promising exceptions are the works of Patera & Orszag (1981) and Rozhdestvensky & Simakin (1984) which simulate turbulent channel flows at Reynolds numbers up to 5000 (see also Deissler 1984).

Parametrization of turbulent mixing processes is incorporated into increasingly-elaborate large-scale computer models of atmospheric and ocean dynamics. Employment of empirical 'eddy diffusion coefficients' to model turbulence is widespread but defensible only *faute de mieux*. Models of global weather patterns, for example, contain many free parameters, and these may be adjusted to reproduce known realistic features with considerable success. But the use of such models as predictive tools, to estimate, say, the effect of melting the polar icecap, is fraught with dangers.

They are very likely to have been constructed to be 'robust', producing small responses to small perturbations; but the real environment is not necessarily so resilient. Various mechanisms, which have been described in previous chapters, produce large responses to small disturbances in relatively simple dynamical systems; is it reasonable to expect complex systems to behave otherwise?

Undoubtedly, the next decade will see a further increase in the use and power of computational methods and a likely decline in the application of analytical, particularly asymptotic, techniques. However, the very success of a direct computational approach carries its own dangers. The user of a 'package' for solving the Navier–Stokes equations need not be so conversant with underlying mathematical structure, nor develop as much physical insight, as one whose work is analytically based. At present, the best numerical work has grown naturally from the (relatively) firm ground of weakly-nonlinear theory, and its practitioners are themselves well-versed in analytical methods. Emphasis on adequate spatial and temporal resolution has enabled reproduction and extension of known results of linear and weakly-nonlinear theory. Inadequate resolution may yield spectacular failures or, worse, results which are plausible but incorrect. The development of 'computational fluid dynamics' as a separate, self-contained, discipline must be resisted.

That the past twenty years have witnessed important theoretical advances should be evident. But they have also seen extravagant claims, engendered by over-enthusiasm for an attractive idea, which mislead the ill-informed and are detrimental to science. The irrational temptation to extrapolate results of rational theories beyond their range of validity is ever-present: so too is uncritical belief in over-crude empirical modelling. The 'key' to understanding nonlinear wave motion and transition to turbulence is not any *one* of solitons, bifurcation theory, catastrophe theory, strange attractors, period-doubling cascades, *et cetera*. Fashionable, and fascinating, theoretical bandwagons add momentum to scientific progress but can also carry the unwary up blind alleys. The richness of fluid mechanics is such that many new surprises and insights still await discovery.

REFERENCES

Ablowitz, M. J. & Haberman, R. (1975) *Phys. Rev. Lett.* **35**, 1185–1188. Nonlinear evolution equations – two and three dimensions.

Ablowitz, M. J. & Segur, H. (1979) *J. Fluid Mech.* **92**, 691–715. On the evolution of packets of water waves.

Ablowitz, M. J. & Segur, H. (1981) *Solitons and the Inverse Scattering Transform.* S.I.A.M. Philadelphia.

Abramowitz, M. & Stegun, I. A. (eds) (1964) *Handbook of Mathematical Functions.* U.S. Govt. Printing Office, Washington, D.C. [also Dover, 1965].

Acheson, D. J. (1976) *J. Fluid Mech.* **77**, 433–472. On over-reflexion.

Acheson, D. J. (1980) *J. Fluid Mech.* **96**, 723–733. 'Stable' density stratification as a catalyst for instability.

Ahlers, G. & Behringer, R. P. (1978) *Phys. Rev. Lett.* **40**, 712–719. Evolution of turbulence for the Rayleigh–Bénard instability.

Akylas, T. R. (1982) *Stud. Appl. Math.* **67**, 1–24. A nonlinear theory for the generation of water waves by wind.

Akylas, T. R. & Benney, D. J. (1980) *Stud. Appl. Math.* **63**, 209–226. Direct resonance in nonlinear wave systems.

Akylas, T. R. & Benney, D. J. (1982) *Stud. Appl. Math.* **67**, 107–123. The evolution of waves near direct-resonance conditions.

Alber, I. E. (1978) *Proc. Roy. Soc. Lond.* **A363**, 525–546. The effects of randomness on the stability of two-dimensional surface wavetrains.

Andereck, C. D., Dickman, R. & Swinney, H. L. (1983) *Phys. Fluids* **26**, 1395–1401. New flows in a circular Couette system with co-rotating cylinders.

Andrews, D. G. & McIntyre, M. E. (1978a) *J. Fluid Mech.* **89**, 609–646. An exact theory of nonlinear waves on a Lagrangian-mean flow.

Andrews, D. G. & McIntyre, M. E. (1978b) *J. Fluid Mech.* **89**, 647–664 (Corrigendum **95**, 796). On wave-action and its relatives.

Anker, D. & Freeman, N. C. (1978a) *Proc. Roy. Soc. Lond.* **A360**, 529–540. On the soliton solutions of the Davey–Stewartson equation for long waves.

Anker, D. & Freeman, N. C. (1978b) *J. Fluid Mech.* **87**, 17–31. Interpretation of three-soliton interactions in terms of resonant triads.

Antar, B. N. & Benek, J. A. (1978) *Phys. Fluids* **21**, 183–189. Temporal eigenvalue spectrum of the Orr–Sommerfeld equation for the Blasius boundary layer.

Antar, B. N. & Collins, F. G. (1975) *Phys. Fluids* **18**, 289–297. Numerical calculation of finite amplitude effects in unstable laminar boundary layers.

289

Armstrong, J. A., Bloembergen, N., Ducuing, J. & Pershan, P. S. (1962) *Phys. Rev.* **127**, 1918–1939. Interactions between light waves in a nonlinear dielectric.

Armstrong, J. A., Sudhanshu, S. J. & Shiren, N. S. (1970) *IEEE J. Quantum Electronics* **QE-6**, 123–129. Some effects of group-velocity dispersion on parametric interactions.

Baldwin, P. & Roberts, P. H. (1970) *Mathematika* **17**, 102–119. The critical layer in stratified shear flow.

Ball, F. K. (1964) *J. Fluid Mech.* **20**, 465–478. Energy transfer between external and internal gravity waves.

Banks, W. H. H., Drazin, P. G. & Zaturska, M. B. (1976) *J. Fluid Mech.* **75**, 149–171. On the normal modes of parallel flow of inviscid stratified fluid.

Bannerjee, P. P. & Korpel, A. (1982) *Phys. Fluids* **25**, 1938–1943. Subharmonic generation by resonant three-wave interaction of deep-water capillary waves.

Barenblatt, G. I., Iooss, G. & Joseph, D. D. (eds) (1983) *Nonlinear Dynamics and Turbulence*, Pitman.

Barnett, T. P. & Kenyon, K. E. (1975) *Rep. Prog. Phys.* **38**, 667–729. Recent advances in the study of wind waves.

Barr, A. D. S. & Ashworth, R. P. (1977) *U.S. Air Force Off. Sci. Res.* Rep. 74-2723. Parametric and nonlinear mode interaction behaviour in the dynamics of structures.

Batchelor, G. K. (1956) *J. Fluid Mech.* **1**, 177–190. On steady laminar flow with closed streamlines at large Reynolds number.

Bekefi, G. (1966) *Radiation processes in plasmas*, Wiley.

Béland, M. (1976) *J. Atmos. Sci.* **33**, 2066–2078. Numerical study of the nonlinear Rossby wave critical level development in a barotropic zonal flow.

Béland, M. (1978) *J. Atmos. Sci.* **35**, 1802–1815. The evolution of a nonlinear Rossby wave critical level: effects of viscosity.

Bénard, H. (1900) *Revue Gén. Sci. Pur. Appl.* **11**, 1261–1271, 1309–1328. Les tourbillons cellulaires dans une nappe liquide.

Bénard, H. (1901) *Ann. Chim. Phys.* **7**, Sér 23, 62–144. Les tourbillons cellulaires dans une nappe liquid transportant de la chaleur par convection en régime permanent.

Benjamin, T. Brooke (1957) *J. Fluid Mech.* **2**, 554–574 (and Corrigendum, **3**, (1958) 657). Wave formation in laminar flow down an inclined plane.

Benjamin, T. Brooke (1959) *J. Fluid Mech.* **6**, 161–205. Shearing flow over a wavy boundary.

Benjamin, T. Brooke (1961) *J. Fluid Mech.* **10**, 401–419. The development of three-dimensional disturbances in an unstable film of liquid flowing down an inclined plane.

Benjamin, T. Brooke (1963) *J. Fluid Mech.* **16**, 436–450. The threefold classification of unstable disturbances in flexible surfaces bounding inviscid flows.

Benjamin, T. Brooke (1967) *Proc. Roy. Soc. Lond.* **A299**, 59–75. Instability of periodic wavetrains in nonlinear dispersive systems.

Benjamin, T. Brooke (1978) *Proc. Roy. Soc. Lond.* **A359**, 1–26. Bifurcation phenomena in steady flows of a viscous fluid. I. Theory.

Benjamin, T. Brooke (1981) In *Transition and Turbulence* (ed. R. E. Meyer), pp. 25–41. Academic. New observations in the Taylor experiment.

Benjamin, T. Brooke & Feir, J. E. (1967) *J. Fluid Mech.* **27**, 417–430. The disintegration of wavetrains on deep water. Part 1. Theory.

Benjamin, T. Brooke & Mullin, T. (1981) *Proc. Roy. Soc. Lond.* **A377**, 221–249. Anomalous modes in the Taylor experiment.

Benjamin, T. Brooke & Mullin, T. (1982) *J. Fluid Mech.* **121**, 219–230. Notes on the multiplicity of flows in the Taylor experiment.

Benney, D. J. (1961) *J. Fluid Mech.* **10**, 209–236. A non-linear theory for oscillations in a parallel flow.

Benney, D. J. (1962) *J. Fluid Mech.* **14**, 577–584. Non-linear gravity wave interactions.

Benney, D. J. (1964) *Phys. Fluids* **7**, 319–326. Finite amplitude effects in an unstable laminar boundary layer.

Benney, D. J. (1966) *J. Math. and Phys.* **45**, 150–155. Long waves on liquid films.

Benney, D. J. (1976) *Stud. Appl. Math.* **55**, 93–106. Significant interactions between small and large scale surface waves.

Benney, D. J. (1984) *Stud. Appl. Math.* **70**, 1–19. The evolution of disturbances in shear flows at high Reynolds numbers.

Benney, D. J. & Bergeron, R. F. (1969) *Stud. Appl. Math.* **48**, 181–204. A new class of nonlinear waves in parallel flows.

Benney, D. J. & Gustavsson, L. H. (1981) *Stud. Appl. Math.* **64**, 185–209. A new mechanism for linear and nonlinear hydrodynamic instability.

Benney, D. J. & Lin, C. C. (1960) *Phys. Fluids* **3**, 656–657. On the secondary motion induced by oscillations in a shear flow.

Benney, D. J. & Maslowe, S. A. (1975) *Stud. Appl. Math.* **54**, 181–205. The evolution in space and time of nonlinear waves in parallel shear flows.

Benney, D. J. & Newell, A. C. (1967) *J. Math. and Phys.* **46**, 133–139. The propagation of nonlinear wave envelopes.

Benney, D. J. & Roskes, G. (1969) *Stud. Appl. Math.* **48**, 377–385. Wave instabilities.

Betchov, R. & Szewczyk, A. (1963) *Phys. Fluids* **6**, 1391–1396. Stability of a shear layer between parallel streams.

Bingham, R. & Lashmore-Davies, C. N. (1979) *J. Plasma Phys.* **21**, 51–59. On the nonlinear development of the Langmuir modulational instability.

Blake, J. R. & Sleigh, M. A. (1975) *Swimming and Flying in Nature* vol. 1, (ed. Wu, T. Y-T., Brokaw, C. J. & Brennen, C.) Plenum, New York. Hydromechanical aspects of ciliary propulsion.

Blennerhassett, P. J. (1980) *Phil. Trans. Roy. Soc. Lond.* **A298**, 451–494. On the generation of waves by wind.

Blumen, W. (1971) *Geophys. Fluid Dyn.* **2**, 189–200. On the stability of plane flow with horizontal shear to three-dimensional nondivergent disturbances.

Bodonyi, R. J. & Smith, F. T. (1981) *Proc. Roy. Soc. Lond.* **A375**, 65–92. The upper branch stability of the Blasius boundary layer, including non-parallel flow effects.

Bodonyi, R. J., Smith, F. T. & Gajjar, J. (1983) *I.M.A. J. Appl. Math.* **30**, 1–19. Amplitude-dependent stability of boundary-layer flow with a strongly nonlinear critical layer.

Booker, J. R. & Bretherton, F. P. (1967) *J. Fluid Mech.* **27**, 513–539. The critical layer for internal gravity waves in a shear flow.

Bouthier, M. (1973) *J. de Mécanique* **12**, 75–95. Stabilité linéaire des écoulements presque parallèles, II. La couche limite de Blasius.

Bouthier, M. (1983) *Comptes rendus Acad. Sci. Paris* **296**, Sér II, 593–596. Sur la stabilité des écoulements non parallèles et le spectre continu.

Breeding, R. J. (1971) *J. Fluid Mech.* **50**, 545–563. A nonlinear investigation of critical levels for internal atmospheric gravity waves.

Breeding, R. J. (1972) *J. Geophys. Res.* **77**, 2681–2692. A nonlinear model of the break-up of internal gravity waves due to their exponential growth with height.

Brennen, C. (1974) *J. Fluid Mech.* **65**, 799–824. An oscillating boundary-layer theory for ciliary propulsion.

Bretherton, F. P. (1964) *J. Fluid Mech.* **20**, 457–479. Resonant interactions between waves. The case of discrete oscillations.

Bretherton, F. P. (1966) *Quart. J. Roy. Met. Soc.* **92**, 466–480. The propagation of groups of internal waves in a shear flow.

Bretherton, F. P. (1971) *Lectures Appl. Math.* **13**, 61–102, Am. Math. Soc. The general linear theory of wave propagation.

Bretherton, F. P. & Garrett, C. J. R. (1968) *Proc. Roy. Soc. Lond.* **A302**, 529–554. Wavetrains in inhomogeneous moving media.

Briggs, R. J., Daugherty, J. D. & Levy, R. H. (1970) *Phys. Fluids* **13**, 421–432. Role of Landau damping in crossed-field electron beams and inviscid shear flow.

Brown, S. N., Rosen, A. S. & Maslowe, S. A. (1981) *Proc. Roy. Soc. Lond.* **A375**, 271–293. The evolution of a quasi-steady critical layer in a stratified viscous shear layer.

Brown, S. N. & Stewartson, K. (1978a) *Geophys. Astrophys. Fluid Dyn.* **10**, 1–24. The evolution of the critical layer of a Rossby wave. Part II.

Brown, S. N. & Stewartson, K. (1978b) *Proc. Roy. Soc. Lond.* **A363**, 175–194. The evolution of a small inviscid disturbance to a marginally unstable stratified shear flow; stage two.

Brown, S. N. & Stewartson, K. (1980) *J. Fluid Mech.* **100**, 811–816. The algebraic decay of disturbances in a stratified linear shear flow.

Brown, S. N. & Stewartson, K. (1982a) *J. Fluid Mech.* **115**, 217–230. On the nonlinear reflection of a gravity wave at a critical level. Part 2.

Brown, S. N. & Stewartson, K. (1982b) *J. Fluid Mech.* **115**, 231–250. On the nonlinear reflection of a gravity wave at a critical level. Part 3.

Bryant, P. J. (1982) *J. Fluid Mech.* **114**, 443–466. Modulation by swell of waves and wave groups on the ocean.

Burgers, J. M. (1946) *Adv. Appl. Mech.* **1**, 171–196. A mathematical model illustrating the theory of turbulence.

Burke, A. T. (1983) 'On gravity induced critical layers'. Ph.D. Thesis, University of London.

Busse, F. H. (1967) *J. Fluid Mech.* **30**, 625–649. The stability of finite amplitude cellular convection and its relation to an extremum principle.

Busse, F. H. (1978) *Rep. Prog. Phys.* **41**, 1930–1967. Non-linear properties of thermal convection.

Busse, F. H. (1981a) In *Transition and Turbulence* (ed. R. E. Meyer), pp. 43–61, Academic. Transition to turbulence in thermal convection with and without rotation.

Busse, F. H. (1981b) In *Hydrodynamic Instabilities and the Transition to Turbulence* (eds H. L. Swinney & J. P. Gollub), pp. 97–137, Springer. Transition to turbulence in Rayleigh–Bénard convection.

Busse, F. H. & Clever, R. M. (1979) In *Recent Developments in Theoretical and Experimental Fluid Mechanics* (eds U. Müller, K. G. Roesner & B. Schmidt), pp. 376–385, Springer. Nonstationary convection in a rotating system.

Busse, F. H. & Heikes, K. E. (1980) *Science* **208**, 173–175. Convection in a rotating layer: a simple case of turbulence.

Busse, F. H. & Whitehead, J. A. (1971) *J. Fluid Mech.* **47**, 305–320. Instabilities of convection rolls in a high Prandtl number fluid.

Busse, F. H. & Whitehead, J. A. (1974) *J. Fluid Mech.* **66**, 67–79. Oscillatory and convective instabilities in large Prandtl number convection.

Cairns, R. A. (1979) *J. Fluid Mech.* **92**, 1–14. The role of negative energy waves in some instabilities of parallel flows.

Cairns, R. A. & Lashmore-Davies, C. N. (1983a) *Proc. 3rd Joint Varenna–Grenoble Internat. Symp. on Heating in Toroidal Plasmas*, Grenoble 1982, Vol. 2, 755–760. A mode conversion interpretation of the absorption of electron cyclotron radiation for perpendicular propagation.

Cairns, R. A. & Lashmore-Davies, C. N. (1983b) *Phys. Fluids* **26**, 1268–1274. A unified theory of a class of mode conversion problems.

Campbell, D. & Rose, H. (eds) (1983) *Physica 7D*. 'Order in Chaos': Proc. Internat. Conf., Los Alamos (1982).

Caponi, E. A., Fornberg, B., Knight, D. D., McLean, J. W., Saffman, P. G. & Yuen, H. C. (1982) *J. Fluid Mech.* **124**, 347–362. Calculations of laminar viscous flow over a moving wavy surface.

Caponi, E. A., Saffman, P. G. & Yuen, H. C. (1982) *Phys. Fluids* **25**, 2159–2166. Instability and confined chaos in a nonlinear dispersive wave system.

Case, K. M. (1961) *J. Fluid Mech.* **10**, 420–429. Hydrodynamic stability and the inviscid limit.

Case, K. M. & Chiu, S. C. (1977) *Phys. Fluids* **20**, 742–745. Three-wave resonant interactions of gravity-capillary waves.

Chabert-d'Hieres, G. (1960) *Houille Blanche* **15**, 153–163. Etude du clapotis.

Chandrasekhar, S. (1961) *Hydrodynamic and Hydromagnetic Stability*, Clarendon, Oxford.

Chen, B. & Saffman, P. G. (1980a) *Stud. Appl. Math.* **62**, 1–21. Numerical evidence for the existence of new types of gravity waves of permanent form on deep water.

Chen, B. & Saffman, P. G. (1980b) *Stud. Appl. Math.* **62**, 95–111. Steady gravity-capillary waves on deep water. II Numerical results for finite amplitude.

Chen, T. S. & Joseph, D. D. (1973) *J. Fluid Mech.* **58**, 337–351. Subcritical bifurcation of plane Poiseuille flow.

Chester, W. (1968) *Proc. Roy. Soc. Lond.* **A306**, 5–22. Resonant oscillations of water waves. I. Theory.

Chester, W. & Bones, J. A. (1968) *Proc. Roy. Soc. Lond.* **A306**, 23–39. Resonant oscillations of water waves. II. Experiment.

Childress, S. (1981) *Mechanics of Swimming and Flying*, Cambridge Univ. Press.

Chu, F. Y. F. (1975a) *Phys. Lett.* **51A**, 129–130. Special solutions for nonlinear wave-particle interaction.

Chu, F. Y. F. (1975b) *Phys. Rev.* **A12**, 2065–2067. Bäcklund transformation for the wave–wave scattering equations.

Chu, F. Y. F. & Karney, C. F. F. (1977) *Phys. Fluids* **20**, 1728–1732. Solution of the three-wave resonant equations with one wave heavily damped.

Chu, F. Y. F. & Scott, A. C. (1975) *Phys. Rev.* **A12**, 2060–2067. Inverse scattering transform for wave–wave scattering.

Clever, R. M. & Busse, F. H. (1974) *J. Fluid Mech.* **65**, 625–645. Transition to time-dependent convection.

Coffee, T. (1977) *J. Fluid Mech.* **83**, 401–413 (and Corrigendum, **88**, 798 (1978)). Finite amplitude stability of plane Couette flow.

Cohen, B. I., Krommes, J. A., Tang, W. M. & Rosenbluth, M. N. (1976) *Nuclear Fusion* **16**, 971–992. Nonlinear saturation of the dissipative trapped-ion mode by mode coupling.

Cohen, L. S. & Hanratty, T. J. (1965) *A.I.Ch.E.J.* **11**, 138–144. Generation of waves in the concurrent flow of air and a liquid.

Cokelet, E. D. (1977) *Phil. Trans. Roy. Soc. Lond.* **A286**, 183–230. Steep gravity waves in water of arbitrary uniform depth.

Cole, J. A. (1976) *J. Fluid Mech.* **75**, 1–15. Taylor-vortex instability and annulus-length effects.

Coles, D. (1965) *J. Fluid Mech.* **21**, 385–425. Transition in circular Couette flow.

Collins, J. I. (1963) *J. Geophys. Res.* **68**, 6007–6014. Inception of turbulence at the bed under periodic gravity waves.

Coppi, B., Rosenbluth, M. N. & Sudan, R. N. (1969) *Annals of Phys.* **55**, 207–247. Nonlinear interactions of positive and negative energy modes in rarified plasmas (I).

Copson, E. T. (1975) *Partial Differential Equations*, Cambridge Univ. Press.

Corner, D., Barry, M. D. J. & Ross, M. A. S. (1974) *Aero. Res. Counc. Rep.* C.P. 1296, London. Nonlinear stability theory of the flat plate boundary layer.

Corner, D., Houston, D. J. R. & Ross, M. A. S. (1976) *J. Fluid Mech.* **77**, 81–103. Higher eigenstates in boundary-layer stability theory.

Cornille, H. (1979) *J. Math. Phys.* **20**, 1653–1666. Solutions of the nonlinear three-wave equations in three spatial dimensions.

Cowley, S. J. & Smith, F. T. (1985) *J. Fluid Mech.* **156**, 83–100. On the stability of Poiseuille–Couette flow: a bifurcation from infinity.

Craik, A. D. D. (1966) *J. Fluid Mech.* **26**, 369–392. Wind-generated waves in thin liquid films.

Craik, A. D. D. (1968) *J. Fluid Mech.* **34**, 531–549. Resonant gravity-wave interactions in a shear flow.

Craik, A. D. D. (1969) *J. Fluid Mech.* **36**, 685–693. The stability of plane Couette flow with viscosity stratification.

Craik, A. D. D. (1970) *J. Fluid Mech.* **41**, 801–821. A wave-interaction model for the generation of windrows.

Craik, A. D. D. (1971) *J. Fluid Mech.* **50**, 393–413. Nonlinear resonant instability in boundary layers.

Craik, A. D. D. (1975) *Proc. Roy. Soc. Lond.* **A343**, 351–362. Second order resonance and subcritical instability.

Craik, A. D. D. (1977) *J. Fluid Mech.* **81**, 209–223. The generation of Langmuir circulations by an instability mechanism.

Craik, A. D. D. (1978) *Proc. Roy. Soc. Lond.* **A363**, 257–269. Evolution in space and time of resonant wave triads. II. A class of exact solutions.

Craik, A. D. D. (1980) *J. Fluid Mech.* **99**, 247–265. Nonlinear evolution and breakdown in unstable boundary layers.

Craik, A. D. D. (1981) *Proc. Roy. Soc. Lond.* **A373**, 457–476. The development of wave packets in unstable flows.

Craik, A. D. D. (1982a) *Trans. A.S.M.E. Jour. Appl. Mech.* **49**, 284–290. The growth of localized disturbances in unstable flows.

Craik, A. D. D. (1982b) *J. Fluid Mech.* **116**, 187–205. The drift velocity of water waves.

Craik, A. D. D. (1982c) *J. Fluid Mech.* **125**, 27–35. The generalized Lagrangian-mean equations and hydrodynamic stability.

Craik, A. D. D. (1982d) *J. Fluid Mech.* **125**, 37–52. Wave-induced longitudinal-vortex instability in shear flows.

Craik, A. D. D. (1983) *Proc. Roy. Soc. Edin.* **94A**, 85–88. Growth of localized disturbances on a vortex sheet.

Craik, A. D. D. & Adam, J. A. (1978) *Proc. Roy. Soc. Lond.* **A363**, 243–255. Evolution in space and time of resonant wave triads. I. The 'pump-wave approximation'.

Craik, A. D. D. & Adam, J. A. (1979) *J. Fluid Mech.* **92**, 15–33. 'Explosive' resonant wave interactions in a three-layer fluid flow.

Craik, A. D. D. & Leibovich, S. (1976) *J. Fluid Mech.* **73**, 401–426. A rational model for Langmuir circulations.

Crampin, D. J. & Dore, B. D. (1979) *Mathematika* **26**, 224–235. Mass transport induced by standing interfacial waves.

Crapper, G. D. (1957) *J. Fluid Mech.* **2**, 532–540. An exact solution for progressive capillary waves of arbitrary amplitude.

Crapper, G. D. (1970) *J. Fluid Mech.* **40**, 149–159. Nonlinear capillary waves generated by steep gravity waves.

Crawford, D. R., Saffman, P. G. & Yuen, H. C. (1980) *Wave Motion* **2**, 1–16. Evolution of a random inhomogeneous field of nonlinear deep-water gravity waves.

Crawford, D. R., Lake, B. N., Saffman, P. G. & Yuen, H. C. (1981) *J. Fluid Mech.* **105**, 177–191. Stability of weakly nonlinear deep-water waves in two and three dimensions.

Criminale, W. O. & Kovasznay, L. S. G. (1962) *J. Fluid Mech.* **14**, 59–80. The growth of localized disturbances in a laminar boundary layer.

Curry, J. H. (1978) *Comm. Math. Phys.* **60**, 193–204. A generalized Lorenz system.

Da Costa, L. N., Knobloch, E. & Weiss, N. O. (1981) *J. Fluid Mech.* **109**, 25–43. Oscillations in double-diffusive convection.·

Daniels, P. (1977) *Proc. Roy. Soc. Lond.* **A358**, 173–197. The effect of distant side walls on the transition to finite amplitude Bénard convection.

Davey, A. (1962) *J. Fluid Mech.* **14**, 336–368. The growth of Taylor vortices in flow between rotating cylinders.

Davey, A. (1972) *J. Fluid Mech.* **53**, 769–781. The propagation of a weak nonlinear wave.

Davey, A. (1978) *J. Fluid Mech.* **86**, 695–703. On Itoh's finite amplitude stability theory for pipe flow.

Davey, A., DiPrima, R. C. & Stuart, J. T. (1968) *J. Fluid Mech.* **31**, 17–52. On the instability of Taylor vortices.

Davey, A., Hocking, L. M. & Stewartson, K. (1974) *J. Fluid Mech.* **63**, 529–536. On the nonlinear evolution of three-dimensional disturbances in plane Poiseuille flow.

Davey, A. & Nguyen, H. P. F. (1971) *J. Fluid Mech.* **45**, 701–720. Finite-amplitude stability of pipe flow.

Davey, A. & Reid, W. H. (1977a) *J. Fluid Mech.* **80**, 509–525. On the stability of stratified viscous plane Couette flow. Part 1: Constant buoyancy frequency.

Davey, A. & Reid, W. H. (1977b) *J. Fluid Mech.* **80**, 527–534. On the stability of stratified viscous plane Couette flow. Part 2: Variable buoyancy frequency.

Davey, A. & Stewartson, K. (1974) *Proc. Roy. Soc. Lond.* **A338**, 101–110. On three-dimensional packets of surface waves.

Davidson, R. C. (1972) *Methods in Nonlinear Plasma Theory*, Academic, New York.

Davies, A. G. (1982) *Dyn. of Atmos. and Oceans* **6**, 207–232. The reflection of wave energy by modulations on a seabed.

Davies, A. G. & Heathershaw, A. D. (1984) *J. Fluid Mech.* **144**, 419–443. Surface wave propagation over sinusoidally varying topography.

Davis, R. E. (1969) *J. Fluid Mech.* **36**, 337–346. On the high Reynolds number flow over a wavy boundary.

Davis, R. E. (1972) *J. Fluid Mech.* **52**, 287–306. On prediction of the turbulent flow over a wavy boundary.

Davis, R. E. (1974) *J. Fluid Mech.* **63**, 673–693. Perturbed turbulent flow, eddy viscosity and the generation of turbulent stresses.

Davis, R. E. & Acrivos, A. (1967) *J. Fluid Mech.* **30**, 723–736. The stability of oscillatory internal waves.

Deissler, R. G. (1984) *Rev. Mod. Phys.* **56**, 223–254. Turbulent solutions of the equations of fluid motion.

Dhanak, M. R. (1983) *Proc. Roy. Soc. Lond.* **A385**, 53–84. On certain aspects of three-dimensional instability of parallel flows.

Dikii, L. A. (1960a) *J. Appl. Math. Mech.* **24**, 357–369. [trs. of *Prikl. Mat. Mekh.* **24**, 249–259.] On the stability of plane parallel flows of an inhomogeneous fluid.

Dikii, L. A. (1960b) *Sov. Phys. Doklady* **5**, 1179–1182. [trs. of *Dokl. Akad. Nauk, SSSR* **135**, 1068–1071.] The stability of plane-parallel flows of an ideal fluid.

DiPrima, R. C. (1981) In *Transition and Turbulence* (ed. R. E. Meyer), pp. 1–23, Academic. Transition in flow between rotating concentric cylinders.

DiPrima, R. C. & Eagles, P. M. (1977) *Phys. Fluids* **20**, 171–175. Amplification rates and torques for Taylor-vortex flows between rotating cylinders.

DiPrima, R. C., Eckhaus, W. & Segel, L. A. (1971) *J. Fluid Mech.* **49**, 705–744. Nonlinear wave-number interaction in near-critical two-dimensional flows.

DiPrima, R. C. & Habetler, G. J. (1969) *Arch. Rat. Mech. Anal.* **34**, 218–227. A completeness theorem for non-selfadjoint eigenvalue problems in hydrodynamic stability.

DiPrima, R. C. & Swinney, H. L. (1981) In *Hydrodynamic Instabilities and the Transition to Turbulence* (eds H. L. Swinney & J. P. Gollub), pp. 139–180, Springer. Instabilities and transition in flow between concentric rotating cylinders.

Djordjevic, V. D. & Redekopp, L. G. (1977) *J. Fluid Mech.* **79**, 703–714. On two-dimensional packets of capillary-gravity waves.

Dodd, R. K., Eilbeck, J. C., Gibbon, J. D. & Morris, H. C. (1982) *Solitons and Nonlinear Wave Equations*, Academic.

Dore, B. D. (1970) *J. Fluid Mech.* **40**, 113–126. Mass transport in layered fluid systems.

Dore, B. D. (1976) *J. Fluid Mech.* **74**, 819–828. Double boundary layers in standing interfacial waves.

Dore, B. D. (1977) *Quart. J. Mech. Appl. Math.* **30**, 157–173. On mass transport velocity due to progressive waves.

Dore, B. D. (1978a) *Geophys. Astrophys. Fluid Dyn.* **10**, 215–230. Some effects of the air–water interface on gravity waves.

Dore, B. D. (1978b) *J. Engng. Math.* **12**, 289–301. A double boundary-layer model of mass transport in progressive interfacial waves.

Dore, B. D. & Al Zanaidi, M. A. (1979) *Quart. Appl. Math.* **37**, 35–50. On secondary vorticity in internal waves.

Drazin, P. G. (1970) *J. Fluid Mech.* **42**, 321–335. Kelvin–Helmholtz instability of finite amplitude.

Drazin, P. G. (1972) *J. Fluid Mech.* **55**, 577–587. Nonlinear baroclinic instability of a continuous zonal flow of viscous fluid.

Drazin, P. G. (1978) In *Rotating Fluids in Geophysics* (eds P. H. Roberts & A. M. Soward), pp. 139–169, Academic. Variations on a theme of Eady.

Drazin, P. G. & Howard, L. N. (1966) *Advances in Appl. Mech.* **9**, 1–89. Hydrodynamic stability of parallel flow of inviscid fluid.

Drazin, P. G. & Reid, W. H. (1981) *Hydrodynamic Stability*, Cambridge Univ. Press.

Drazin, P. G., Zaturska, M. B. & Banks, W. H. H. (1979) *J. Fluid Mech.* **95**, 681–705. On the normal modes of parallel flow of inviscid stratified fluid. Part 2: Unbounded flow with propagation at infinity.

Dryden, H. L., Murnahan, F. P. & Bateman, H. (1956) *Hydrodynamics*, Dover, New York.

Dysthe, K. B. (1979) *Proc. Roy. Soc. Lond.* **A369**, 105–114. Note on a modification to the nonlinear Schrödinger equation for application to deep water waves.

Eagles, P. M. (1971) *J. Fluid Mech.* **49**, 529–550. On the stability of Taylor vortices by fifth-order amplitude expansions.

Eckart, C. (1963) *Phys. Fluids* **6**, 1042–1047. Extension of Howard's circle theorem to adiabatic jets.

Eckhaus, W. (1963) *J. Mécanique* **2**, 153–172. Problèmes non-linéaires de stabilité dans un espace à deux dimensions. Deuxième partie: Stabilité des solutions périodiques.

Eckhaus, W. (1965) *Studies in Nonlinear Stability Theory*, Springer, Berlin.

Edge, R. D. & Walters, G. (1964) *J. Geophys. Res.* **69**, 1674–1675. The period of standing gravity waves of largest amplitude on water.

Einaudi, F. & Lalas, D. P. (1976) *Trans. A.S.M.E. J. Appl. Mech.* **E98**, 243–248. The effect of boundaries on the stability of inviscid stratified shear flows.

Ellingsen, T., Gjevik, B. & Palm, E. (1970) *J. Fluid Mech.* **40**, 97–112. On the nonlinear stability of plane Couette flow.

Eltayeb, I. A. & McKenzie, J. F. (1975) *J. Fluid Mech.* **72**, 661–671. Critical-level behaviour and wave amplification of a gravity wave incident upon a shear layer.

Engevik, L. (1982) *J. Fluid Mech.* **117**, 457–471. An amplitude-evolution equation for linearly unstable modes in stratified shear flows.

Euler, L. (1765) *Theoria Motus Corporum Solidorum seu Rigidorum*, Griefswald. [L. Euleri Opera Omnia Ser. 2, Vols 3–4.]

Faller, A. J. & Caponi, E. A. (1978) *J. Geophys. Res.* **83**, 3617–3633. Laboratory studies of wind-driven Langmuir circulations.

Faller, A. J. & Cartwright, R. W. (1982) (University of Maryland, Technical Rep. BN-985). Laboratory studies of Langmuir circulations.

Fasel, H. (1976) *J. Fluid Mech.* **78**, 355–383. Investigation of the stability of boundary layers by a finite-difference model of the Navier–Stokes equations.

Fasel, H., Bestek, H. & Schefenacker, R. (1977) *Laminar-Turbulent Transition: Proc. AGARD Conf. No. 224, Lyngby, 1977*, paper 14. Numerical simulation studies of transition phenomena in incompressible, two-dimensional flows.

Fasel, H. & Bestek, H. (1980) In *Laminar-Turbulent Transition: Proc. IUTAM Symp. Stuttgart 1979* (eds R. Eppler & H. Fasel), pp. 173–185. Investigation of nonlinear, spatial disturbance amplification in plane Poiseuille flow.

Fenstermacher, P. R., Swinney, H. L. & Gollub, J. P. (1979) *J. Fluid Mech.* **94**, 103–128. Dynamical instabilities and the transition to chaotic Taylor vortex flow.

Fenton, J. D. & Rienecker, M. M. (1982) *J. Fluid Mech.* **118**, 411–442. A Fourier method for solving nonlinear water-wave problems: application to solitary wave interactions.

Fermi, E., Pasta, J. & Ulam, S. (1955) Studies of nonlinear problems. *Collected Papers of Enrico Fermi*, Vol. 2, 978–988. Chicago.

Fowler, A. C., Gibbon, J. D. & McGuinness, M. J. (1983) *Physica* **7D**, 126–134. The real and complex Lorenz equations and their relevance to physical systems.

Franklin, R. E., Price, M. & Williams, D. C. (1973) *J. Fluid Mech.* **57**, 257–268. Acoustically driven water waves.

Freeman, N. C. & Davey, A. (1975) *Proc. Roy. Soc. Lond.* **A344**, 427–433. On the evolution of packets of long surface waves.

Fritts, D. C. (1984) *Rev. Geophys. Space Phys.* **22**, 275–308. Gravity wave saturation in the middle atmosphere: a review of theory and observations.

Fuchs, V. & Beaudry, G. (1975) *J. Math. Phys.* **16**, 616–619. Effects of damping on nonlinear three-wave interaction.

Gajjar, J. & Smith, F. T. (1985) *J. Fluid Mech.* **157**, 53–77. On the global instability of free disturbances with a time-dependent nonlinear viscous critical layer.

Gargett, A. E. & Hughes, B. A. (1972) *J. Fluid Mech.* **52**, 179–191. On the interaction of surface and internal waves.

Garrett, C. J. R. (1976) *J. Mar. Res.* **34**, 117–130. Generation of Langmuir circulations by surface waves – a feedback mechanism.

Gaster, M. (1968) *J. Fluid Mech.* **32**, 173–184. The development of three-dimensional wave packets in a boundary layer.

Gaster, M. (1974) *J. Fluid Mech.* **66**, 465–480. On the effects of boundary-layer growth on flow stability.

Gaster, M. (1975) *Proc. Roy. Soc. Lond.* **A347**, 271–289. A theoretical model of a wave packet in the boundary layer on a flat plate.

Gaster, M. (1978) *Proc. 12th Symp. Naval Hydrodyn., Washington*. The physical processes causing breakdown to turbulence.

Gaster, M. (1979) *A.I.A.A. Paper* 79-1492 (presented at AIAA 12th Fluid and Plasma Dyn. Conf., Williamsburg, Va.). The propagation of linear wave packets in laminar boundary layers. Asymptotic theory for nonconservative wave systems.

Gaster, M. (1982a) *J. Fluid Mech.* **121**, 365–377. Estimates of the errors incurred in various asymptotic representations of wave packets.

Gaster, M. (1982b) *Proc. Roy. Soc. Lond.* **A384**, 317–332. The development of a two-dimensional wavepacket in a growing boundary layer.

Gaster, M. & Davey, A. (1968) *J. Fluid Mech.* **32**, 801–808. The development of three-dimensional wave packets in unbounded parallel flows.

Gaster, M. & Grant, I. (1975) *Proc. Roy. Soc. Lond.* **A347**, 253–269. An experimental investigation of the formation and development of a wave packet in a laminar boundary layer.

George, W. D., Hellums, J. D. & Martin, B. (1974) *J. Fluid Mech.* **63**, 765–771. Finite-amplitude neutral disturbances in plane Poiseuille flow.

Gertsenshtein, S. Y. & Shtempler, Y. M. (1977) *Sov. Phys. Dokl.* **22**, 300–302. Nonlinear growth of perturbations in boundary layers and their stability.

Gibbon, J. D., James, I. N. & Moroz, I. M. (1979) *Proc. Roy. Soc. Lond.* **A367**, 219–237. An example of soliton behaviour in a rotating baroclinic fluid.

Gibbon, J. D. & McGuinness, M. J. (1981) *Proc. Roy. Soc. Lond.* **A377**, 185–219. Amplitude equations at the critical points of unstable dispersive physical systems.

Gibbon, J. D. & McGuinness, M. J. (1982) *Physica* **5D**, 108–122. The real and complex Lorenz equations in rotating fluids and lasers.

Gilev, V. M., Kachanov, Y. S. & Kozlov, V. V. (1982) Development of three-dimensional wave packets in a boundary layer (in Russian); Publication of Institute of Theoretical and Applied Mechanics, Novosibirsk.

Goldstein, S. (1931) *Proc. Roy. Soc. Lond.* **A132**, 524–548. On the stability of superposed streams of fluid of different densities.

Gollub, J. P. & Benson, S. V. (1980) *J. Fluid Mech.* **100**, 449–470. Many routes to turbulent convection.

Gollub, J. P., McCarriar, A. R. & Steinman, J. F. (1982) *J. Fluid Mech.* **125**, 259–281. Convective pattern evolution and secondary instabilities.

Goncharov, V. V. (1981) *Izv., Atmos. & Oceanic Phys.* **17**, 65–69. (trs. of *Izv. Akad. Nauk. SSSR Fiz. Atmos. i Okeana*). The influence of high-order nonlinearity on three-wave resonance processes.

Gor'kov, L. P. (1957) *Sov. Phys. J.E.T.P.* **6**, 311–315 (1958). [trs. of *J. Eksp. Teor. Fiz.* **33**, 402–407.] Stationary convection in a plane liquid layer near the critical heat transfer point.

Gorman, M. & Swinney, H. L. (1982) *J. Fluid Mech.* **117**, 123–142. Spatial and temporal characteristics of modulated waves in the circular Couette flow.

Gotoh, K. (1965) *J. Phys. Soc. Japan* **20**, 164–169. The damping of the large wavenumber disturbances in a free boundary layer flow.

Greenspan, H. P. & Benney, D. J. (1963) *J. Fluid Mech.* **15**, 133–153. On shear-layer instability, breakdown and transition.

Grimshaw, R. H. J. (1976) *J. Fluid Mech.* **76**, 65–83. Nonlinear aspects of an internal gravity wave co-existing with an unstable mode associated with a Helmholtz velocity profile.

Grimshaw, R. H. J. (1977) *Stud. Appl. Math.* **56**, 241–266. The modulation of an internal gravity-wave packet and the resonance with the mean motion.

Grimshaw, R. H. J. (1979a) *Phil. Trans. Roy. Soc. Lond.* **A292**, 391–417. Mean flows induced by internal gravity wave packets propagating in a shear flow.

Grimshaw, R. H. J. (1979b) *J. Fluid Mech.* **90**, 161–178. On resonant over-reflexion of internal gravity waves from a Helmholtz velocity profile.

Grimshaw, R. H. J. (1981a) *J. Fluid Mech.* **109**, 349–365. Resonant over-reflection of internal gravity waves from a thin shear layer.

Grimshaw, R. H. J. (1981b) *J. Austr. Math. Soc.* **B22**, 318–347. Mean flow generated by a progressing water wave packet.

Grimshaw, R. H. J. (1982) *J. Fluid Mech.* **115**, 347–377. The effect of dissipative processes on mean flows induced by internal gravity-wave packets.

Grimshaw, R. H. J. (1984) *Ann. Rev. Fluid Mech.* **16**, 11–44. Wave action and wave-mean flow interaction, with application to stratified shear flows.

Grosch, C. E. & Salwen, H. (1968) *J. Fluid Mech.* **34**, 177–205. The stability of steady and time-dependent plane Poiseuille flow.

Grosch, C. E. & Salwen, H. (1978) *J. Fluid Mech.* **87**, 33–54. The continuous spectrum of the Orr–Sommerfeld equation. Part 1. The spectrum and the eigenfunctions.

Gustavsson, L. H. (1979) *Phys. Fluids* **22**, 1602–1605. Initial value problem for boundary layer flows.

Gustavsson, L. H. (1981) *J. Fluid Mech.* **112**, 253–264. Resonant growth of three-dimensional disturbances in plane Poiseuille flow.

Gustavsson, L. H. & Hultgren, L. S. (1980) *J. Fluid Mech.* **98**, 149–159. A resonance mechanism in plane Couette flow.

Haberman, R. (1972) *Stud. Appl. Math.* **51**, 139–161. Critical layers in parallel flows.

Haberman, R. (1973a) *J. Fluid Mech.* **58**, 129–142. (and Corrigendum, **61**, 829.) Note on slightly unstable nonlinear wave systems.

Haberman, R. (1973b) *J. Fluid Mech.* **58**, 727–735. Wave-induced distortions of a slightly stratified shear flow: a nonlinear critical layer effect.

Haberman, R. (1976) *S.I.A.M. J. Math. Anal.* **7**, 70–81. Nonlinear perturbations of the Orr–Sommerfeld equation – asymptotic expansion of the logarithmic phase shift across the critical layer.

Haberman, R. (1977) *S.I.A.M. J. Appl. Math.* **32**, 154–163. On the singular behaviour of spatially dependent nonlinear wave envelopes.

Hall, P. (1980a) *Proc. Roy. Soc. Lond.* **A372**, 317–356. Centrifugal instabilities of circumferential flow in finite cylinders: nonlinear theory.

Hall, P. (1980b) *J. Fluid Mech.* **99**, 575–596. Centrifugal instabilities in finite containers: a periodic model.

Hall, P. (1982) *Proc. Roy. Soc. Lond.* **A384**, 359–379. Centrifugal instabilities of circumferential flows in finite cylinders: the wide gap problem.

Hall, P. & Smith, F. T. (1982) *Stud. Appl. Math.* **66**, 241–265. A suggested mechanism for nonlinear wall roughness effects on high Reynolds number flow stability.

Hall, P. & Smith, F. T. (1984) *Stud. Appl. Math.* **70**, 91–120. On the effects of nonparallelism, three-dimensionality, and mode interaction in nonlinear boundary-layer stability.

Hall, P. & Walton, I. C. (1977) *Proc. Roy. Soc. Lond.* **A358**, 199–221. The smooth transition to a convective regime in a two dimensional box.

Hall, P. & Walton, I. C. (1979) *J. Fluid Mech.* **90**, 377–395. Bénard convection in a finite box: secondary and imperfect bifurcations.

Hame, W. & Muller, U. (1975) *Acta Mechanica* **23**, 75–89. Uber die Stabilität einer ebenen Zweischichten Poiseuille Strömung.

Hammack, J. L. & Segur, H. (1974) *J. Fluid Mech.* **65**, 289–314. The Korteweg–de Vries equation and water waves. Part 2. Comparison with experiments.

Hammack, J. L. & Segur, H. (1978) *J. Fluid Mech.* **84**, 337–358. The Korteweg–de Vries equation and water waves. Part 3. Oscillatory waves.

Hanratty, T. J. (1983) In *Waves on Fluid Interfaces* (ed. R. E. Meyer), Academic, pp. 221–259. Interfacial instabilities caused by air flow over a thin liquid layer.

Hart, J. E. (1973) *J. Atmos. Sci.* **30**, 1017–1034. On the behaviour of large-amplitude baroclinic waves.

Hart, J. E. (1979) *Ann. Rev. Fluid Mech.* **11**, 147–172. Finite amplitude baroclinic instability.

Hasimoto, H. (1974) *Proc. Japan Acad.* **50**, 623–627. Exact solution of a certain semi-linear system of partial differential equations related to a migrating predation problem.

Hasimoto, H. & Ono, H. (1972) *J. Phys. Soc. Japan* **33**, 805–811. Nonlinear modulation of gravity waves.

Hasselmann, D. E. (1979) *J. Fluid Mech.* **93**, 491–500. The high wavenumber instabilities of a Stokes wave.

Hasselmann, K. (1962) *J. Fluid Mech.* **12**, 481–500. On the nonlinear energy transfer in a gravity wave spectrum. Part 1: General theory.

Hasselmann, K. (1963) *J. Fluid Mech.* **15**, 273–281. On the nonlinear energy transfer in a gravity wave spectrum. Part 2: Conservation theorems, wave-particle analogy, irreversibility.

Hasselmann, K. (1967a) *Proc. Roy. Soc. Lond.* **A299**, 77–100. Nonlinear interactions treated by the methods of theoretical physics (with application to the generation of waves by wind).

Hasselmann, K. (1967b) *J. Fluid Mech.* **30**, 737–739. A criterion for nonlinear wave stability.

Haurwitz, B. (1931) *Veröff. Geophys. Inst. Univ. Leipzig* **6**, No. 1. Zur Theorie der Wellenbewegungen in Luft und Wasser.

Hayes, W. D. (1970) *Proc. Roy. Soc. Lond.* **A320**, 187–208. Conservation of action and modal wave action.

Hayes, W. D. (1973) *Proc. Roy. Soc. Lond.* **A332**, 199–221. Group velocity and nonlinear dispersive wave propagation.

Hazel, P. (1972) *J. Fluid Mech.* **51**, 39–61. Numerical studies of the stability of inviscid stratified shear flows.

Herbert, T. (1975) *Proc. 4th Internat. Conf. Num. Methods in Fluid Dyn.* (ed. R. D. Richtmyer), 212–217. *Springer Lecture Notes in Physics* **35**. On finite amplitudes of periodic disturbances of the boundary layer along a flat plate.

Herbert, T. (1977) *Laminar-Turbulent Transition: Proc. AGARD Conf. No. 224, Lyngby*, paper 3. Finite amplitude stability of plane parallel flows.

Herbert, T. (1980) *AIAA Jour.* **18**, 243–248. Nonlinear stability of parallel flows by high-order amplitude expansions.

Herbert, T. (1981) In *Proc. 7th Internat. Conf. Num. Methods in Fluid Dyn.* (eds W. C. Reynolds & R. W. MacCormack). *Springer Lecture Notes in Physics*, **141**, p. 200. Numerical studies on nonlinear hydrodynamic stability by computer extended perturbation series

Herbert, T. (1982) *Polish Acad. Sci. Fluid Dynamics Trans.* **11**, 77–126. Stability of plane Poiseuille flow – theory and experiment.

Herbert, T. (1983a) *J. Fluid Mech.* **126**, 167–186. On perturbation methods in nonlinear stability theory.

Herbert, T. (1983b) *Phys. Fluids* **26**, 871–874. Secondary instability of plane channel flow to subharmonic three-dimensional disturbances.

Herbert, T. (1983c) *AIAA Paper 83-1759* (Presented at 16th Fluid and Plasma Dyn. Conf.; Danvers, Mass.) Subharmonic three-dimensional disturbances in unstable plane shear flows.

Herbert, T. (1984) In *Turbulence and Chaotic Phenomena in Fluids. Proc. IUTAM Symp., Kyoto* (ed. T. Tatsumi); North Holland. Modes of secondary instability in plane Poiseuille flow.

Herbert, T. & Morkovin, M. V. (1980) In *Laminar-Turbulent Transition: Proc. IUTAM*

Symp., Stuttgart (ed. R. Eppler & H. Fasel); Springer, pp. 47–72. Dialogue on bridging some gaps in stability and transition research.

Hide, R. & Mason, P. J. (1975) *Adv. Phys.* **24**, 47–100. Sloping convection in a rotating fluid.

Ho, C.-H. & Huerre, P. (1984). *Ann. Rev. Fluid Mech.* **16**, 365–424. Perturbed free shear layers.

Hocking, L. M. (1975) *Quart. J. Mech. Appl. Math.* **28**, 341–353. Nonlinear instability of the asymptotic suction velocity profile.

Hocking, L. M. & Stewartson, K. (1971) *Mathematika* **18**, 219–239. On the nonlinear response of a marginally unstable plane parallel flow to a three-dimensional disturbance.

Hocking, L. M. & Stewartson, K. (1972) *Proc. Roy. Soc. Lond.* **A326**, 289–313. On the nonlinear response of a marginally unstable plane parallel flow to a two-dimensional disturbance.

Hocking, L. M., Stewartson, K. & Stuart, J. T. (1972) *J. Fluid Mech.* **51**, 707–735. A nonlinear instability burst in plane parallel flow.

Hogan, S. J. (1980) *J. Fluid Mech.* **96**, 417–445. Some effects of surface tension on steep water waves. Part 2.

Hogan, S. J. (1981) *J. Fluid Mech.* **110**, 381–410. Some effects of surface tension on steep water waves. Part 3.

Hogan, S. J. (1984a) *J. Fluid Mech.* **143**, 243–252. Particle trajectories in nonlinear capillary waves.

Hogan, S. J. (1984b) *Phys. Fluids* **27**, 42–45. Subharmonic generation of deep-water capillary waves.

Holton, J. R. & Lindzen, R. S. (1972) *J. Atmos. Sci.* **29**, 1076–1080. An updated theory for the quasi-biennial oscillation of the tropical stratosphere.

Holyer, J. Y. (1979) *J. Fluid Mech.* **93**, 433–448. Large amplitude progressive interfacial waves.

Hooper, A. P. & Boyd, W. G. C. (1983) *J. Fluid Mech.* **128**, 507–529. Shear-flow instability at the interface between two viscous fluids.

Hooper, A. P. & Grimshaw, R. (1985) *Phys. Fluids* **28**, 37–45. Nonlinear instability at the interface between two viscous fluids.

Hopman, H. J. (1971) *Proc. 10th Internat. Conf. Phenomena in Ionized Gases, Oxford.* Evidence for explosive three-wave interaction.

Howard, L. N. (1961) *J. Fluid Mech.* **10**, 509–512. Note on a paper of John W. Miles.

Howard, L. N. & Maslowe, S. A. (1973) *Boundary-Layer Meteorol.* **4**, 511–523. Stability of stratified shear flows.

Huerre, P. (1980) *Phil. Trans. Roy. Soc. Lond.* **A293**, 643–675. The nonlinear stability of a free shear layer in the viscous critical layer regime.

Huerre, P. & Scott, J. F. (1980) *Proc. Roy. Soc. Lond.* **A371**, 509–524. Effects of critical layer structure on the nonlinear evolution of waves in free shear layers.

Hui, W. H. & Hamilton, J. (1979) *J. Fluid Mech.* **93**, 117–134. Exact solutions of a three-dimensional nonlinear Schrödinger equation applied to gravity waves.

Hultgren, L. S. & Gustavsson, L. H. (1981) *Phys. Fluids* **24**, 1000–1004. Algebraic growth of disturbances in a laminar boundary layer.

Huntley, I. (1972) *J. Fluid Mech.* **53**, 209–216. Observations on a spatial-resonance phenomenon.

Huntley, I. & Smith, R. (1973) *J. Fluid Mech.* **61**, 401–413. Hysteresis and nonlinear detuning in a spatial-resonance phenomenon.

Huppert, H. E. & Moore, D. R. (1976) *J. Fluid Mech.* **78**, 821–854. Nonlinear double-diffusive convection.

Huppert, H. E. & Turner, J. S. (1981) *J. Fluid Mech.* **106**, 299–329. Double-diffusive convection.

Hussain, A. K. M. F. (1983) *Phys. Fluids* **26**, 2816–2850. Coherent structures – reality and myth.

Ibrahim, R. A. & Barr, A. D. S. (1975a) *J. Sound and Vibration* **42**, 159–179. Autoparametric resonance in a structure containing a liquid. Part I: two mode interaction.

Ibrahim, R. A. & Barr, A. D. S. (1975b) *J. Sound and Vibration* **42**, 181–200. Autoparametric resonance in a structure containing a liquid. Part II: three mode interaction.

Ikeda, M. (1977) *J. Phys. Soc. Japan* **42**, 1764–1771. Nonlinearity and non-periodicity of a two-dimensional disturbance in plane Poiseuille flow.

Ikeda, M. (1978) *J. Phys. Soc. Japan* **44**, 667–675. Instability of plane Poiseuille flow caused by a nonlinear, nonperiodic and three-dimensional disturbance.

Itoh, N. (1974a) *Trans. Japan Soc. Aero. Space Sci.* **17**, 65–75. A power series method for the numerical treatment of the Orr–Sommerfeld equation.

Itoh, N. (1974b) *Trans. Japan Soc. Aero. Space Sci.* **17**, 160–174. Spatial growth of finite wave disturbances in parallel and nearly parallel flows. Part 1. The theoretical analysis and the numerical results for plane Poiseuille flow.

Itoh, N. (1974c) *Trans. Japan Soc. Aero. Space Sci.* **17**, 175–186. Spatial growth of finite wave disturbances in parallel and nearly parallel flows. Part 2. The numerical results for the flatplate boundary layer.

Itoh, N. (1977a) *J. Fluid Mech.* **82**, 455–467. Nonlinear stability of parallel flows with subcritical Reynolds numbers. Part 1. An asymptotic theory valid for small amplitude disturbances.

Itoh, N. (1977b) *J. Fluid Mech.* **82**, 469–479. Nonlinear stability of parallel flows with subcritical Reynolds numbers. Part 2. Stability of pipe Poiseuille flow to finite axisymmetric disturbances.

Itoh, N. (1980a) In *Laminar-Turbulent Transition: Proc. IUTAM Symp. Stuttgart* (ed. R. Eppler & H. Fasel), pp. 86–95. Springer. Linear stability theory for wave packet disturbances in parallel and nearly parallel flows.

Itoh, N. (1980b) *Trans. Japan Soc. Aero. Space Sci.* **23**, 91–103. Three-dimensional growth of finite wave disturbances in plane Poiseuille flow.

Itoh, N. (1981) *Proc. Roy. Soc. Lond.* **A375**, 565–578. Secondary instability of laminar flow.

Janssen, P. A. E. M. (1983) *J. Fluid Mech.* **126**, 1–11. On a fourth-order envelope equation for deep-water waves.

Jeffreys, H. (1925) *Proc. Roy. Soc. Lond.* **A107**, 189–206. On the formation of water waves by wind.

Jimenez, J. & Whitham, G. B. (1976) *Proc. Roy. Soc. Lond.* **A349**, 277–287. An averaged Lagrangian method for dissipative wavetrains.

Jones, C. S. (1981) *J. Fluid Mech.* **102**, 249–261. Nonlinear Taylor vortices and their stability.

Jones, D. S. & Morgan, J. D. (1972) *Proc. Camb. Phil. Soc.* **72**, 465–488. The instability of a vortex sheet on a subsonic stream under acoustic radiation.

Jones, W. L. (1968) *J. Fluid Mech.* **34**, 609–624. Reflexion and stability of waves in stably stratified fluids with shear flow: a numerical study.

Jones, W. L. & Houghton, D. D. (1972) *J. Atmos. Sci.* **29**, 844–849. The self-destructing internal gravity wave.

Jordinson, R. (1970) *J. Fluid Mech.* **43**, 801–811. The flat-plate boundary layer. Part 1. Numerical integration of the Orr–Sommerfeld equation.

Jordinson, R. (1971) *Phys. Fluids* **14**, 2535–2537. Spectrum of eigenvalues of the Orr–Sommerfeld equation for Blasius flow.

Joseph, D. D. (1976) *Stability of Fluid Motions.* 2 vols. Springer Tracts in Natural Philosophy, vols 27 and 28. Springer, Berlin.

Joseph, D. D. (1981) In *Hydrodynamic Instabilities and the Transition to Turbulence* (eds H. L. Swinney & J. P. Gollub), pp. 27–76. Academic. Hydrodynamic stability and bifurcation.

Joseph, D. D. & Carmi, S. (1969) *Quart. Appl. Math.* **26**, 575–599. Stability of Poiseuille flow in pipes, annuli and channels.

Jurkus, A. & Robson, P. N. (1960) *Proc. I.E.E.* **107b**, 119–122. Saturation effects in a travelling-wave parametric amplifier.

Kachanov, Yu. S., Kozlov, V. V. & Levchenko, V. Ya. (1977) *Fluid Dyn.* **3**, 383–390. [trs. of *Mekh. Zhid. i Gaza* **3**, 49–53.] Nonlinear development of a wave in a boundary layer.

Kachanov, Yu. S., Kozlov, V. V., Levchenko, V. Ya. & Ramazanov, M. P. (1985) In *Laminar-Turbulent Transition: Proc. IUTAM Symp. Novosibirsk* (ed. V. V. Kozlov), pp. 61–73. On the nature of K-breakdown of a laminar boundary layer. New experimental data.

Kachanov, Yu. S. & Levchenko, V. Ya. (1984) *J. Fluid Mech.* **138**, 209–247. The resonant interaction of disturbances at laminar-turbulent transition in a boundary layer.

Kadomtsev, B. B. & Petviashvilli, V. I. (1970) *Sov. Phys. Dokl.* **15**, 539–541 (Engl. trs.). On the stability of solitary waves in weakly dispersive media.

Karpman, V. I. & Maslov, E. M. (1977) *Sov. Phys. – JETP* **46**, 281–291, [trs. of *Zh. Eksp. Teor. Fiz.* **73**, 537–559]. Perturbation theory for solitons.

Kaup, D. J. (1976) *S.I.A.M. J. Appl. Math.* **31**, 121–133. A perturbation expansion for the Zakharov-Shabat inverse scattering transform.

Kaup, D. J. (1980) *Stud. Appl. Math.* **62**, 75–83. A method for solving the separable initial value problem of the full three-dimensional three-wave interaction.

Kaup, D. J. (1981a) *J. Math. Phys.* **22**, 1176–1181. The lump solutions and the Bäcklund transformation for the three-dimensional three-wave resonant interaction.

Kaup, D. J. (1981b) *Nonlinear phenomena in physics and biology. Proc. NATO Advanced Study Institute,* 1980, *Banff, Canada,* Plenum, pp. 95–123. The linearity of nonlinear soliton equations and the three wave resonance interaction.

Kaup, D. J., Reiman, A. & Bers, A. (1979) *Rev. Mod. Phys.* **51**, 275–310. [Errata, in *Rev. Mod. Phys.* **51**, 915, are corrected in reprints.] Space-time evolution of nonlinear three-wave interactions. I. Interactions in a homogeneous medium.

Kawahara, T. (1983) *Phys. Rev. Lett.* **51**, 381–383. Formation of saturated solutions in a nonlinear dispersive system with instability and dissipation.

Keady, G. & Norbury, J. (1978) *Math. Proc. Camb. Phil. Soc.* **83**, 137–157. On the existence theory for irrotational water waves.

Keener, T. P. & McLaughlin, D. W. (1977) *Phys. Rev.* **A16**, 777–790. Solitons under perturbations.

Kelly, R. E. (1967) *J. Fluid Mech.* **27**, 657–689. On the stability of an inviscid shear layer which is periodic in space and time.

Kelly, R. E. (1968) *J. Fluid Mech.* **31**, 789–799. On the resonant interaction of neutral disturbances in two inviscid shear flows.

Kelly, R. E. (1970) *J. Fluid Mech.* **42**, 139–150. Wave-induced boundary layers in a stratified fluid.

Kelly, R. E. & Maslowe, S. A. (1970) *Stud. Appl. Math.* **49**, 301–326. The nonlinear critical layer in a slightly stratified shear flow.

Kelly, R. E. & Pal, D. (1976) *Proc. Heat Transf. Fluid Mech. Inst.* 1–17. Thermal convection induced between non-uniformly heated horizontal surfaces.

Kelly, R. E. & Pal, D. (1978) *J. Fluid Mech.* **86**, 433–456. Thermal convection with spatially periodic boundary conditions: resonant wavelength excitation.

Kim, Y. Y. & Hanratty, T. J. (1971) *J. Fluid Mech.* **50**, 107–132. Weak quadratic interactions of two-dimensional waves.

King, G. P., Li, Y., Lee, W. & Swinney, H. L. (1984) *J. Fluid Mech.* **141**, 365–390. Wave speeds in wavy Taylor-vortex flow.

Kinnersley, W. (1976) *J. Fluid Mech.* **77**, 229–241. Exact large amplitude capillary waves on sheets of fluid.

Kinsman, B. (1969) *Wind Waves: Their Generation and Propagation on the Ocean Surface.* Prentice-Hall, Englewood Cliffs, N.J.

Kirby, J. T. & Dalrymple, R. A. (1983) *Phys. Fluids* **26**, 2916–2918. Oblique envelope solutions of the Davey–Stewartson equations in intermediate water depth.

Klebanoff, P. S., Tidstrom, K. D. & Sargent, L. M. (1962) *J. Fluid Mech.* **12**, 1–34. The three-dimensional nature of boundary-layer instability.

Knops, R. J. & Wilkes, E. W. (1966) *Internat. J. Engng. Sci.* **4**, 303–329. On Movchan's theorems for stability of continuous systems.

Kogelman, S. & DiPrima, R. C. (1970) *Phys. Fluids* **13**, 1–11. Stability of spatially periodic supercritical flows in hydrodynamics.

Komoda, H. (1967) *Phys. Fluids* Suppl. **10**, 587–594. Nonlinear development of disturbances in a laminar boundary layer.

Koop, C. G. (1981) *J. Fluid Mech.* **113**, 347–386. A preliminary investigation of the interaction of internal gravity waves with a steady shearing motion.

Koop, C. G. & Butler, G. (1981) *J. Fluid Mech.* **112**, 225–251. An investigation of internal solitary waves in a two-fluid system.

Koop, C. G. & Redekopp, L. G. (1981) *J. Fluid Mech.* **111**, 367–409. The interaction of long and short internal gravity waves: theory and experiment.

Koschmieder, E. L. (1967) *J. Fluid Mech.* **30**, 9–15. On convection under an air surface.

Koschmieder, E. L. (1980) In *Laminar-Turbulent Transition: Proc. IUTAM Symp., Stuttgart,* (ed. R. Eppler & H. Fasel), pp. 396–404. Springer. Transition from laminar to turbulent Taylor vortex flow.

Kovasznay, L. S. G., Komoda, H. & Vasudeva, B. R. (1962) *Proc. 1962 Heat Transf. & Fluid Mech. Inst.,* 1–26. Detailed flow field in transition.

Kozlov, V. V. (ed.) (1985) *Laminar-Turbulent Transition: Proc. IUTAM Symp. Novosibirsk,* Springer.

Kozlov, V. V. & Ramazanov, M. P. (1984) *J. Fluid Mech.* **147**, 149–157. Development of finite-amplitude disturbances in Poiseuille flow.

Krishnamurti, R. (1970) *J. Fluid Mech.* **42**, 309–320. On the transition to turbulent convection. Part 2. The transition to time-dependent flow.

Krishnamurti, R. (1973) *J. Fluid Mech.* **60**, 285–303. Some further studies on the transition to turbulent convection.

Kuppers, G. & Lortz, D. (1969) *J. Fluid Mech.* **35**, 609–620. Transition from laminar convection to thermal turbulence in a rotating fluid layer.

Lake, B. M., Yuen, H. C., Rungalder, H. & Ferguson, W. E. Jr. (1977) *J. Fluid Mech.* **83**, 49–74. Nonlinear deep-water waves: Theory and experiment. Part 2, Evolution of a continuous wave train.

Lalas, D. P. & Einaudi, F. (1976) *J. Atmos. Sci.* **33**, 1248–1259. On the characteristics of gravity waves generated by atmospheric shear layers.

Lalas, D. P., Einaudi, F. & Fua, D. (1976) *J. Atmos. Sci.* **33**, 59–69. The destabilizing effect of the ground on Kelvin–Helmholtz waves in the atmosphere.

Lamb. H. (1932) *Hydrodynamics* 6th edn., Cambridge Univ. Press.

Landahl, M. T. (1962) *J. Fluid Mech.* **13**, 609–632. On the stability of a laminar incompressible boundary layer over a flexible surface.

Landahl, M. T. (1972) *J. Fluid Mech.* **56**, 775–802. Wave mechanics of breakdown.

Landahl, M. T. (1980) *J. Fluid Mech.* **98**, 243–251. A note on an algebraic instability of inviscid parallel shear flows.

Landahl, M. T. (1982) *Phys. Fluids* **25**, 1512–1516. The application of kinematic wave theory to wave trains and packets with small dissipation.

Landau, L. (1944) *C.R. Acad. Sci. U.R.S.S.* **44**, 311–314. (Also *Collected Papers* (1965), pp. 387–391.) On the problem of turbulence.

Lange, C. G. & Newell, A. C. (1971) *S.I.A.M. J. Appl. Math.* **21**, 605–629. The post-buckling problem for thin elastic shells.

Leblond, P. H. & Mysak, L. A. (1978) *Waves in the Ocean*, Elsevier; Amsterdam.

Lees, L. & Lin, C. C. (1946) *N.A.C.A. Tech. Note* TN 1115. Investigation of the stability of the laminar boundary layer in a compressible fluid.

Leibovich, S. (1977a) *J. Fluid Mech.* **79**, 715–743. On the evolution of the system of wind drift currents and Langmuir circulations in the ocean. Part 1. Theory and averaged current.

Leibovich, S. (1977b) *J. Fluid Mech.* **82**, 561–585. Convective instability of stably stratified water in the ocean.

Leibovich, S. (1980) *J. Fluid Mech.* **99**, 715–724. On wave–current interaction theories of Langmuir circulations.

Leibovich, S. (1983) *Ann. Rev. Fluid Mech.* **15**, 391–427. On wave–current interaction theories of Langmuir circulations.

Leibovich, S. & Paolucci, S. (1981) *J. Fluid Mech.* **102**, 141–167. The instability of the ocean to Langmuir circulation.

Leibovich, S. & Radhakrishnan, K. (1977) *J. Fluid Mech.* **80**, 481–507. On the evolution of the system of wind drift currents and Langmuir circulations in the ocean. Part 2. Structure of the Langmuir vortices.

Leibovich, S. & Seebass, A. R. (eds) (1974) *Nonlinear Waves*, Cornell Univ. Press.

Leibovich, S. & Ulrich, D. (1972) *J. Geophys. Res.* **77**, 1683–1688. A note on the growth of small-scale Langmuir circulations.

Lekoudis, S. G. (1980) *AIAA J.* **18**, 122–124. Resonant wave interactions on a swept wing.

Levi-Civita, T. (1925) *Math. Ann.* **93**, 264–314. Détermination rigoureuse des ondes permanentes d'ampleur finie.

Lewis, J. E., Lake, B. M. & Ko, D. R. S. (1974) *J. Fluid Mech.* **63**, 773–800. On the interaction of internal waves and surface gravity waves.

Libchaber, A. & Maurer, J. (1978) *J. Phys. Lett.* (*Paris*) **39**, L369. Local probe in a Rayleigh–Bénard experiment in liquid helium.

Libchaber, A., Fauve, S. & Laroche, C. (1983) *Physica* **7D**, 73–84. Two-parameter study of the routes to chaos.

Lighthill, M. J. (1962) *J. Fluid Mech.* **14**, 385–398. Physical interpretation of the mathematical theory of wave generation by wind.

Lighthill, M. J. (1965) *J. Inst. Math. Applic.* **1**, 269–306. Contributions to the theory of waves in nonlinear dispersive systems.

Lighthill, M. J. (1978) *Waves in Fluids*, Cambridge Univ. Press.

Lin, C. C. (1955) *The Theory of Hydrodynamic Stability*, Cambridge Univ. Press.

Lin, C. C. (1961) *J. Fluid Mech.* **10**, 430–438. Some mathematical problems in the theory of the stability of parallel flows.

Lin, S. P. & Krishna, M. V. G. (1977) *Phys. Fluids* **20**, 2005–2011. Stability of a liquid film with respect to initially finite three-dimensional disturbances.

Lindzen, R. S. (1973) *Boundary-layer Meteor.* **4**, 327–343. Wave-mean flow interactions in the upper atmosphere.

Lindzen, R. S. (1974) *J. Atmos. Sci.* **31**, 1507–1514. Stability of a Helmholtz velocity profile in a continuously stratified infinite Boussinesq fluid – Applications to clear-air turbulence.

Lindzen, R. S. & Barker, J. W. (1985) *J. Fluid Mech.* **151**, 189–217. Instability and wave over-reflection in stably stratified shear flow.

Lipps, F. B. (1976) *J. Fluid Mech.* **75**, 113–148. Numerical simulation of three-dimensional Bénard convection in air.

Liu, J. T. C. (1969) *Phys. Fluids* **12**, 1763–1774. Finite-amplitude instability of the compressible laminar wake. Weakly nonlinear theory.

Liu, A-K. & Davis, S. H. (1977) *J. Fluid Mech.* **81**, 63–84. Viscous attenuation of mean drift in water waves.

Lo, E. & Mei, C. C. (1985) *J. Fluid Mech.* **150**, 395–416. A numberical study of water-wave modulation based on a higher-order nonlinear Schrödinger equation.

Loesch, A. Z. (1974) *J. Atmos. Sci.* **31**, 1177–1217. Resonant interactions between unstable and neutral baroclinic waves. Parts I & II.

Loesch, A. Z. & Domaracki, A. (1977) *J. Atmos. Sci.* **34**, 22–35. Dynamics of N resonantly interacting baroclinic waves.

Longuet-Higgins, M. S. (1953) *Phil. Trans. Roy. Soc. Lond.* **A245**, 535–581. Mass transport in water waves.

Longuet-Higgins, M. S. (1962) *J. Fluid Mech.* **12**, 321–332. Resonant interactions between two trains of gravity waves.

Longuet-Higgins, M. S. (1975) *Proc. Roy. Soc. Lond.* **A342**, 157–174. Integral properties of periodic gravity waves of finite amplitude.

Longuet-Higgins, M. S. (1976) *Proc. Roy. Soc. Lond.* **A347**, 311–328. On the nonlinear transfer of energy in the peak of a gravity wave spectrum: a simplified model.

Longuet-Higgins, M. S. (1978a) *Proc. Roy. Soc. Lond.* **A360**, 471–488. The instabilities of gravity waves of finite amplitude in deep water. I. Superharmonics.

Longuet-Higgins, M. S. (1978b) *Proc. Roy. Soc. Lond.* **A360**, 489–505. The instabilities of gravity waves of finite amplitude in deep water. II. Subharmonics.

Longuet-Higgins, M. S. (1981) *Proc. Roy. Soc. Lond.* **A376**, 377–400. On the overturning of gravity waves.

Longuet-Higgins, M. S. & Cokelet, E. D. (1976) *Proc. Roy. Soc. Lond.* **A350**, 1–26. The deformation of steep surface waves on water. I: A numerical method of computation.

Longuet-Higgins, M. S. & Cokelet, E. D. (1978) *Proc. Roy. Soc. Lond.* **A364**, 1–28. The deformation of steep surface waves on water. II: Growth of normal-mode instabilities.

Longuet-Higgins, M. S. & Gill, A. E. (1967) *Proc. Roy. Soc. Lond.* **A299**, 120–140. Resonant interactions between planetary waves.

Longuet-Higgins, M. S. & Smith, N. D. (1966) *J. Fluid Mech.* **25**, 417–435. An experiment on third order resonant wave interactions.

Longuet-Higgins, M. S. & Stewart, R. W. (1960) *J. Fluid Mech.* **8**, 565–583. Changes in the form of short gravity waves on long waves and tidal streams.

Longuet-Higgins, M. S. & Stewart, R. W. (1961) *J. Fluid Mech.* **10**, 529–549. The changes in amplitude of short gravity waves on steady non-uniform currents.

Longuet-Higgins, M. S. & Stewart, R. W. (1962) *J. Fluid Mech.* **13**, 481–504. Radiation stress and mass transport in gravity waves, with application to 'surf beats'.

Lorenz, E. N. (1963) *J. Atmos. Sci.* **20**, 130–141. Deterministic non-periodic flow.

Love, A. E. H. (1927) *A Treatise on the Mathematical Theory of Elasticity* 4th edn. Cambridge. [Reprinted Dover, New York, 1944.]

Luke, J. C. (1967) *J. Fluid Mech.* **27**, 395–397. A variational principle for a fluid with a free surface.

L'vov, V. S., Predtechensky, A. A. & Chernykh, A. I. (1983) In *Nonlinear Dynamics and Turbulence* (eds G. I. Barenblatt, G. Iooss & D. D. Joseph), pp. 238–280, Pitman. Bifurcations and chaos in the system of Taylor vortices – laboratory and numerical experiment.

McEwan, A. D. (1971) *J. Fluid Mech.* **50**, 431–448. Degeneration of resonantly-excited standing internal gravity waves.

McEwan, A. D., Mander, D. W. & Smith, R. K. (1972) *J. Fluid Mech.* **55**, 589–608. Forced resonant second-order interactions between damped internal waves.

McEwan, A. D. & Robinson, R. M. (1975) *J. Fluid Mech.* **67**, 667–687. Parametric instability of internal gravity waves.

McGoldrick, L. F. (1965) *J. Fluid Mech.* **21**, 305–331. Resonant interactions among capillary–gravity waves.

McGoldrick, L. F. (1970a) *J. Fluid Mech.* **40**, 251–271. An experiment on second-order capillary–gravity resonant wave interactions.

McGoldrick, L. F. (1970b) *J. Fluid Mech.* **42**, 193–200. On Wilton's ripples: a special case of resonant interactions.

McGoldrick, L. F. (1972) *J. Fluid Mech.* **52**, 725–751. On the rippling of small waves: a harmonic nonlinear nearly resonant interaction.

McGoldrick, L. F., Phillips, O. M., Huang, N. & Hodgson, T. (1966) *J. Fluid Mech.* **25**, 437–456. Measurements on resonant wave interactions.

McIntyre, M. E. (1981) *J. Fluid Mech.* **106**, 331–347. On the 'wave momentum' myth.

McLean, J. W. (1982a) *J. Fluid Mech.* **114**, 315–330. Instabilities of finite-amplitude water waves.

McLean, J. W. (1982b) *J. Fluid Mech.* **114**, 331–341. Instabilities of finite-amplitude gravity waves on water of finite depth.

McLean, J. W., Ma, Y. C., Martin, D. U., Saffman, P. G. & Yuen, H. C. (1981) *Phys. Rev. Lett.* **46**, 817–820. Three-dimensional instability of finite-amplitude water waves.

Ma, Y.-C. (1978) *Stud. Appl. Math.* **59**, 201–221. The complete solution of the long-wave-short-wave resonance equations.

Ma, Y.-C. (1981) *Wave Motion* **3**, 257–267. The resonant interaction among long and short waves.

Ma, Y.-C. (1982) *Wave Motion* **4**, 113–125. On steady three-dimensional deep water weakly nonlinear gravity waves.

Ma, Y.-C. (1984a) *Phys. Fluids* **27**, 571–578. Resonant triads and direct resonance for Kelvin–Helmholtz waves.

Ma, Y.-C. (1984b) *Stud. Appl. Math.* **70**, 201–213. On nonlinear Kelvin–Helmholtz waves near direct resonance.

Ma, Y.-C. & Redekopp, L. G. (1979) *Phys. Fluids* **22**, 1872–1876. Some solutions pertaining to the resonant interaction of long and short waves.

Mack, L. M. (1976) *J. Fluid Mech.* **73**, 497–520. A numerical study of the temporal eigenvalue spectrum of the Blasius boundary layer.

Mack, L. M. (1977) In *Laminar-Turbulent Transition. Proc. AGARD Conf. No. 224, Lyngby, Denmark*, Paper 1. Transition prediction and linear stability theory.

Mack, L. M. & Kendall, J. M. (1983) *AIAA Paper* 83-0046. Wave patterns produced by a localized harmonic source in a Blasius boundary layer.

Madsen, O. S. (1978) *J. Phys. Oceanog.* **8**, 1009–1015. Mass transport in deep-water waves.

Mahony, J. J. & Smith, R. (1972) *J. Fluid Mech.* **53**, 193–208. On a model representation for certain spatial-resonance phenomena.

Malkus, W. V. R. & Veronis, G. (1958) *J. Fluid Mech.* **4**, 225–260. Finite amplitude cellular convection.

Martin, S., Simmons, W. & Wunsch, C. (1972) *J. Fluid Mech.* **53**, 17–44. The excitation of resonant triads by single internal waves.

Martin, D. U., Saffman, P. G. & Yuen, H. C. (1980) *Wave Motion* **2**, 215–229. Stability of plane wave solutions of the two-space-dimensional nonlinear Schrödinger equation.

Maseev, L. M. (1968) *Fluid Dyn.* **3**, No. 6, 23–24. [trs. of *Mekh. Zhid. i Gaza* **3**, No. 6, 42–45.] Occurrence of three-dimensional perturbations in a boundary layer.

Maslowe, S. A. (1972) *Stud. Appl. Math.* **51**, 1–16. The generation of clear-air turbulence by nonlinear waves.

Maslowe, S. A. (1977a) *J. Fluid Mech.* **79**, 689–702. Weakly nonlinear stability of a viscous free shear layer.

Maslowe, S. A. (1977b) *Quart J. Roy. Met. Soc.* **103**, 769–783. Weakly nonlinear stability theory of stratified shear flows.

Maslowe, S. A. (1981) In *Hydrodynamic Instabilities and the Transition to Turbulence* (eds H. L. Swinney & J. P. Gollub), pp. 181–228. Springer. Shear flow instabilities and transition.

May, R. M. & Leonard, W. J. (1975) *S.I.A.M. J. Appl. Math.* **29**, 243–253. Nonlinear aspects of competition between three species.

Mei, C. C. (1985) *J. Fluid Mech.* **152**, 315–335. Resonant reflection of surface water waves by periodic sandbars.

Meiron, D. I., Saffman, P. G. & Yuen, H. C. (1982) *J. Fluid Mech.* **124**, 109–121. Calculation of steady three-dimensional deep-water waves.

Meiron, D. I. & Saffman, P. G. (1983) *J. Fluid Mech.* **129**, 213–218. Overhanging interfacial gravity waves of large amplitude.

Melville, W. K. (1980) *J. Fluid Mech.* **98**, 285–297. On the Mach reflexion of a solitary wave.

Melville, W. K. (1982) *J. Fluid Mech.* **115**, 165–185. The instability and breaking of deep-water waves.

Merkine, L.-O. & Shtilman, L. (1984) *Proc. Roy. Soc. Lond.* **A395**, 313–340. Explosive instability of baroclinic waves.

Meyer, R. E. (ed.) (1981) *Transition and Turbulence.* Academic.

Michell, J. H. (1893) *Philos. Mag.* [5] **36**, 430–437. The highest waves in water.

Miksad, R. W. (1972) *J. Fluid Mech.* **56**, 695–719. Experiments on the nonlinear stages of free-shear-layer transition.

Miksad, R. W. (1973) *J. Fluid Mech.* **59**, 1–21. Experiments on nonlinear interactions in the transition of a free shear layer.

Miksad, R. W., Jones, F. L., Powers, E. J., Kim, Y. C. & Khadra, L. (1982) *J. Fluid Mech.* **123**, 1–29. Experiments on the role of amplitude and phase modulation during transition to turbulence.

Miksad, R. W., Jones, F. L. & Powers, E. J. (1983) *Phys. Fluids* **26**, 1402–1409. Measurements of nonlinear interactions during natural transition of a symmetric wake.

Miles, J. W. (1957a) *J. Fluid Mech.* **3**, 185–204 (and Corrigenda, **6**, (1959) 582). On the generation of surface waves by shear flows.

Miles, J. W. (1957b) *J. Acoust. Soc. Am.* **29**, 226–228. On the reflection of sound at an interface of relative motion.

Miles, J. W. (1959) *J. Fluid Mech.* **6**, 583–598. On the generation of surface waves by shear flows. Part 3: Kelvin–Helmholtz instability.

Miles, J. W. (1960) *J. Fluid Mech.* **8**, 593–610. The hydrodynamic stability of a thin film of liquid in uniform shearing motion.

Miles, J. W. (1962) *J. Fluid Mech.* **13**, 433–448. On the generation of surface waves by shear flows, Part 4.

Miles, J. W. (1967) *Proc. Roy. Soc. Lond.* **A297**, 459–475. Surface wave damping in closed basins.

Miles, J. W. (1976a) *J. Fluid Mech.* **75**, 419–448. Nonlinear surface waves in closed basins.

Miles, J. W. (1976b) *Stud. Appl. Math.* **55**, 351–359. On internal resonance of two damped oscillators.

Miles, J. W. (1977a) *J. Fluid Mech.* **79**, 157–169. Obliquely interacting solitary waves.

Miles, J. W. (1977b) *J. Fluid Mech.* **79**, 171–179. Resonantly interacting solitary waves.

Miles, J. W. (1980) *Ann. Rev. Fluid Mech.* **12**, 11–43. Solitary waves.

Moon, H. T., Huerre, P. & Redekopp, L. G. (1983) *Physica* **7D**, 135–150. Transitions to chaos in the Ginzburg–Landau equation.

Moore, D. R., Toomre, J., Knobloch, E. & Weiss, N. O. (1983) *Nature* **303**, 663–667. Period doubling and chaos in partial differential equations for thermosolutal convection.

Moore, D. R. & Weiss, N. O. (1973) *J. Fluid Mech.* **58**, 289–312. Two-dimensional Rayleigh–Bénard convection.

Morkovin, M. (1986?) *Guide to Experiments on Instabilities and Laminar Turbulent Transition in Shear Layers.* AIAA, New York. (forthcoming).

Moroz, I. M. & Brindley, J. (1981) *Proc. Roy. Soc. Lond.* A**377**, 379–404. Evolution of baroclinic wave packets in a flow with continuous shear and stratification.

Moroz, I. M. & Brindley, J. (1984) *Stud. Appl. Math.* **70**, 21–61. Nonlinear amplitude evolution of baroclinic wave trains and wave packets.

Mullin, T. (1982) *J. Fluid Mech.* **121**, 207–218. Mutations of steady cellular flows in the Taylor experiment.

Mullin, T. & Benjamin, T. Brooke (1980) *Nature* **288**, 567–569. Transition to oscillatory motion in the Taylor experiment.

Mullin, T., Benjamin T. Brooke, Schatzel, K. & Pike, E. R. (1981) *Phys. Lett.* **83A**, 333–336. New aspects of unsteady Couette flow.

Murdock, J. W. (1977) *AIAA Jour.* **15**, 1167–1173. A numerical study of nonlinear effects on boundary-layer stability.

Murdock, J. W. & Stewartson, K. (1977). *Phys. Fluids* **20**, 1404–1411. Spectrum of the Orr–Sommerfeld equation.

Murdock, J. W. & Taylor, T. D. (1977) *Laminar-Turbulent Transition: Proc. AGARD Conf. No. 224, Lyngby, 1977,* paper 4. Numerical investigation of nonlinear wave interaction in a two-dimensional boundary layer.

Nakaya, C. (1974) *J. Phys. Soc. Japan* **36**, 1164–1173. Domain of stable periodic vortex flows in a viscous fluid between concentric circular cylinders.

Nakaya, C. (1975) *J. Phys. Soc. Japan* **38**, 576–585. The second stability boundary for circular Couette flow.

Nayfeh, A. H. (1970) *J. Fluid Mech.* **40**, 671–684. Finite amplitude surface waves in a liquid layer.

Nayfeh, A. H. (1973) *Perturbation Methods.* Wiley-Interscience.

Nayfeh, A. H. (1981) *J. Fluid Mech.* **107**, 441–453. Effect of streamwise vortices on Tollmien–Schlichting waves.

Nayfeh, A. H. & Bozatli, A. N. (1979a) *Phys. Fluids* **22**, 805–813. Secondary instability in boundary-layer flows.

Nayfeh, A. H. & Bozatli, A. N. (1979b) *Proc. AIAA 12th Fluid & Plasma Dyn. Conf.* Williamsburg, Va.: AIAA Paper 79-1456. Nonlinear wave interactions in boundary layers.

Nayfeh, A. H. & Bozatli, A. N. (1980) *Phys. Fluids* **23**, 448–458. Nonlinear interaction of two waves in boundary-layer flows.

Nayfeh, A. H. & Saric, W. S. (1972) *J. Fluid Mech.* **55**, 311–327. Nonlinear waves in a Kelvin–Helmholtz flow.

Nelson, G. & Craik, A. D. D. (1977) *Phys. Fluids* **20**, 698–700. Growth of streamwise vorticity in unstable boundary layers.

Newell, A. C. (1972) *J. Atmos. Sci.* **29**, 64–76. The post bifurcation stage of baroclinic instability.

Newell, A. C. (1978) *S.I.A.M. J. Appl. Math.* **35**, 650–664. Long waves–short waves; a solvable model.

Newell, A. C. & Whitehead, J. A. (1969) *J. Fluid Mech.* **38**, 279–304. Finite bandwidth, finite amplitude convection.

Nishioka, M., Asai, M. & Iida, S. (1980) In *Laminar-Turbulent Transition: Proc. IUTAM Symp. Stuttgart* (eds R. Eppler & H. Fasel), pp. 37–46, Springer. An experimental investigation of the secondary instability.

Nishioka, M. & Asai, M. (1985) In *Laminar-Turbulent Transition: Proc. IUTAM Symp. Novosibirsk* (ed. V. V. Kozlov) pp. 173–182. Three-dimensional wave-disturbances in plane Poiseuille flow.

Nishioka, M., Iida, S. & Ichikawa, Y. (1975) *J. Fluid Mech.* **72**, 731–751. An experimental investigation of the stability of plane Poiseuille flow.

Normand, C., Pomeau, Y. & Velarde, M. G. (1977) *Rev. Mod. Phys.* **49**, 581–624. Convective instability: a physicist's approach.

Olfe, D. B. & Rottman, J. W. (1980) *J. Fluid Mech.* **100**, 801–810. Some new highest-wave solutions for deep-water waves of permanent form.

Orszag, S. A. (1971) *J. Fluid Mech.* **50**, 689–703. Accurate solution of the Orr–Sommerfeld stability equation.

Orszag, S. A. & Kells, L. C. (1980) *J. Fluid Mech.* **96**, 159–205. Transition to turbulence in plane Poiseuille and plane Couette flow.

Orszag, S. A. & Patera, A. T. (1981) In *Transition and Turbulence* (ed. R. E. Meyer), pp. 127–146. Academic. Subcritical transition to turbulence in planar shear flows.

Orszag, S. A. & Patera, A. T. (1983) *J. Fluid Mech.* **128**, 347–385. Secondary instability of wall-bounded shear flows.

Ostrovsky, L. A. & Stepanyants, Yu. A. (1983) *Fluid Dyn.* **17**, 540–546. [trs. of *Izv. Akad. Nauk. SSSR Mekh. Zhidk. i Gaza.*] Nonlinear stage of the shearing instability in a stratified liquid of finite depth.

Ostrovsky, L. A., Stepanyants, Yu. A. & Tsimring, L. Sh. (1983) *Int. J. Nonlinear Mech.* **19**, 151–161. Radiation instability in a stratified shear flow.

Ostrovsky, L. A. & Tsimring, L. Sh. (1981) *Atmos. Ocean Phys.* **17**, 564–565. [trs. of *Izv. Akad. Nauk. SSSR Fiz. Atmos. i Okeana*, **17**, 766.] Radiation instability of shear flows in a stratified fluid.

Palm, E. (1960) *J. Fluid Mech.* **8**, 183–192. On the tendency towards hexagonal cells in steady convection.

Palm, E. (1975) *Ann. Rev. Fluid Mech.* **77**, 39–61. Nonlinear thermal convection.

Palm, E., Ellingsen, T. & Gjevik, B. (1967) *J. Fluid Mech.* **30**, 651–661. On the occurrence of cellular motion in Bénard convection.

Patel, V. C. & Head, M. R. (1969) *J. Fluid Mech.* **38**, 181–201. Some observations on skin friction and velocity profiles in fully developed pipe and channel flows.

Patera, A. T. & Orszag, S. A. (1981) *J. Fluid Mech.* **112**, 467–474. Finite-amplitude stability of axisymmetric pipe flow.

Patera, A. T. & Orszag, S. A. (1981) In *Proc. 7th Internat. Conf. on Numerical Methods in Fluid Dynamics* (eds R. W. MacCormack & W. C. Reynolds); Springer. Transition and turbulence in planar channel flows.

Patnaik, P. C., Sherman, F. S. & Corcos, G. M. (1976) *J. Fluid Mech.* **73**, 215–240. A numerical simulation of Kelvin–Helmholtz waves of finite amplitude.

Pearson, J. R. A. (1958) *J. Fluid Mech.* **4**, 489–500. On convection cells induced by surface tension.

Pedlosky, J. (1970) *J. Atmos. Sci.* **27**, 15–30. Finite-amplitude baroclinic waves.

Pedlosky, J. (1971) *J. Atmos. Sci.* **28**, 587–597. Finite-amplitude baroclinic waves with small dissipation.

Pedlosky, J. (1972) *J. Atmos. Sci.* **29**, 680–686. Finite-amplitude baroclinic wave packets.

Pedlosky, J. (1975) *J. Phys. Oceanog.* **5**, 608–614. The amplitude of baroclinic wave triads and mesoscale motion in the ocean.

Pedlosky, J. (1976) *J. Fluid Mech.* **78**, 621–637. On the dynamics of finite amplitude baroclinic waves as a function of supercriticality.

Pedlosky, J. (1977) *J. Atmos. Sci.* **34**, 1898–1912. A model of wave-amplitude vacillation.

Pedlosky, J. (1979) *J. Atmos. Sci.* **36**, 1908–1924. Finite-amplitude baroclinic waves in a continuous model of the atmosphere.

Pekeris, C. L. & Shkoller, B. (1967) *J. Fluid Mech.* **29**, 31–38. Stability of plane Poiseuille flow to periodic disturbances of finite amplitude in the vicinity of the neutral curve.

Peltier, W. R., Hallé, J. & Clark, T. L. (1978) *Geophys. Astrophys. Fluid Dyn.* **10**, 53–87. The evolution of finite amplitude Kelvin–Helmholtz billows.

Penney, W. G. & Price, A. T. (1952) *Phil. Trans. Roy. Soc. Lond.* **A244**, 251–284. Some gravity wave problems in the motion of perfect liquids. Part II. Finite periodic stationary gravity waves in a perfect fluid.

Peregrine, D. H. (1972) In *Waves on Beaches and Resulting Sediment Transport* (ed. R. E. Meyer), pp. 95–121. Academic. Equations for water waves and the approximations behind them.

Peregrine, D. H. (1976) *Advances in Appl. Mech.* **16**, 9–117. Interaction of water waves and currents.

Peregrine, D. H. (1983) *J. Austral. Math. Soc.* **B25**, 16–43. Water waves, nonlinear Schrödinger equations and their solutions.

Pereira, N. R. (1977) *Phys. Fluids* **20**, 1735–1743. Solitons in the damped nonlinear Schrödinger equation.

Pereira, N. R. & Chu, F. Y. F. (1979) *Phys. Fluids* **22**, 874–881. Damped double solitons in the nonlinear Schrödinger equation.

Pereira, N. R. & Stenflo, L. (1977) *Phys. Fluids* **20**, 1733–1734. Nonlinear Schrödinger equation including growth and damping.

Phillips, O. M. (1957) *J. Fluid Mech.* **2**, 417–445. On the generation of waves by turbulent wind.

Phillips, O. M. (1960) *J. Fluid Mech.* **9**, 193–217. On the dynamics of unsteady gravity waves of finite amplitude. Part 1. The elementary interactions.

Phillips, O. M. (1961) *J. Fluid Mech.* **11**, 143–155. On the dynamics of unsteady gravity waves of finite amplitude. Part 2. Local properties of a random wave field.

Phillips, O. M. (1966) *Proc. 6th Naval Hydrodyn. Symp., Washington D.C.*, Paper 21. On internal wave interactions.

Phillips, O. M. (1974) In *Nonlinear Waves* (ed. S. Leibovich & A. R. Seebass), Cornell Univ. Press, pp. 186–211. Wave interactions.

Phillips, O. M. (1977) *The Dynamics of the Upper Ocean*, 2nd edn. Cambridge Univ. Press.

Phillips, O. M. (1981a) *J. Fluid Mech.* **106**, 215–227. Wave interactions – the evolution of an idea.

Phillips, O. M. (1981b) *J. Fluid Mech.* **107**, 465–485. The dispersion of short wavelets in the presence of a dominant long wave.

Pinsker, Z. G. (1978) *Dynamical scattering of X-rays in crystals*. Springer.

Plant, W. J. & Wright, J. W. (1977) *J. Fluid Mech.* **82**, 767–793. Growth and equilibrium of short gravity waves in a wind-wave tank.

Plumb, R. A. (1977) *J. Atmos. Sci.* **34**, 1847–1858. The interaction of two internal waves with the mean flow: implications for the theory of the quasi-biennial oscillation.

Plumb, R. A. & McEwan, A. D. (1978) *J. Atmos. Sci.* **35**, 1827–1839. The instability of a forced standing wave in a viscous stratified fluid: a laboratory analogue of the quasi-biennial oscillation.

Prasad, P. & Krishnan, E. V. (1978) *J. Phys. Soc. Japan* 1028–1032. On multi-dimensional packets of surface waves.

Raetz, G. S. (1959) *Norair Rep.* NOR-59-383. Hawthorne, Calif. A new theory of the cause of transition in fluid flows.

Raetz, G. S. (1964) *Norair Rep.* NOR-64-111. Hawthorne, Calif. Current status of resonance theory of transition.

Ramamonjiarisoa, A. & Coantic, M. (1976) *C.R. Acad. Sci. Paris* **B282**, 111–113. Loi expérimental de dispersion des vagues produites par le vent sur une faible longueur d'action.

Ramamonjiarisoa, A. & Giovanangeli, J.-P. (1978) *C.R. Acad. Sci. Paris* **B287**, 133–136. Observations de la vitesse de propagation des vagues engendrées par le vent au large.

Ramazanov, M. P. (1985) In *Laminar-Turbulent Transition: Proc. IUTAM Symp. Novosibirsk* (ed. V. V. Kozlov) pp. 183–190. Development of finite-amplitude disturbances in Poiseuille flow.

Rand, D. (1981) *Arch. Rat. Mech. Anal.* **79**, 1–37. Dynamics and symmetry. Predictions for modulated waves in rotating fluids.

Rayleigh, Lord (J. W. Strutt) (1896) *The Theory of Sound* 2 vols. 2nd edn. Macmillan, London. (reprinted Dover, New York, 1945).

Reid, W. H. (1965) *Basic Developments in Fluid Dynamics* 1, (ed. M. Holt), pp. 249–307. Academic, New York. The stability of parallel flows.

Reiman, A. (1979) *Rev. Mod. Phys.* **51**, 311–330. Space-time evolution of nonlinear three-wave interactions. II: Interactions in an inhomogeneous medium.

Reshotko, E. (1976) *Ann. Rev. Fluid Mech.* **8**, 311–349. Boundary-layer stability and transition.

Reynolds, W. C. & Potter, M. C. (1967) *J. Fluid Mech.* **27**, 465–492. Finite amplitude instability of parallel shear flows.

Ribner, H. S. (1957) *J. Acoust. Soc. Amer.* **29**, 435–441. Reflection, transmission and amplification of sound by a moving medium.

Riley, N. (1967) *J. Inst. Math. Applic.* **3**, 419–434. Oscillatory viscous flows. Review and extension.

Ripa, P. (1981) *J. Fluid Mech.* **103**, 87–115. On the theory of nonlinear wave–wave interactions among geophysical waves.

Roberts, A. J. (1983) *J. Fluid Mech.* **135**, 301–321. Highly nonlinear short-crested water waves.

Robinson, J. L. (1974) *J. Fluid Mech.* **63**, 723–752. The inviscid nonlinear instability of parallel shear flows.

Romanov, V. A. (1973) *Functional Anal. & its Applics.* **7**, 137–146. [trs. of *Funkc. Anal. i Prolozen* 7, (2), 62–73.] Stability of plane-parallel Couette flow.

Ross, J. A., Barnes, F. H., Burns, J. G. & Ross, M. A. S. (1970) *J. Fluid Mech.* **43**, 819–832. The flat plate boundary layer. Part 3. Comparison of theory with experiment.

Rott, N. (1970) *Z.A.M.P.* **21**, 570–582. A multiple pendulum for the demonstration of nonlinear coupling.

Rottman, J. W. (1982) *J. Fluid Mech.* **124**, 283–306. Steep standing waves at a fluid interface.

Rottman, J. W. & Olfe, D. B. (1979) *J. Fluid Mech.* **94**, 777–793. Numerical calculations of steady gravity-capillary waves using an integro-differential formulation.

Roux, J.-C., Simoyi, R. H. & Swinney, H. L. (1983) *Physica* **8D**, 257–266. Observations of a strange attractor.

Rowlands, G. (1974) *J. Inst. Math. Applic.* **13**, 367–377. On the stability of solutions of the nonlinear Schrödinger equation.

Rozhdestvensky, B. L. & Simakin, I. N. (1984) *J. Fluid Mech.* **147**, 261–289. Secondary flows in a plane channel: their relationship and comparison with turbulent flows.

Ruelle, D. & Takens, F. (1971) *Comm. Math. Phys.* **20**, 167–192. On the nature of a strange attractor.

Russell, R. C. H. & Osorio, J. D. C. (1958) *Proc. 6th Conf. Coastal Engng.* (*Miami* 1957) pp. 171–193, Counc. Wave Res., Univ. California, Berkeley. An experimental investigation of drift profiles in closed channels.

Saffman, P. G. & Yuen, H. C. (1978) *Phys. Fluids* **21**, 1450–1451. Stability of a plane soliton to infinitesimal two-dimensional perturbations.

Saffman, P. G. & Yuen, H. C. (1979) *J. Fluid Mech.* **95**, 707–715. A note on numerical computations of large amplitude standing waves.

Saffman, P. G. & Yuen, H. C. (1982) *J. Fluid Mech.* **123**, 459–476. Finite-amplitude interfacial waves in the presence of a current.

Salwen, H., Cotton, F. W. & Grosch, C. E. (1980) *J. Fluid Mech.* **98**, 273–284. Linear stability of Poiseuille flow in a circular pipe.

Salwen, H. & Grosch, C. E. (1972) *J. Fluid Mech.* **54**, 93–112. The stability of Poiseuille flow in a pipe of circular cross-section.

Salwen, H. & Grosch, C. E. (1981) *J. Fluid Mech.* **104**, 445–465. The continuous spectrum of the Orr–Sommerfeld equation. Part 2. Eigenfunction expansions.

Saric, W. S. & Nayfeh, A. H. (1975) *Phys. Fluids* **18**, 945–950. Non-parallel stability of boundary layer flows.

Saric, W. S. & Thomas, A. S. W. (1984) In *Turbulence and Chaotic Phenomena in Fluids, Proc. IUTAM Symp., Kyoto* (ed. T. Tatsumi). North Holland. Experiments on the subharmonic route to turbulence in boundary layers.

Sato, H. (1970) *J. Fluid Mech.* **44**, 741–765. An experimental study of nonlinear interaction of velocity fluctuations in the transition region of a two-dimensional wake.

Sato, H. & Kuriki, K. (1961) *J. Fluid Mech.* **11**, 321–352. The mechanism of transition in the wake of a thin flat plate placed parallel to a uniform flow.

Satsuma, J. & Ablowitz, M. J. (1979) *J. Math. Phys.* **20**, 1496–1503. Two dimensional lumps in nonlinear dispersive systems.

Scanlon, J. W. & Segel, L. A. (1967) *J. Fluid Mech.* **30**, 149–162. Finite amplitude cellular convection induced by surface tension.

Schade, H. (1964) *Phys. Fluids* **7**, 623–628. Contribution to the nonlinear stability theory of inviscid shear layers.

Schaeffer, D. G. (1980) *Math. Proc. Camb. Philos. Soc.* **87**, 307–337. Qualitative analysis of a model for boundary effects in the Taylor problem.

Schluter, A., Lortz, D. & Busse, F. (1965) *J. Fluid Mech.* **23**, 129–144. On the stability of steady finite amplitude convection.

Schubauer, G. B. & Skramstad, H. K. (1947) *J. Res. Nat. Bur. Stand.* **38**, 251–292. Laminar boundary-layer oscillations and transition on a flat plate.

Schwartz, L. W. (1974) *J. Fluid Mech.* **62**, 553–578. Computer extension and analytic continuation of Stokes' expansion for gravity waves.

Schwartz, L. W., Springett, B. E. & Donnelly, R. J. (1964) *J. Fluid Mech.* **20**, 281–289. Modes of instability in spiral flow between rotating cylinders.

Schwartz, L. W. & Fenton, J. D. (1982) *Ann. Rev. Fluid Mech.* **14**, 39–60. Strongly nonlinear waves.

Schwartz, L. W. & Vanden-Broeck, J.-M. (1979) *J. Fluid Mech.* **95**, 119–139. Numerical solution of the exact equations for capillary-gravity waves.

Schwartz, L. W. & Whitney, A. K. (1981) *J. Fluid Mech.* **107**, 147–171. A semianalytic solution for nonlinear standing waves in deep water.

Scott, J. C. (1979) *Oil on Troubled Waters. A bibliography on the effects of surface-active films on surface-wave motions.* Multiscience Publ. Co., London.

Scott, J. C. (1981) *J. Fluid Mech.* **108**, 127–131. The propagation of capillary-gravity waves on a clean water surface.

Segel, L. A. (1962) *J. Fluid Mech.* **14**, 97–114. The nonlinear interaction of two disturbances in the thermal convection problem.

Segel, L. A. (1965) *J. Fluid Mech.* **21**, 359–384. The nonlinear interaction of a finite number of disturbances to a layer of fluid heated from below.

Segel, L. A. & Stuart, J. T. (1962) *J. Fluid Mech.* **13**, 289–306. On the question of the preferred mode in cellular thermal convection.

Segur, H. (1981) *Phys. Fluids* **24**, 2372–2374. Viscous decay of envelope solitons in water waves.

Segur, H. & Hammack, J. L. (1982) *J. Fluid Mech.* **118**, 285–304. Soliton models of long waves.

Sen, P. K. & Venkateswarlu, D. (1983) *J. Fluid Mech.* **133**, 179–206. On the stability of plane Poiseuille flow to finite-amplitude disturbances, considering the higher-order Landau constants.

Shen, S. F. (1964) In *Theory of Laminar Flows* (ed. F. K. Moore), pp. 719–853, Princeton. Stability of laminar flows.

Sherman, J. & McLaughlin, J. B. (1978) *Comm. Math. Phys.* **58**, 9–17. Power spectra of nonlinearly coupled waves.

Shnol', E. E. (1974) *J. Appl. Math. Mech.* **38**, 464–468. [trs. of *Prikl. Mat. Mekh.* **38**, 502–506.] On the instability of plane-parallel flows of perfect fluid.

Simmons, W. F. (1969) *Proc. Roy. Soc. Lond.* **A309**, 551–575. A variational method for weak resonant wave interactions.

Smith, F. I. P. & Craik, A. D. D. (1971) *J. Fluid Mech.* **45**, 527–544. Wind-generated waves in thin liquid films with soluble contaminant.

Smith. F. T. (1979a) *Proc. Roy. Soc. Lond.* **A366**, 91–109. On the non-parallel flow stability of the Blasius boundary layer.

Smith, F. T. (1979b) *Proc. Roy. Soc. Lond.* **A368**, 573–589 (and Corrigenda: **A371**, (1980) 439–440). Nonlinear stability of boundary layers for disturbances of various sizes.

Smith, F. T. & Bodonyi, R. J. (1982a) *J. Fluid Mech.* **118**, 165–185. Nonlinear critical layers and their development in streaming-flow stability.

Smith, F. T. & Bodonyi, R. J. (1982b) *Proc. Roy. Soc. Lond.* **A384**, 463–489. Amplitude-dependent neutral modes in the Hagen–Poiseuille flow through a circular pipe.

Smith, M. K. & Davis, S. H. (1982) *J. Fluid Mech.* **121**, 187–206. The instability of sheared liquid layers.

Spooner, G. F. & Criminale, W. O. (1982) *J. Fluid Mech.* **115**, 327–346. The evolution of disturbances in an Ekman boundary layer.

Stewartson, K. (1975) *Polish Acad. Sci. Fluid Dyn. Trans.* **7**, 101–128. Some aspects of nonlinear stability theory.

Stewartson, K. (1978) *Geophys. Astrophys. Fluid Dyn.* **9**, 185–200. The evolution of the critical layer of a Rossby wave.

Stewartson, K. (1981) *IMA Jour. Appl. Math.* **27**, 133–175. Marginally stable inviscid flows with critical layers.

Stewartson, K. & Stuart, J. T. (1971) *J. Fluid Mech.* **48**, 529–545. A nonlinear instability theory for a wave system in plane Poiseuille flow.

Stiassnie, M. & Shemer, L. (1984) *J. Fluid Mech.* **143**, 47–67. On modifications of the Zakharov equation for surface gravity waves.

Stoker, J. J. (1957) *Water Waves.* Interscience, New York.

Stokes, G. G. (1847) *Trans. Camb. Phil. Soc.* **8**, 441–455. (Reprinted in *Math. Phys. Papers* 1, 197–219.) On the theory of oscillatory waves.

Stokes, G. G. (1880a) *Mathematical and Physical Papers*, Vol. 1, 219–229. Appendix to 'On the theory of oscillatory waves'.

Stokes, G. G. (1880b) *Mathematical and Physical Papers*, Vol. 1, 314–326. Cambridge. Supplement to a paper on the theory of oscillatory waves.

Stommel, H. (1951) *Weather*, **6**, 72–74. Streaks on natural water surfaces.

Struminskii, V. V. & Skobelev, B. Yu. (1980) *Sov. Phys. Doklady.* **25**, 345–347. [trs. of *Dokl. Akad. Nauk. SSSR* **252**, 566–570.] Nonlinear neutral curve for Poiseuille flow.

Stuart, J. T. (1958) *J. Fluid Mech.* **4**, 1–21. On the nonlinear mechanics of hydrodynamic stability.

Stuart, J. T. (1960) *J. Fluid Mech.* **9**, 353–370. On the nonlinear mechanics of wave disturbances in stable and unstable parallel flows, Part 1. The basic behaviour in plane Poiseuille flow.

Stuart, J. T. (1962a) *Proc. 10th Internat. Congr. Appl. Mech.* Stresa, Italy 1960, (eds F. Rolla & W. T. Koiter), pp. 63–97. Elsevier, Amsterdam. Nonlinear effects in hydrodynamic stability.

Stuart, J. T. (1962b) *Adv. Aero. Sci.* **3–4**, 121–142. On three-dimensional nonlinear effects in the stability of parallel flows.

Stuart, J. T. (1963) In *Laminar Boundary Layers* (ed. L. Rosenhead), pp. 492–579. Oxford University Press. Hydrodynamic stability.

Stuart, J. T. (1966) *J. Fluid Mech.* **24**, 673–687. Double boundary layers in oscillatory viscous flow.

Stuart, J. T. (1967) *J. Fluid Mech.* **29**, 417–440. On finite amplitude oscillations in laminar mixing layers.

Stuart, J. T. & DiPrima, R. C. (1978) *Proc. Roy. Soc. Lond.* **A362**, 27–41. The Eckhaus and Benjamin–Feir resonance mechanisms.

Stuart, J. T. & DiPrima, R. C. (1980) *Proc. Roy. Soc. Lond.* **A372**, 357–365. On the mathematics of Taylor-vortex flows in cylinders of finite length.

Su, M.-Y. (1982) *J. Fluid Mech.* **124**, 73–108. Three-dimensional deep-water waves. Part 1. Experimental measurement of skew and symmetric wave patterns.

Su, M.-Y., Bergin, M., Marler, P. & Myrick, R. (1982) *J. Fluid Mech.* **124**, 45–72. Experiments on nonlinear instabilities and evolution of steep gravity wave trains.

Swinney, H. L. (1983) *Physica* **7D**, 3–15. Observations of order and chaos in nonlinear systems.

Swinney, H. L. & Gollub, J. P. (eds) (1981) *Hydrodynamic Instabilities and the Transition to Turbulence.* Springer.

Synge, J. L. (1938) *Semicentenn. Publ. Amer. Math. Soc.* **2**, 227–269. Hydrodynamical stability.

Tadjbaksh, I. & Keller, J. B. (1960) *J. Fluid Mech.* **8**, 442–451. Standing surface waves of finite amplitude.

Taneda, S. (1983) *Phys. Fluids* **26**, 2801–2806. Visual observations on the amplification of artificial disturbances in turbulent shear flows.

Tani, I. (1969) *Ann. Rev. Fluid Mech.* **1**, 169–196. Boundary-layer transition.

Tani, I. (1980) In *Laminar-Turbulent Transition: Proc. IUTAM Symp.* Stuttgart (eds. R. Eppler & H. Fasel). Springer, pp. 263–276. Some thoughts on boundary layer transition.

Tani, I. (1981) *Proc. Indian Acad. Sci. (Engg. Sci.)* **4**, 219–238. Three-dimensional aspects of boundary layer transition.

Taniuti, T. & Washimi, H. (1968) *Phys. Rev. Lett.* **21**, 209–212. Self-trapping and instability of hydromagnetic waves along the magnetic field in a cold plasma.

Taylor, G. I. (1931) *Proc. Roy. Soc. Lond.* **A132**, 499–523. [Reprinted in *Scientific Papers* (1960), Vol. II, 219–239.] Effect of variation in density on the stability of superposed streams of fluid.

Taylor, G. I. (1951) *Proc. Roy. Soc. Lond.* **A209**, 447–461. [Reprinted in *Scientific Papers*, **IV**, 147–162.] Analysis of the swimming of microscopic organisms.

Taylor, G. I. (1953) *Proc. Roy. Soc. Lond.* **A218**, 44–59. An experimental study of standing waves.

Thom, R. (1975) *Structural Stability and Morphogenesis*, Benjamin; Reading, Mass.

Thomas, A. S. W. & Saric, W. S. (1981) *Bull. Amer. Phys. Soc.* **26**, 1252. Harmonic and subharmonic waves during boundary-layer transition. (Abstract.)

Thorpe, S. A. (1966) *J. Fluid Mech.* **24**, 737–751. On wave interactions in a stratified fluid.

Thorpe, S. A. (1968a) *Phil. Trans. Roy. Soc. Lond.* **A263**, 563–614. On the shape of progressive internal waves.

Thorpe, S. A. (1968b) *J. Fluid Mech.* **32**, 489–528. On standing internal gravity waves of finite amplitude.

Thorpe, S. A. (1971) *J. Fluid Mech.* **46**, 299–319. Experiments on the instability of stratified shear flows: miscible fluids.

Thorpe, S. A. (1973) *J. Fluid Mech.* **61**, 731–751. Experiments on instability and turbulence in a stratified shear flow.

Thorpe, S. A. (1981) *J. Fluid Mech.* **103**, 321–344. An experimental study of critical layers.

Thorpe, S. A. & Hall, A. J. (1982) *J. Fluid Mech.* **114**, 237–250. Observations of the thermal structure of Langmuir circulation.

Thyagaraja, A. (1983) In *Nonlinear Waves* (ed. L. Debnath) pp. 308–325, Cambridge. Recurrence phenomena and the number of effective degrees of freedom in nonlinear wave motions.

Tietjens, O. (1925) *Z. angew. Math. Mech.* **5**, 200–217. Beiträge zur Entstehung der Turbulenz.

Toland, J. F. (1978) *Proc. Roy. Soc. Lond.* **A363**, 469–485. On the existence of a wave of greatest height and Stokes' conjecture.

Toomre, J., Gough, D. O. & Spiegel, E. A. (1982) *J. Fluid Mech.* **125**, 99–122. Time-dependent solutions of multimode convection equations.

Tuck, E. O. (1968) *J. Fluid Mech.* **31**, 305–308. A note on a swimming problem.

Turner, J. S. (1973) *Buoyancy Effects in Fluids*, Cambridge Univ. Press.

Turner, J. G. & Boyd, T. J. M. (1978) *J. Math. Phys.* **19**, 1403–1413. Three- and four-wave interactions in plasmas.

Turpin, F.-M., Benmoussa, C. & Mei, C. C. (1983) *J. Fluid Mech.* **132**, 1–23. Effects of slowly varying depth and current on the evolution of a Stokes wave packet.

Usher, J. R. & Craik, A. D. D. (1974) *J. Fluid Mech.* **66**, 209–221. Nonlinear wave interactions in shear flows. Part 1. A variational formulation.

Usher, J. R. & Craik, A. D. D. (1975) *J. Fluid Mech.* **70**, 437–461. Nonlinear wave interactions in shear flows. Part 2. Third-order theory.

Vakhitov, N. G. & Kolokolov, A. A. (1973) *Radiophys. Quant. Electr.* **16**, 745–753. [trs. of *Izv. Vys. Uch. Zav. Radiofiz.* **16**, 1020–1028.] Stationary solutions of the wave equation in the medium with nonlinearity saturation.

Vanden-Broeck, J. M. & Keller, J. B. (1980) *J. Fluid Mech.* **98**, 161–169. A new family of capillary waves.

Van Duin, C. A. & Kelder, H. (1982) *J. Fluid Mech.* **120**, 505–521. Reflection properties of internal gravity waves incident upon a hyperbolic tangent shear layer.

Vlasov, V. N., Petrishchev, I. A. & Talanov, V. I. (1974) *Quant. Electron. Radiophys.* **14**, 1002.

Volodin, A. G. & Zel'man, M. B. (1977) *Fluid Dyn.* **12**, 192–196. [trs. of *Izv. Akad. Nauk. SSSR Mekh. Zhid. i Gaza.*] Pairwise nonlinear interactions of Tollmien–Schlichting waves in flows of the boundary-layer type.

Volodin, A. G. & Zel'man, M. B. (1979) *Fluid Dyn.* **13**, 698–703 [trs. of *Mekh. Zhid. i Gaza* **5**, 78–84.] Three-wave resonance interaction of disturbances in a boundary layer.

Walden, R. W. & Donnelly, R. J. (1979) *Phys. Rev. Lett.* **42**, 301–304. Re-emergent order of chaotic circular Couette flow.

Wang, P. K. C. (1972) *J. Math. Phys.* **13**, 943–947. Bounds for solution of nonlinear wave–wave interacting systems with well-defined phase description.

Warn, T. & Warn, H. (1978) *Stud. Appl. Math.* **59**, 37–71. The evolution of a nonlinear critical level.

Watanabe, T. (1969) *J. Phys. Soc. Japan* **27**, 1341–1350. A nonlinear theory of two-stream instability.

Watson, J. (1960) *J. Fluid Mech.* **9**, 371–389. On the nonlinear mechanics of wave disturbances in stable and unstable parallel flows. Part 2. The development of a solution for plane Poiseuille and for plane Couette flow.

Watson, J. (1962) *J. Fluid Mech.* **14**, 211–221. On spatially-growing finite disturbances in plane Poiseuille flow.

Weber, J. E. (1983) *J. Fluid Mech.* **137**, 115–129. Attenuated wave-induced drift in a viscous rotating ocean.

Weidman, P. D. & Maxworthy, T. (1978) *J. Fluid Mech.* **85**, 417–431. Experiments on strong interactions between solitary waves.

Weiland, J. & Wilhelmsson, H. (1977) *Coherent Nonlinear Interaction of Waves in Plasmas*. Oxford: Pergamon.

Weinstein, M. (1981) *Proc. Roy. Soc. Lond.* **A375**, 155–167. Nonlinear instability in plane Poiseuille flow: a quantitative comparison between the method of amplitude expansions and the method of multiple scales.

Weissman, M. A. (1979) *Phil. Trans. Roy. Soc. Lond.* **A290**, 639–685. Nonlinear wave packets in the Kelvin–Helmholtz instability.

Wersinger, J-M., Finn, J. M. & Ott, E. (1980a) *Phys. Rev. Lett.* **44**, 453–456. Bifurcations and strange behaviour in instability saturation by nonlinear mode coupling.

Wersinger, J-M., Finn, J. M. & Ott, E. (1980b) *Phys. Fluids* **23**, 1142–1154. Bifurcation and 'strange' behaviour in instability saturation by nonlinear three wave mode coupling.

West, B. J. (1981) *Deep Water Gravity Waves*. Lecture Notes in Physics No.146, Springer.

Whitehead, J. A. (1976) *J. Fluid Mech.* **75**, 715–720. The propagation of dislocations in Rayleigh–Bénard rolls and bimodal flow.

Whitham, G. B. (1962) *J. Fluid Mech.* **12**, 135–147. Mass, momentum and energy flux in water waves.

Whitham, G. B. (1965) *J. Fluid Mech.* **22**, 273–283. A general approach to linear and nonlinear dispersive waves using a Lagrangian.

Whitham, G. B. (1967) *J. Fluid Mech.* **27**, 399–412. Nonlinear dispersion of water waves.

Whitham, G. B. (1970) *J. Fluid Mech.* **44**, 373–395. Two-timing, variational principles and waves.

Whitham, G. B. (1974) *Linear and Nonlinear Waves*. Wiley, New York.

Wilhelmsson, H. & Pavlenko, V. P. (1973) *Physica Scripta* **7**, 213–216. Five wave interaction – a possibility for enhancement of optical or microwave radiation by nonlinear coupling of explosively unstable plasma waves.

Wilhelmsson, H., Watanabe, M. & Nishikawa, K. (1977) *Phys. Lett.* **60A**, 311–313. Theory for space–time evolution of explosive-type instability.

Williams, J. M. (1981) *Phil. Trans. Roy. Soc. Lond.* **302**, 139–188. Limiting gravity waves in water of finite depth.

Wilton, J. R. (1913) *Phil. Mag.* **26**, (Ser. 6) 1053–1058. On the highest wave in deep water.

Wilton, J. R. (1915) *Phil. Mag.* **29**, (Ser. 6) 688–700. On ripples.

Winant, C. D. & Browand, F. K. (1974) *J. Fluid Mech.* **63**, 237–255. Vortex pairing: the mechanism of turbulent mixing-layer growth at moderate Reynolds number.

Wortmann, F. X. (1977) In *Laminar-Turbulent Transition: Proc. AGARD Conf. No. 224, Lyngby, Denmark*, Paper 12. The incompressible fluid motion downstream of two-dimensional Tollmien–Schlichting waves.

Wygnanski, I., Haritonidis, J. H. & Kaplan, R. E. (1979) *J. Fluid Mech.* **92**, 505–528. On a Tollmien–Schlichting wave packet produced by a turbulent spot.

Yahata, H. (1981) *Prog. Theor. Phys.* **66**, 879–891. Temporal development of the Taylor vortices in a rotating fluid: IV.

Yahata, H. (1982) *Prog. Theor. Phys.* **68**, 1070–1081. Transition to turbulence in the Rayleigh–Bénard convection.

Yahata, H. (1983a) *Prog. Theor. Phys.* **69**, 396–402. Temporal development of the Taylor vortices in a rotating fluid: V.

Yahata, H. (1983b) *Prog. Theor. Phys.* **69**, 1802–1805. Period-doubling cascade in the Rayleigh–Bénard convection.

Yih, C.-S. (1967) *J. Fluid Mech.* **27**, 337–352. Instability due to viscosity stratification.

Yorke, J. A. & Yorke, E. D. (1981) In *Hydrodynamic Instabilities and the Transition to Turbulence* (eds H. L. Swinney & J. P. Gollub), Springer, pp. 77–95. Chaotic behaviour and fluid dynamics.

Yuen, H. C. (1983) In *Waves on Fluid Interfaces* (ed. R. E. Meyer) Academic, pp. 17–40. Instability of finite-amplitude interfacial waves.

Yuen, H. C. & Lake, B. M. (1975) *Phys. Fluids.* **18**, 956–960. Nonlinear deep water waves: theory and experiment.

Yuen, H. C. & Lake, B. M. (1980) *Ann. Rev. Fluid Mech.* **12**, 303–334. Instabilities of waves on deep water.

Yuen, H. C. & Lake, B. M. (1982) *Advances in Appl. Mech.* **22**, 67–229. Nonlinear dynamics of deep-water gravity waves.

Zabusky, N. J. & Deem, G. S. (1971) *J. Fluid Mech.* **47**, 353–379. Dynamical evolution of two-dimensional unstable shear flows.

Zahn, J.-P., Toomre, J., Spiegel, E. A. & Gough, D. O. (1974) *J. Fluid Mech.* **64**, 319–345. Nonlinear cellular motions in Poiseuille channel flow.

Zakharov, V. E. (1966) *Sov. Phys. – J.E.T.P.* **24**, 740–744. [trs. of *Zh. Eksp. Teor. Fiz.* **51**, 1107–1114.] The instability of waves in nonlinear dispersive media.

Zakharov, V. E. (1968) *J. Appl. Mech. Tech. Phys.* **2**, 190–194. [trs. of *Zh. Prikl. Mekh. Tekh. Fiz.* **9** (2), 86–94.] Stability of periodic waves of finite amplitude on the surface of a deep fluid.

Zakharov, V. E. (1972) *Sov. Phys. – J.E.T.P.* **35**, 908–914. [trs. of *Zh. Eksp. Teor. Fiz.* **62**, 1745–1751.] Collapse of Langmuir waves.

Zakharov, V. E. (1976) *Soviet Phys. Dokl.* **21**, 322–323. [trs. of *Dokl. Akad. Nauk. S.S.S.R.* **228**, 1314–1316.] Exact solutions to the problem of the parametric interaction of three-dimensional wave packets.

Zakharov, V. E. & Manakov, S. V. (1973) *Soviet Phys. – J.E.T.P. Lett.* **18**, 243–247. [trs. of *Zh. Eksp. Teor. Fiz. Pis'ma Red.* **18**, 413–417.] Resonant interaction of wave packets in nonlinear media.

Zakharov, V. E. & Manakov, S. V. (1975) *Soviet Phys. – J.E.T.P.* **42**, 842–850 (1976). [trs. of *Zh. Eksp. Teor. Fiz.* **69**, 1654–1673.] The theory of resonance interaction of wave packets in nonlinear media.

Zakharov, V. E. & Rubenchik, A. M. (1974) *Sov. Phys. – J.E.T.P.* **38**, 494–500. [trs. of *Zh. Eksp. Teor. Fiz.* **65**, 997–1011.] Instability of waveguides and solitons in nonlinear media.

Zakharov, V. E. & Shabat, A. B. (1972) *Sov. Phys. – J.E.T.P.* **34**, 62–69. [trs. of *Zh. Eksp. Teor. Fiz.* **61**, 118–134.] Exact theory of two-dimensional self-focusing and one-dimensional self-modulating waves in nonlinear media.

Zhou, H. (1982) *Proc. Roy. Soc. Lond.* **A381**, 407–418. On the nonlinear theory of stability of plane Poiseuille flow in the subcritical range.

Subject index